MOLECULAR BREEDING OF FORAGE AND TURF

Developments in Plant Breeding

VOLUME 11

The titles published in this series are listed at the end of this volume.

Molecular Breeding of Forage and Turf

Proceedings of the 3rd International Symposium, Molecular Breeding of Forage and Turf, Dallas, Texas, and Ardmore, Oklahoma, U.S.A., May, 18–22, 2003

Edited by

ANDREW HOPKINS

ZENG-YU WANG

ROUF MIAN

MARY SLEDGE
Forage Improvement Division,
The Samuel Roberts Noble Foundation,
Ardmore, Oklahoma, U.S.A.

and

REED E. BARKER
USDA-ARS, Narional, Forage Seed Production Center,
Corvallis, Oregon, U.S.A.

KLUWER ACADEMIC PUBLISHERS

DORDRECHT / BOSTON / LONDON

A C.I.P. Catalogue record for this book is available from the Library of Congress.

ISBN 1-4020-1867-3

Published by Kluwer Academic Publishers,
P.O. Box 17, 3300 AA Dordrecht, The Netherlands.

Sold and distributed in North, Central and South America
by Kluwer Academic Publishers,
101 Philip Drive, Norwell, MA 02061, U.S.A.

In all other countries, sold and distributed
by Kluwer Academic Publishers,
P.O. Box 322, 3300 AH Dordrecht, The Netherlands.

Printed on acid-free paper

Printed in the Netherlands.

Molecular Breeding *of* Forage *and* Turf

TABLE OF CONTENTS

Flowering and Reproductive Development

Genomics of Plant-Symbiont Relations

Improvement for Animal, Human and Environmental Welfare

Development and Application of Molecular Technologies in Forage and Turf Improvement

Bioinformatics

Population and Quantitative Genetics

Functional Genomics and Genome Sequencing of *Medicago truncatula*

Field Test, Risk Assessment and Biosafety

Intellectual Property Rights

Preface

Forage grasses and forage legumes are critical to livestock industries throughout the world. They play a major role in providing high quality roughage for the economical production of meat, milk and fiber products, and are important in soil conservation and environmental protection. Besides being used as forage, some of the grasses are grown specifically for turf or amenity purposes on sports fields, golf courses, parks, lawns and roadsides. Turf grasses contribute considerably to our environment by adding beauty to the surroundings, providing a safe playing surface for sports and recreation, and preventing erosion. Genetic improvement is one of the most effective ways to increase productivity of forage and turf. Due to the biological complexity of forage and turf species and the associated difficulties encountered by traditional breeding methods, the potential of molecular breeding for the development of improved cultivars is evident. The joint efforts of molecular biologists, plant breeders and scientists in related disciplines will make the available biotechnological methods useful for accelerating forage and turf improvement.

The 3rd International Symposium on Molecular Breeding of Forage and Turf was held May 18-22, 2003, in Dallas, Texas and Ardmore, Oklahoma. The Conference was hosted by The Samuel Roberts Noble Foundation in cooperation with Texas A&M University. Attendees included breeders, molecular biologists, geneticists, agronomists and biochemists from sixteen countries. The program featured plenary addresses from leading international speakers, selected oral presentations, volunteered poster presentations, as well as tours of the Noble Foundation in Ardmore, Oklahoma, and the Texas A&M Agricultural Research and Extension Center in Dallas.

This book includes papers from the plenary lectures and selected oral presentations of the Conference. A wide variety of themes are included and a collection of authoritative reports provided on the recent progress and understanding of molecular technologies and their application in plant improvement. Almost all relevant areas in molecular breeding of forage and turf, from gene discovery to the development of improved cultivars, are discussed in the proceedings.

The 3rd International Symposium on Molecular Breeding of Forage and Turf and the publication of this volume, *Molecular Breeding of Forage and Turf*, have been supported by the Noble Foundation, Texas A&M University, National Science Foundation, USDA-ARS, U.S. Golf Association, Texas

Turfgrass Association, Controlled Environments Ltd. (Conviron), PhytoTechnology Laboratories L.L.C., The Scotts Co., Qiagen Inc., Monsanto Co., Forage and Grassland Foundation, Grass Breeders Conference, American Forage and Grassland Council, Crop Science Society of America and North American Alfalfa Improvement Conference. We express our sincere thanks for their sponsorship and support.

We thank Mervyn Humphreys, Odd Arne Rognli, German Spangenberg and Hotoshi Nakagawa of the International Organizing Committee, as well as Milt Engelke, Mark Hussey, Shan Ingram, LIoyd Sumner and Brian Unruh of the Local Organizing Committee for their contributions to the success of the conference. We are also grateful to Michael Cawley, Joe Bouton, Steven Rhines, Carol Resz, Ronnie Bloomfield, Scott McNeil and Darla Snelson of the Noble Foundation for their enthusiastic support and help. We thank Dallas Meeting Management Inc. for detailed planning of the conference. We thank Noeline Gibson and Jacco Flipsen of Kluwer Academic Publishers for their assistance and cooperation in the publication of this volume. Finally, we express our gratitude to the authors whose dedication and work made this book possible. Without any doubt, the use of the technologies and tools outlined in this book will expand our knowledge in designing better forage and turf for the future.

Andrew Hopkins
Rouf Mian
Reed E. Barker
Zeng-Yu Wang
Mary Sledge

September 2003

Molecular Improvement of Forages — from Genomics to GMOs

Richard A. Dixon
Plant Biology Division, The Samuel Roberts Noble Foundation, 2510 Sam Noble Parkway, Ardmore, Oklahoma 73401, USA. (Email: radixon@noble.org).

Key words: genomics, *Medicago truncatula*, digestibility, lignification, pasture bloat, condensed tannins, saponins

Abstract:

Quality traits are major targets for the biotechnological improvement of forage crops. Many years of research have identified lignin as an impediment to digestibility, lack of condensed tannins as promoting pasture bloat and limiting nitrogen nutrition, and triterpene saponins as anti-palatability factors. Recent progress on understanding and manipulating the biosynthetic pathways leading to lignin, condensed tannins and saponins will facilitate engineering of alfalfa and other forage crops for reduced bloating potential and improved digestibility and palatability. Developments in genomics technology, centered on selected model species, have accelerated the pace of gene discovery in secondary metabolism and other complex pathways in plants. The rapidly emerging genomics resources for model species such as the legume *Medicago truncatula* will impact many aspects of forage improvement, with, in the case of *M. truncatula*, direct and immediate relevance for alfalfa. The first biotech products to reach the farm have been modified for input traits such as insect or herbicide resistance. Millions of acres in the US are now planted with genetically modified corn, soybean and cotton, but such products have met with resistance from environmentalists and the public in Europe and elsewhere. Forage crops with genetically improved quality (output) traits will benefit both the health of the animals that consume them and the environment through reductions in waste excretion and greenhouse gas emission. Policymakers should be made aware of these attributes. Genetically improved forage crops represent a unique opportunity for demonstrating the global benefits of biotechnology.

A. Hopkins, Z.Y. Wang, R. Mian, M. Sledge and R.E. Barker (eds.), Molecular Breeding of Forage and Turf, 1-19.

1. INTRODUCTION

Most of the research funding for plant improvement in the United States has, historically, gone to the major commodity crops such as corn, wheat and soybean. More recently, the worldwide revolution in plant genomics has been centered primarily on just two species, *Arabidopsis thaliana* and rice (Bevan et al. 1999; Delseny et al. 2001; Ausubel 2002; Goff et al. 2002). A vigorous debate continues as to how the findings made in these model systems will translate to other economically important species.

Molecular improvement of forage crops presents both new challenges and clear opportunities for the application of biotechnology. In the past several years, researchers worldwide have begun to develop genetic model systems for forage legumes and grasses. Forage quality traits such as digestibility, nutritional quality and palatability present the molecular biologist with interesting new targets for gene discovery. Genetic modification of these traits should enhance economics, animal health and the environment, and presents a case study for explaining the potential benefits of GMOs. This article briefly reviews the science behind molecular approaches to forage quality improvement.

2. GENE DISCOVERY PROGRAMS FOR FORAGE CROP IMPROVEMENT

Classical breeding approaches have, with few exceptions, been the mainstay of forage improvement over the past half century. More recently, molecular tools such as QTL analysis (Paterson et al. 1988) and marker assisted selection have facilitated these endeavors. Clearly, knowing the functions of all the genes within a plant would provide an invaluable resource for molecular breeding. However, the genomes of most forage crops (grasses and legumes) are complex and unlikely to be subjects of large scale sequencing projects in the foreseeable future.

Medicago truncatula (also known as barrel medic because of the shape of its seed pods) is a forage legume commonly grown in Australia. It originates from Mediterranean regions, and has recently been introduced as a warm season annual legume to the Gulf Coast States in the US. *M. truncatula* is very closely related to the world's major forage legume, alfalfa (*Medicago sativa*). However, whereas alfalfa is an outcrossing autotetraploid with four copies of each of its eight chromosomes, *M. truncatula* has a simple diploid genome and can be readily self-pollinated, facilitating genetic analysis.

In addition to its small genome, *M. truncatula* has a fast generation time and can be transformed genetically using relatively standard protocols, and has thus been adopted as a model for legume genomics (Cook 1999; Oldroyd and Geurts 2001). Genes from *M. truncatula* share very high sequence identity to their counterparts from alfalfa, and also appear to be arranged in a similar order on the chromosomes, making *M. truncatula* an excellent model for understanding the molecular biology of alfalfa.

The *Medicago* gene index at the National Center for Genome Resources (Bell et al. 2001) and the TIGR *Medicago* gene index (Quackenbush et al. 2000), provide information on approximately 200,000 expressed sequence tags (ESTs) from *M. truncatula*, and a whole genome sequencing project for *M. truncatula* is in progress at the University of Oklahoma (Trends in Plant Science 7: 101, 2002). It is possible to view and analyze sequences of all the expressed genes sequenced to date, and to compute their expression patterns *in silico*, by simple search and query commands with various Plant Gene Index databases, such as those available at the TIGR website (http://www.tigr.org/tdb/tgi.shtml). An example of this, showing differential expression patterns of the *M. truncatula* genes encoding L-phenylalanine ammonia-lyase, the first enzyme in the phenylpropanoid pathway leading to lignin, condensed tannins and antimicrobial isoflavonoid phytoalexins, is shown in Figure 1. An important feature of the *M. truncatula* EST data is that nearly 40 different cDNA libraries, representing a range of tissues and biological conditions, have been sequenced, greatly facilitating *in silico* analysis of gene expression patterns.

In addition to the *Medicago* genomics resources outlined above, a new collaborative program between the Forage Biotechnology Group and Plant Biology Division at the Samuel Roberts Noble Foundation aims to produce in excess of 60,000 ESTs for tall fescue (*Festuca arundinacea*). Similar programs in perennial ryegrass (*Lolium perenne*) and white clover (*Trifolium repens*) are in progress at Agriculture Victoria-DNRE (Australia) and AgResearch Limited (New Zealand), and a project to sequence the gene-rich regions of the legume *Lotus japonicus* is underway at the Kasuza DNA Institute in Japan. Such genomics resources provide a basis for gene discovery in biochemical pathways that affect forage quality traits, as well as providing a rich source of genetic markers, such as simple sequence repeat (SSR) motifs, for marker-assisted classical breeding.

DNA sequence information alone is only one part of an integrated genomics program. Having a "unigene set" of all the expressed genes in a plant allows the researcher to analyze responses to biotic and abiotic stresses, and developmental programs, on a global level using DNA array techniques (Wu et al. 2001). However, such an approach is essentially correlative, and does not provide the final proof of gene function. For this reason, it is important to develop both forward and reverse genetic approaches to examine plant gene function. This has been successfully done in Arabidopsis, where T-DNA insertion mutants (Azpiroz-Leehan and Feldmann 1997) now exist in essentially every gene in the organism, and T-DNA activation tagging can be employed to generate dominant, gain of function mutations (Weigel et al. 2000). Development of such resources, which might also include alternative mutational approaches such as transposon tagging (Fitzmaurice et al. 1992), fast neutron bombardment to generate deletions, or virus-induced gene silencing for rapid transient analysis of target gene function (Baulcombe 1999), will be a rate limiting factor for the full exploitation of genomics approaches to forage crops.

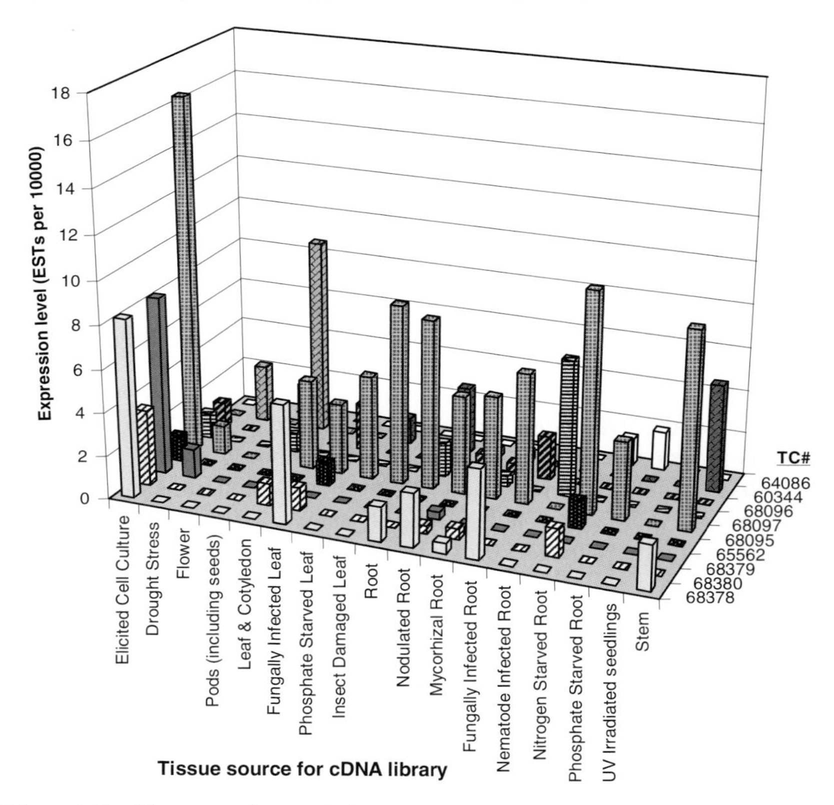

Figure 1. In silico expression analysis of PAL genes in *M. truncatula*. EST sequences can be clustered into tentative consensus (TC) sequences, representing transcripts originating from one specific gene. The database can be queried to determine the number of times an EST corresponding to a particular TC has been sequenced in a particular library or group of libraries. This number is normalized to the total number of ESTs sequenced in that library(s) to give an approximate value for the expression level of the gene represented by that TC. Note that only three of the nine putative TCs appear to be expressed in stems, and therefore be candidates for involvement in stem lignification.

It is possible that knowledge derived from the Arabidopsis resources will in some cases be of value for forage crop improvement, the understanding of new pathways for lignin biosynthesis being an example (Humphreys and Chapple 2002). Furthermore, it will not be feasible to develop genetic systems for each individual forage crop. *M. truncatula* provides an excellent model for alfalfa, except for the fact that it is an annual whereas alfalfa is a perennial. Among the monocots, rice has the advantage of a sequenced genome and quite good genetic resources (Ronald et al. 1992; Matsumura et al. 1999; Goff et al. 2002). Corn has had excellent genetic resources for many years (Gierl and Saedler 1989), but has a very large genome. Comparative mapping studies (Ahn et al. 1993) should allow translation of genetic data from the more tractable systems to genetically complex forage grasses. The beauty of the EST approach is that its success is essentially independent of genomic complexity.

3. TRANSGENIC ALTERATION OF FORAGE QUALITY TRAITS

The targeted modification of biochemical pathways for forage crop improvement requires knowledge of the pathways themselves at the enzymatic and underlying genetic levels. In some cases, such as that of the lignin pathway, this knowledge is available, and successes have already been reported. In other cases, such as the tannins and saponins, there is still a need for basic gene discovery, making these pathways prime candidates for the genomics approach. The following sections briefly review the current status of research aimed at forage improvement by modifying lignin, tannins or triterpene saponins.

3.1 Lignin

Lignin is a major structural component of secondarily thickened plant cell walls. It is a complex polymer of hydroxylated and methoxylated phenylpropane units, linked via oxidative coupling (Boudet et al. 1995). Because of the negative effects of lignin on forage quality, there is considerable interest in genetic manipulation to alter the quantity and/or quality of the lignin polymer (Dixon et al. 1996). At the same time, lignin is important for stem rigidity and hydrophobicity of vascular elements, and, particularly in cereal crops, may be an important inducible defensive barrier against fungal pathogen attack (Beardmore et al. 1983). Thus, lignin modification must not compromise basic functions for the plant and thereby result in negative traits such as lodging or disease susceptibility. Potential improvements to forage quality associated with a reduction in lignin content, or changes in lignin quality, are summarized in Table 1.

Table 1. Potential benefits of transgenic alfalfa with improved cell wall digestibility

- Increased energy from forage
  - Dietary fiber is required for rumen health; increased digestibility of this fiber will result in more energy for milk/beef production.
  - Fiber digestibility will likely become a major limiting factor in further increasing milk production in the U.S.
- Increased milk/beef production potential
  - USDFRC estimates that a 10% increase in fiber digestibility would result in an annual $350 million increase in milk/beef production.
- Decreased generation of manure
  - USDFRC estimates that a 10% increase in fiber digestibility = 2.8 million tons decrease in manure solids produced each year.

Dicotyledonous angiosperm lignins contain two major monomer species. Guaiacyl (G) units have single hydroxyl and methoxyl substituents on the aromatic ring and arise from coniferyl alcohol, whereas syringyl (S) units are di-methoxylated and arise from sinapyl alcohol (Figure 2). Lignin from grasses also contains a significant proportion of H units derived from coumaryl alcohol (single hydroxyl substitiuent). The monomeric units in lignin are joined through more than five different types of linkages (Davin and Lewis 1992), and polymerization proceeds via free radical reactions believed to be initiated by the enzymes peroxidase and laccase. Suprisingly, the mechanisms that determine the relative proportions of the linkage types in a particular lignin polymer are currently unknown, as is the degree of ordered structure within the polymer. Equally important for the lignin:digestibility relationship may be the extent of cross-linking of the lignin to cell wall polysaccharides. Again, this process is not fully understood.

Current models for the biosynthetic pathways leading to formation of the H, G and S monolignols have been reviewed elsewhere (Humphreys and Chapple 2002). A summary of the currently accepted view of the potential routes to monolignols is given in Figure 2. There is significant debate as to whether the pathways are linear or proceed via a complex "metabolic grid" (Dixon et al. 2001).

To date, attempts to genetically modify lignin in forage crops have targeted only three enzymes of the monolignol pathway, caffeic acid 3-*O*-methyltransferase (COMT), caffeoyl CoA 3-*O*-methyltransferase (CCoAOMT) and cinnamyl alcohol dehydrogenase (CAD). These reactions are shown in Figure 2.

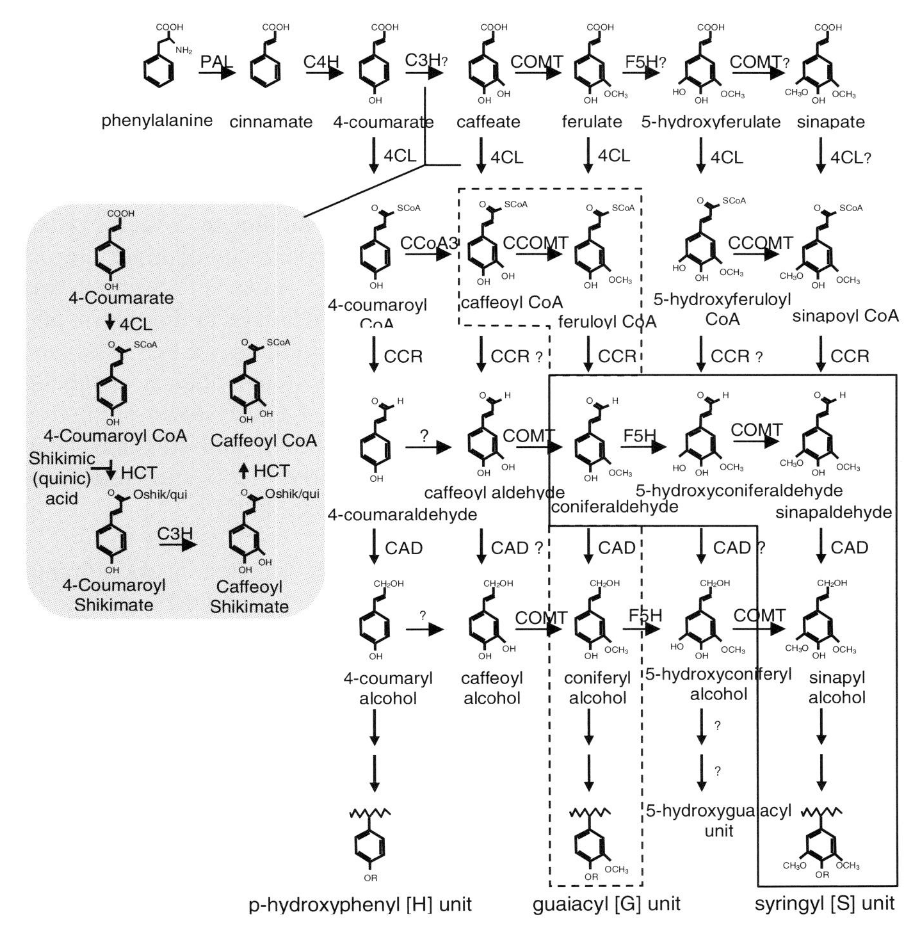

Figure 2. Schematic pathway for the biosynthesis of monomeric constituents of lignin. 4-Hydroxyphenyl units are present at higher levels in lignin from monocots than from dicots.

Constitutive cauliflower mosaic virus 35S promoter-driven antisense reduction of COMT to less than 5% of wild-type values in the tropical pasture legume *Stylosanthes humilis* resulted in no apparent reduction in lignin levels but in a strong reduction in S lignin based on histochemical analysis (Rae et al. 2001). *In vitro* digestibility of stem material in rumen fluid was increased by up to 10% in the transgenic plants exhibiting strongest COMT down-regulation. Up to 30% decreases in Klason lignin levels were observed in transgenic alfalfa in which COMT down-regulation was targeted using the vascular-tissue specific bean *PAL2* promoter, although acetyl bromide soluble lignin was not reduced. Use of this promoter resulted in near total down-regulation of COMT transcripts and protein (Guo et al. 2000) whereas earlier attempts at COMT down-regulation in alfalfa using the 35S promoter were less effective (V. Sewalt and RAD, unpublished results). COMT down-regulation in alfalfa was shown to lead to a loss of S residues in both the ß-*O*-4-linked uncondensed fraction (the major fraction in most lignins) and in the condensed fraction resolved as a range of differently linked dimers (Guo et al. 2000). The effect of COMT down-regulation on S lignin therefore likely reflects a true metabolic reduction in S units, rather than a change in lignin composition resulting in appearance of more S units in the non-condensed fraction of the polymer. Loss of S-units was accompanied by appearance of 5-hydroxyguaiacyl residues in the lignin, and the presence of these residues, and their linkage to yield novel benzodioxane units, has been confirmed by the use of 2-dimensional nuclear magnetic resonance techniques (Marita et al. 2002). Thus, COMT down-regulation results in a striking alteration in lignin composition, and this has been confirmed in several different species, including both dicots and monocots (Jouanin et al. 2000; Ralph et al. 2001; Piquemal et al. 2002).

It has been proposed that the first methylation reaction in monolignol biosynthesis is catalyzed by CCoAOMT (Figure 2), with COMT catalyzing the second methylation to yield syringyl units (Zhong et al. 1998). However, near elimination of CCoAOMT activity reduced G lignin by up to 50% in some alfalfa lines, but had no effect on S lignin (Guo et al. 2000). This suggests that the 3-*O*-methylation reactions in G and S lignin biosynthesis in alfalfa might occur by different routes.

Analysis of in rumen digestibility of transgenic alfalfa in fistulated steers revealed that down-regulation of either COMT or CCoAOMT resulted in significant improvements in digestibility (Guo et al. 2001). Particularly striking was the observation that the digestion kinetics of forage from CCoAOMT down-regulated plants were biphasic, with digestion continuing beyond the time when it had ceased for forage from wild-type and COMT down-regulated plants. The OMT down-regulated lines have been crossed with an elite commercial alfalfa cultivar, and the improved digestibility trait has been shown to hold up in large-scale field trials in Idaho, Wisconsin and Indiana. Currently, attempts are in progress to introduce the improved digestibility trait into a "Roundup-Ready" background for commercialization.

Antisense down-regulation of CAD in transgenic alfalfa to approximately 30% of wild-type level leads to a red coloration of the stem and a reduction in S/G ratio primarily due to a decrease in S units (Baucher et al. 1999). CAD down-regulated alfalfa was fed to cannulated sheep, and it was shown that the digestibility of one line grown in the greenhouse, and in two lines grown in the field, was slightly increased.

Although lignin composition and carbohydrate cross-linking differ between dicots and monocots, the above approaches for improving forage digestibility would appear to be equally effective in monocots. This could be predicted, because brown midrib mutants of corn have been known for many years, and one, the *bm3* mutant, has altered lignin content and composition as a result of a mutation in the *COMT* gene (Vignols et al. 1995). Antisense down-regulation of COMT in corn reproduces the brown midrib phenotype (Piquemal et al. 2002). Recent studies in transgenic tall fescue have indicated that digestibility can be significantly improved by down-regulation of either CAD or COMT (Chen et al. 2003). Interestingly, it has been shown that Bt corn plants expressing the *Bacillus thuringensis* toxin for insect control have a small but significant increase in lignin content (Saxena and Stotzky 2001). The molecular basis for this observation is not clear at present, nor is it known whether this effect would be seen in other forage crops harboring *Bt* genes.

3.2 Condensed Tannins

Condensed tannins (CTs, also known as proanthocyanidins) are phenolic polymers that bind to protein. They are synthesized by a branch of the flavonoid pathway (Figure 3). Although CTs occur in the fruits and seeds of many plants, they are either absent or present in very low amounts in many major forage sources such as alfalfa, white clover, corn silage, corn grain and soybean. The presence of CTs in the leaves of forage plants protects ruminant animals against pasture bloat and improves their nitrogen nutrition by increasing the amount of by-pass protein (dietary protein entering the small intestine from the rumen) (Broderick 1995; Aerts et al. 1999; Barry and McNabb 1999; Coulman et al. 2000; McMahon et al. 2000). In laboratory studies, treatment of feed proteins with modest amounts of tannins (around 2-4% of dry matter) reduced both proteolysis during ensiling and rumen fermentation. In studies performed with sheep in New Zealand (Douglas et al. 1999), increasing dietary tannin from trace amounts to 4% of dry matter increased by-pass protein, and a diet containing only 2% tannin strongly increased absorption of essential amino acids by the small intestine by up to 60%. Milk production of non-supplemented Holstein cows is significantly increased by tannins in birdsfoot trefoil (J. Grabber, USDFRC web site). These, and other advantages of the presence of low concentrations of tannins in forage crops, are listed in Table 2. At the same time, high concentrations of tannins can decrease palatability of forages, and can negatively impact nutritive value (Smulikowska et al. 2001).

Table 2. Potential benefits of transgenic forage crops with low (2-4% dry weight) levels of condensed tannins.

- Reduced rumen fermentation leading to a reduction in incidence of pasture bloat
- Reduction in methane gas emissions from ruminants
- Reduced protein degradation during ensiling
- Improved absorption of essential amino acids, leading to increased meat, milk and wool production
- Reduced excretion of soluble nitrogen in the urine
- Reduced mineralization of carbon and nitrogen in the soil

The problem for engineering tannins into tissues of a plant that do not normally make them is that the biosynthetic pathways specific for the formation of CTs are still poorly understood. Most progress on tannin biosynthesis and its regulation has been made in non-forage species, by using genetic approaches in barley (which accumulates low molecular weight proanthocyanidin polymers lacking (-)-epicatechin, and in *Arabidopsis thaliana*, where mutants impaired in CT production can be readily scored by their transparent seed testa (Shirley et al. 1995).

Mutations in the *BANYULS* (*BAN*) gene (named after the color of a French red wine) result in precocious accumulation of red anthocyanins (flower pigments) and loss of CTs in the Arabidopsis seed coat (Devic et al. 1999). On this basis, and the amino acid sequence similarity of BAN to that of dihydroflavonol reductase, an enzyme that catalyzes an earlier step in the flavonoid pathway (Figure 3), it was suggested that *BAN* encodes leucoanthocyanidin reductase (LAR) (Devic et al. 1999), an enzyme proposed to convert flavan-3,4-diols to 2,3-*trans*-flavan-3-ols such as (+)-catechin (Stafford and Lester 1984; Tanner and Kristiansen 1993), a "starter unit" for CT condensation (Figure 3). However, it has recently been shown that BANYULS from both Arabidopsis and *Medicago truncatula* is a novel anthocyanidin reductase that converts anthocyanin to the corresponding 2,3-*cis*-flavan-3-ol such as (-)-epicatechin (Xie et al. 2003). The CT from *Medicago* seed coat consists of 4→8 linked (-)-epicatechin residues with a (+)-catechin residue as “starter” (Koupai-Abyazani et al. 1993), a common structure among CTs. Thus, BAN activity may be involved in the biosynthesis of the repeating units in many CTs.

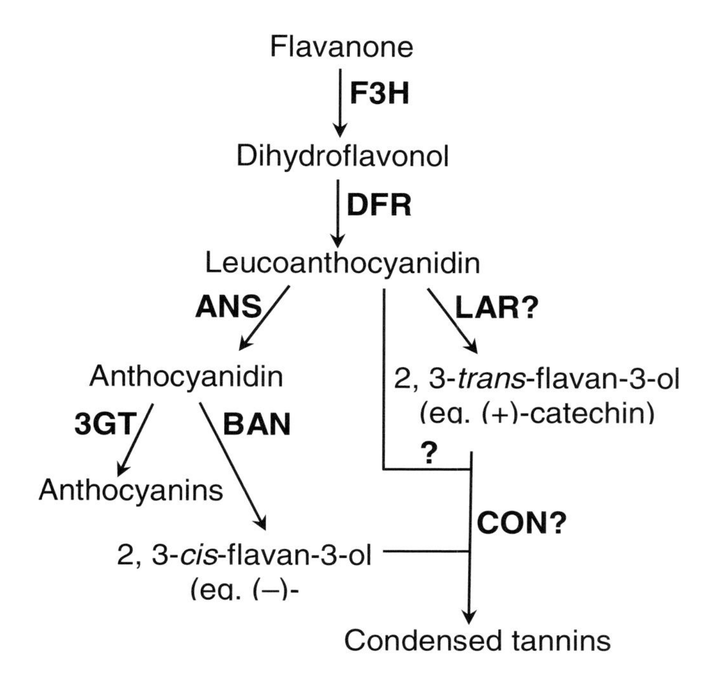

Figure 3. Schematic pathway for the biosynthesis of condensed tannins.

Introduction of *Medicago* BAN into transgenic tobacco resulted in a depletion of the pink anthocyanin pigmentation in the flowers, and accumulation of material that stained with the tannin-specific reagents dimethylaminocinnamaldehyde and butanol-HCl (Xie et al. 2002). This material appears to be a polymeric CT based on its behavior on cellulose and Sephadex LH20 chromatography. Thus, it appears possible to produce CTs in tobacco flowers by simple ectopic expression of the *BAN* gene. Preliminary evidence indicates that tobacco flowers may naturally produce CTs, although at very low levels, and it has proven possible, by ectopic expression of transcription factors, to increase CT production in leaves of species that naturally accumulate these compounds (Robbins et al. 2003). However, it is unlikely that engineering CTs into leaves of forage legumes such as alfalfa and white clover will be quite as simple. Formation of the two monomer types typical of alfalfa seed coat CT will require both BAN (for production of (-)-epicatechin) and a second enzyme for production of the (+)-catechin starter. This second enzyme might be a leucoanthocyanidin reductase (Tanner and Kristiansen 1993), or perhaps a form of BAN that produces the flavan-3-ol with the 2,3-*trans* stereochemistry of (+)-catechin. In Arabidopsis, a number of transcription factors (Nesi et al. 2001), as well as a multidrug resistance type transporter (Debeaujon et al. 2001), are required for accumulation of CTs in the seed coat. It is likely that tissues that do not naturally accumulate CTs will, at minimum, require a source of anthocyanin, enzymes for formation of flavan-3-ols such as catechin and epicatechin, and transporter proteins to move the monomeric units into the vacuole. Whether a specific enzyme is required for polymerization of the monomers has been debated for many years, but is yet to be resolved.

Production of a source of anthocyanin for CT synthesis in leaves is less of a problem than would at first sight appear. In fact, several forage legumes, such as white clover and barrel medic, contain an anthocyanin "spot" on the leaves, and the size of this spot appears to be under both genetic and environmental control. Furthermore, although anthocyanin biosynthesis requires many enzymes, it is possible to coordinately induce the pathway by ectopic expression of certain transcription factors such as the *PAP-1* gene of Arabidopsis (Borevitz et al. 2001) or *MYB* and *MYC* genes from corn (Grotewold et al. 1998). Availability of transgenic plants accumulating the monomers necessary for CT biosynthesis will provide a basis for discovery of the downstream genes necessary for CT assembly. Although over 35 years of biochemical studies have failed to provide an answer as to how CT assembly is regulated, genomics/bioinformatic approaches can provide sets of candidate genes that can be evaluated by either stable or transient expression in a genetic background producing the monomers.

Likely products of this technology in agriculture will include bloat-safe alfalfa (Coulman et al. 2000), which will also significantly reduce greenhouse gas emission from cattle [J. Lee, AgResearch New Zealand, media release, May 2002], have improved silage quality (Albrecht and Muck, 1991), and increase the efficiency of alfalfa protein utilization by dairy cows, leading to reduced urine-N losses to the environment and a decreased requirement for feeding of supplemental protein (Broderick 1995). CTs are also of considerable importance for human health and have been implicated in improving cardiovascular health and preventing urinary tract infections (Bagchi et al. 2000; Foo et al. 2000). They are also critical for flavor and astringency in wine, tea and other beverages.

3.3 Triterpene Saponins

Triterpene glycoside saponins are attracting increasing interest in view of their multiple biological activities (Table 3). These both positively and negatively impact plant traits, and can be divided into properties beneficial for plant protection, negatively impacting forage quality (Cheeke 1976; Oleszek 1996; Small 1996; Oleszek et al. 1999), or of biomedical significance. Poultry are particularly sensitive to triterpene saponins, and this limits the use of alfalfa as a poultry feed. Alfalfa could otherwise be a feed of choice, because it results in eggs with a rich golden yolk. Despite the obvious interest in facilitating or inhibiting production of triterpene saponins for crop improvement or development of pharmacological agents, most of the steps in their biosynthesis remain uncharacterized at the molecular level.

In many plant species, the triterpene saponins form a relatively complex class of molecules. Those from alfalfa (and soybean) have been studied for many years (Pedersen et al. 1967; Oleszek 1996). We have chosen barrel medic as a model in which to understand the biosynthesis of triterpene saponins, making use of the extensive genomics resources for discovery of both biosynthetic and regulatory genes. The idea is that a better understanding of the biosynthetic pathways and their control points will facilitate engineering to alter the content of saponins such that forage quality will be improved but the defensive functions of saponins for the plant will be in large part maintained. Time will tell whether this can be achieved.

Table 3. Biological activities of triterpene saponins

A. Functions in plant defense
- Allelopathic
- Antifungal
- Anti-insect

B. Properties impacting forage quality
- Toxic to monogastrics
- Anti-palatability
- Reduce forage digestibility

C. Pharmacological/biomedical activities
- Anti-cholesterol
- Anti-cancer (eg. avicins from *Acacia victoriae*)
- Adjuvant
- Hemolytic

The saponins of *M. truncatula* are glycosides of at least five different triterpene aglycones, soyasapogenol B, soyasapogenol E, medicagenic acid, hederagenin and bayogenin (Huhman and Sumner 2002). These aglycones are most likely derived from β-amyrin, a product of the cyclization of 2, 3-oxidosqualene (Figure 4). Oxidosqualene is a common intermediate in the biosynthesis of triterpenes and sterols. Mining of *M. truncatula* EST datasets, DNA array analyses, and use of clustering algorithms has led to the identification of genes encoding the first three enzymes of triterpene aglycone formation, squalene synthase (SS), squalene epoxidase (SE), and β-amyrin synthase (β-AS) (Suzuki et al., 2002), as well as a series of cytochrome P450 and glycosyltransferase genes that may be involved in the later stages of the pathway (L. Achnine and R.A. Dixon, unpublished results). The first three enzymes were functionally characterized by expression in *E. coli* or yeast (Suzuki et al. 2002). Important areas for future research on triterpene saponins for legume improvement and commercial exploitation include: obtaining an in depth understanding of their biosynthesis from initial cyclization to final conjugation; discovering regulatory genes for co-ordinated up-regulation of triterpene pathways; and utilizing transgenic approaches to learn more about triterpene function as a basis for genetic modification studies. As with lignin modification, it is critical that transgenic plants with altered saponin profiles are not compromised in disease or pest resistance. This is particularly important in view of the clear role of triterpene saponins as pre-formed inhibitors of fungal infection in oats (Papadopoulou et al. 1999).

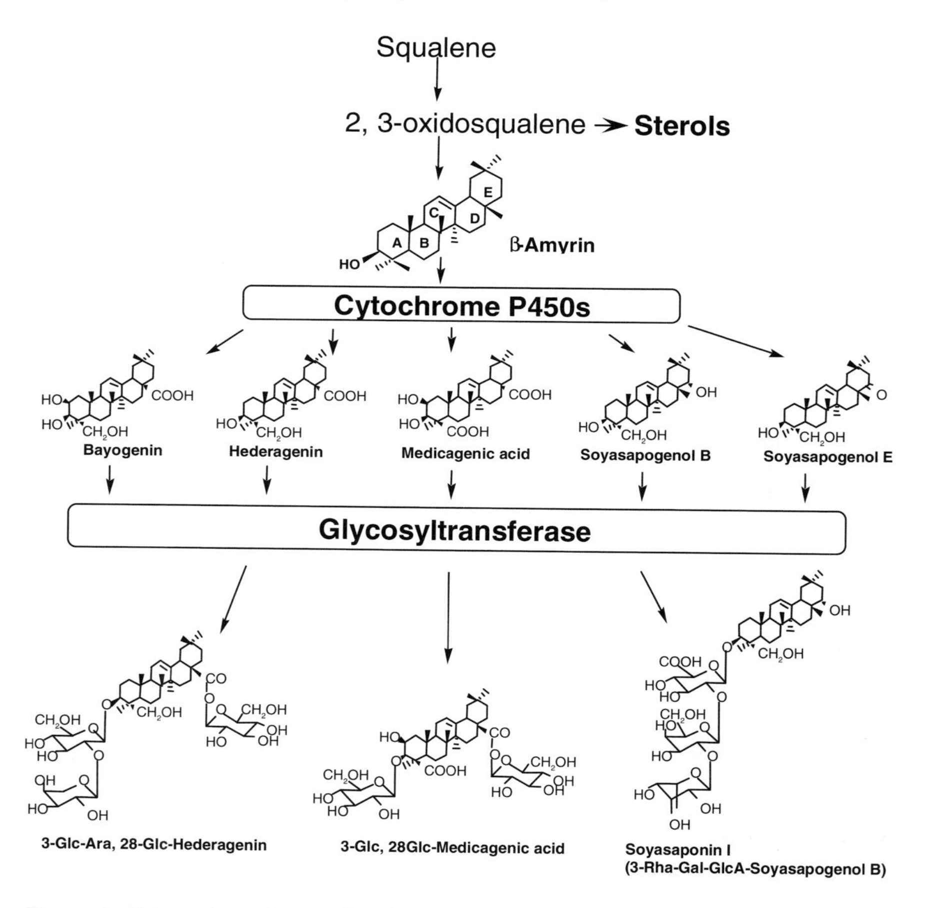

Figure 4. Schematic pathways for the biosynthesis of triterpene saponins in *Medicago truncatula*. Only three of the approximately 35 different saponin glycosides present in *M. truncatula* are shown.

4. IMPROVED FORAGES AND THE GMO DEBATE

The past 10 years has seen a major revolution in agriculture in the United States. It is remarkable that over 70 % of the soybean and cotton, and over 30% of the corn, grown in the US is now transgenic. This adoption of new technology has been faster than that associated with the "green revolution" in the latter half of the last century. However, the types of traits that have been introduced have been primarily limited to weed control using Roundup™ resistance, and insect control, using *Bacillus thuringensis* (BT) toxins, input traits that benefit the farmer and Biotech company, but not the consumer. Although this huge "field experiment" has yet to provide any evidence for negative health or serious environmental impacts, attitudes differ enormously between the US and Europe as to the acceptability of GMOs in agriculture. US citizens put considerable trust in their government with respect to its ability to regulate food safety, whereas, in Europe, with its recent BSE and foot and mouth disease outbreaks, government competency is less taken for granted, and the opinions of environmental pressure groups may be favored over those of government agencies.

Whatever the logic behind the above opinions, it is certainly true that the impact of agriculture on the everyday life of citizens is, overall, somewhat different in the US and Europe. With some exceptions, large-scale production agriculture generally takes place in the US in areas of relatively low population whereas, in Europe, agriculture IS the environment, taking place in the "countryside" in parallel with the recreational activities (hiking, camping) of a large percentage of the population. It is therefore perhaps ironical that the use of genetically modified crops can be so opposed in Europe when even the limited input traits developed so far, and tested so successfully in the US, bring environmental benefits. The factors outlined in Tables 1 and 2 present additional environmental benefits of genetically modified forage crops that are easy to understand by non-scientists (eg. reductions in manure, urine nitrogen and greenhouse gasses) and could be used to make a stronger case for agricultural biotechnology. Forage biotechnologists should lead the fight for acceptance of genetically modified crops by communicating the above described benefits to educators and policymakers worldwide.

ACKNOWLEDGEMENTS

The author thanks Dr Mark McCaslin for helpful discussions and the information in Table 1, and Dr Joe Bouton for advice and critical reading of the manuscript. The work described from the author's laboratory was conducted by an excellent group of postdoctoral fellows and research support staff including Dianjing Guo, Fang Chen, Parvathi Kota, De-Yu Xie, Shashi Sharma, Lahoucine Achnine, Hideyuki Suzuki and Jack Blount, and was supported by the Samuel Roberts Noble Foundation, the National Science Foundation, and Forage Genetics International.

REFERENCES

Aerts RJ, Barry TN, McNabb WC (1999) Polyphenols and agriculture: beneficial effects of proanthocyanidins in forages. Agric. Ecosystems Environ. 75: 1-12.

Ahn S, Anderson JA, Sorrells ME, Tanksley SD (1993) Homoeologous relationships of rice, wheat and maize chromosomes. Mol. Gen. Genet. 241: 483-490.

Albrecht KA, Muck RE (1991) Proteolysis in ensiled forage legumes that vary in tannin concentration. Crop Sci. 31: 464-469.

Ausubel FM (2002) Summaries of National Science Foundation-sponsored *Arabidopsis* 2010 projects and National Science Foundation-sponsored plant genome projects that are generating *Arabidopsis* resources for the communigy. Plant Physiol. 129: 394-437.

Azpiroz-Leehan R, Feldmann KA (1997) T-DNA insertion mutagenesis in Arabidopsis: going back and forth. Trends Genet. 13: 152-156.

Bagchi D, Bagchi M, Stohs SJ, Das DK, Ray SD, Kuszynski CA, Joshi SS, Pruess HG (2000) Free radicals and grape seed proanthocyanidn extract: importance in human health and disease prevention. Toxicology 148: 187-197.

Barry TN, McNabb WC (1999) The implications of condensed tannins on the nutritive value of temperate forages fed to ruminants. Brit. J. Nutr. 81: 263-272.

Baucher M, BernardVailhe MA, Chabbert B, Besle JM, Opsomer C, VanMontagu M, Botterman J (1999) Down-regulation of cinnamyl alcohol dehydrogenase in transgenic alfalfa (*Medicago sativa* L.) and the effect on lignin composition and digestibility. Plant Mol. Biol. 39: 437-447.

Baulcombe DC (1999) Fast forward genetics based on virus-induced gene silencing. Curr. Opinion Plant Biol. 2: 109-113.

Beardmore J, Ride JP, Granger JW (1983) Cellular lignification as a factor in the hypersensitive resistance of wheat to stem rust. Physiol. Plant Pathol. 22: 209-220.

Bell C, Dixon RA, Farmer AD, Flores R, Inman J, Gonzales RA, Harrison MJ, Paiva NL, Scott AD, Weller JW, May GD (2001) The *Medicago* genome initiative: a model legume database. Nucleic Acids Res. 29: 114-117.

Bevan M, Bancroft I, Mewes HW, Martienssen R, McCombie R (1999) Clearing a path through the jungle: progress in Arabidopsis genomics. Bioessays 21: 110-120.

Borevitz J, Xia Y, Blount JW, Dixon RA, Lamb C (2001) Activation tagging identifies a conserved MYB regulator of phenylpropanoid biosynthesis. Plant Cell 12: 2383-2393.

Boudet AM, Lapierre C, Grima-Pettenati J (1995) Tansley review No. 80. Biochemistry and molecular biology of lignification. New Phytologist 129: 203-236.

Broderick GA (1995) Desirable characteristics of forage legumes for improving protein utilization in ruminants. J. Animal Sci. 73: 2760-2773.

Cheeke PR (1976) Nutritional and physiological properties of saponins. Nutr. Rep. Int. 13: 315-324.

Chen L, Auh C, Dowling P, Bell J, Chen F, Hopkins A, Dixon RA, Wang ZY (2003) Improved forage digestibility of tall fescue (*Festuca arundinacea*) by transgenic down-regulation of cinnamyl alcohol dehydrogenase. Plant Biotechnol. J. (in press).

Cook D, R. (1999) *Medicago truncatula* - a model in the making! Curr. Opinion Plant Biol. 2: 301-304.

Coulman B, Goplen B, Majak W, McAllister T, Cheng KJ, Berg B, Hall J, McCartney D, Acharya S (2000) A review of the development of a bloat-reduced alfalfa cultivar. Can. J. Plant Sci. 80: 487-491.

Davin LB, Lewis NG (1992) Phenylpropanoid metabolism: biosynthesis of monolignols, lignans and neolignans, lignins and suberins. Rec. Adv. Phytochem. 26: 325-375.

Debeaujon I, Peeters AJM, Leon-Kloosterziel KM, Korneef M (2001) The *TRANSPARENT TESTA 12* gene of Arabidopsis encodes a multidrug secondary transporter-like protein required for flavonoid sequestration in vacuoles of the seed coat endothelium. Plant Cell 13: 853-871.

Delseny M, Salses J, Cooke R, Sallaud C, Regad F, Lagoda P, Guiderdoni E, Ventelon M, Brugidou C, Bhesquière (2001) Rice genomics: Present and future. Plant Physiol. Biochem. 39: 323-334.

Devic M, Guilleminot J, Debeaujon I, Bechtold N, Bensaude E, Koornneef M, Pelletier G, Delseny M (1999) The BANYULS gene encodes a DFR-like protein and is a marker of early seed coat development. Plant J. 19: 387-398.

Dixon RA, Chen F, Guo D, Parvathi K (2001) The biosynthesis of monolignols: a "metabolic grid", or independent pathways to guaiacyl and syringyl units? Phytochemistry 57: 1069-1084.

Dixon RA, Lamb CJ, Masoud S, Sewalt VJH, Paiva NL (1996) Metabolic engineering: prospects for crop improvement through the genetic manipulation of phenylpropanoid biosynthesis and defense responses- a review. Gene 179: 61-71.

Douglas GB, Stienezen M, Waghorn GC, Foote AG, Purchas RW (1999) Effect of condensed tannins in birdsfoot trefoil (*Lotus corniculatus*) and sulla (*Hedysarum coronarium*) on body weight, carcass fat depth, and wool growth of lambs in New Zealand. New Zealand J. Agric. Res. 42: 55-64.

Fitzmaurice WP, Lehman LJ, Nguyen LV, Thompson WF, Wernsman EA, Conkling MA (1992) Development and characterization of a generalized gene tagging system for higher plants using an engineered maize transposon *Ac*. Plant Mol. Biol. 20: 177-198.

Foo LY, Lu Y, Howell AB, Vorsa N (2000) The structure of cranberry proanthocyanidins which inhibit adherence of uropathogenic P-fimbriated *Escherichia coli* in vitro. Phytochemistry 54: 173-181.

Gierl A, Saedler H (1989) Maize transposable elements. Ann. Rev. Genet. 23: 71-85.

Goff SA, al e (2002) A draft sequence of the rice genome (*Oryza sativa* L. ssp *japonica*). Science 296: 92-100.

Grotewold E, Chamberlin M, Snook M, Siame B, Butler L, Swenson J, Maddock S, St. Clair G, Bowen B (1998) Engineering secondary metabolism in maize cells by ectopic expression of transcription factors. Plant Cell 10: 721-749.

Guo D, Chen F, Inoue K, Blount JW, Dixon RA (2000) Down-regulation of caffeic acid 3-*O*-methyltransferase and caffeoyl CoA 3-*O*-methyltransferase in transgenic alfalfa (*Medicago sativa* L.): impacts on lignin structure and implications for the biosynthesis of G and S lignin. Plant Cell 13: 73-88.

Guo D, Chen F, Wheeler J, Winder J, Selman S, Peterson M, Dixon RA (2001) Improvement of in-rumen digestibility of alfalfa forage by genetic manipulation of lignin O-methyltransferases. Transgenic Res. 10: 457-464.

Huhman DV, Sumner LW (2002) Metabolic profiling of saponin glycosides in *Medicago sativa* and *Medicago truncatula* using HPLC coupled to an electrospary ion-trap mass spectrometer. Phytochemistry 59: 347-360.

Humphreys JM, Chapple C (2002) Rewriting the lignin roadmap. Curr. Opinion Plant Biol. 5: 224-229.

Jouanin L, Goujon T, deNadai V, Martin MT, Mila I, Vallet C, Pollet B, Yoshinaga A, Chabbert B, PetitConil M, Lapierre C (2000) Lignification in transgenic poplars with extremely reduced caffeic acid O-methyltransferase activity. Plant Physiol. 123: 1363-1373.

Koupai-Abyazani MR, McCallum J, Muir AD, Lees GL, Bohm BA, Towers GHN, Gruber MY (1993) Purification and characterization of a proanthocyanidin polymer from seed of alfalfa (*Medicago sativa* cv. beaver). J. Agric. Food Chem. 41: 565-569.

Marita JM, Ralph J, Hatfield RD, Guo D, Chen F, Dixon RA (2002) Structural and compositional modifications in lignin of transgenic alfalfa down-regulated in caffeic acid 3-O-methyltransferase and caffeoyl CoA 3-O-methyltransferase. Phytochemistry 62: 53-65.

Matsumura H, Nirasawa S, Terauchi R (1999) Transcript profiling in rice (*Oryza sativa* L.) seedlings using serial analysis of gene expression (SAGE). Plant J. 20: 719-726.

McMahon LR, McAllister TA, Berg BP, Majak W, Acharya SN, Popp JD, Coulman BE, Wang Y, Cheng KJ (2000) A review of the effects of forage condensed tannins on ruminal fermentation and bloat in grazing cattle. Can. J. Plant Sci. 80: 469-485.

Nesi N, Jond C, Debeaujon I, Caboche M, Lepiniec L (2001) The Arabidopsis TT2 gene encodes an R2R3 MYB domain protein that acts as a key determinant for proanthocyanidin accumulation in developing seed. Plant Cell 13: 2099-2114.

Oldroyd GE, Geurts R (2001) *Medicago truncatula*, going where no plant has gone before. Trends Plant Sci. 6: 552-554.

Oleszek W, ed (1996) Alfalfa saponins: structure, biological activity, and chemotaxonomy. Plenum Press, New York.

Oleszek W, Junkuszew M, Stochmal A (1999) Determination and toxicity of saponins from *Amaranthus cruentus* seeds. J. Agric. Food Chem. 47: 3685-3687.

Papadopoulou K, Melton RE, Leggett M, Daniels MJ, Osbourn AE (1999) Compromised disease resistance in saponin-deficient plants. Proc. Natl. Acad. Sci. USA 96: 12923-12928

Paterson AH, Lander ES, Hewitt JD, Peterson S, Lincoln SE, Tanksley SD (1988) Resolution of quantitative traits into mendelian factors by using a complete linkage map of restriction fragment length polymorphisms. Nature 335: 721-726.

Pedersen MW, Zimmer DE, McAllister DR, Anderson JO (1967) Comparative stduies of saponin of several alfalfa varieties using chemical and biochemical assays. Crop Sci. 7: 349-352.

Piquemal J, Chamayou S, Nadaud I, Beckert M, Barriere Y, Mila I, Lapierre C, Rigau J, Puigdomenech P, Jauneau A, Digonnet C, Boudet A-M, Goffner D, Pichon M (2002) Down-regulation of caffeic acid *O*-methyltransferase in maize revisited using a transgenic approach. Plant Physiol. 130: 1675-1685.

Quackenbush J, Liang F, Holt I, Pertea G, Upton J (2000) The TIGR gene indices: reconstruction and representation of expressed gene sequences. Nucleic Acids Res. 28: 141-145.

Rae AL, Manners JM, Jones RJ, McIntyre CL, Lu DY (2001) Antisense suppression of the lignin biosynthetic enzyme, caffeate O-methyltransferase, improves in vitro digestibility of the tropical pasture legume, *Stylosanthes humilis*. Australian J. Plant Physiol. 28: 289-297.

Ralph J, Lapierre C, Lu FC, Marita JM, Pilate G, VanDoorsselaere J, Boerjan W, Jouanin L (2001) NMR evidence for benzodioxane structures resulting from incorporation of 5-hydroxyconiferyl alcohol into lignins of O-methyltransferase-deficient poplars. J. Agric. Food Chem. 49: 86-91.

Robbins MP, Paolocci F, Hughes J-W, Turchetti V, Allison G, Arcioni S, Morris P, Damiani F (2003) *Sn,* a maize bHLH gene, modulates anthocyanin and condensed tannin pathways in *Lotus corniculatus*. J. Exp. Bot. 54: 239-248.

Ronald PC, Albano B, Tabien R, Abenes L, Wu KS, McCouch S, Tanksley SD (1992) Genetic and physical analysis of the rice blight disease resistance locus, Xa21. Mol. Gen. Genet. 236: 113-120.

Saxena D, Stotzky G (2001) Bt corn has a higher lignin content than non-Bt corn. Amer. J. Bot. 88: 1704-1706.

Shirley BW, Kubasek WL, Storz G, Bruggemann E, Koornneef M, Ausubel FM, Goodman HM (1995) Analysis of Arabidopsis mutants deficient in flavonoid biosynthesis. Plant J. 8: 659-671.

Small E (1996) Adaptations to herbivory in alfalfa (*Medicago sativa*). Can. J. Bot. 74: 807-822.

Smulikowska S, Pastuszewska B, Swiech E, Ochtabinska A, Mieczkowska A, Nguyen VC, Buraczewska L (2001) Tannin content affects negatively nutritive value of pea for monogastrics. J. Animal Feed Sci. 10: 511-523.

Stafford HA, Lester HH (1984) Flavan-3-ol biosynthesis. The conversion of (+)-dihydroquercetin and flavan-3,4-*cis*-diol (leucocyanidin) to (+)-catechin by reductases extracted from cell suspension cultures of Douglas fir. Plant Physiol. 76: 184-186.

Suzuki H, Achnine L, Xu R, Matsuda SPT, Dixon RA (2002) A genomics approach to the early stages of triterpene saponin biosynthesis in *Medicago truncatula*. Plant J. 32: 1033-1048.

Tanner GJ, Kristiansen KN (1993) Synthesis of 3,4-cis[3H]Leucocyanidin and enzymatic reduction to catechin. Anal. Biochem. 209: 274-277.

Vignols F, Rigau J, Torres MA, Capellades M, Puigdoménech P (1995) The *brown midrib3 (bm3)* mutation in maize occurs in the gene encoding caffeic acid O-methyltransferase. Plant Cell 7: 407-416.

Weigel D, Ahn JH, Blazquez MA, Borewitz J, Christensen SK, Fankhauser C, Ferrandiz C, Kardailsky I, Malanchaurvil EJ, Neff MM, Nguyen JT, Sato S, Wang Z-H, Xia Y, Dison RA, Harrison MJ, Lamb CJ, Yanofsky MF, Chory J (2000) Activation tagging in Arabidopsis. Plant Physiol. 122: 1003-1013.

Wu SH, Ramonell K, Gollub J, Somerville S (2001) Plant gene expression profiling with DNA microarrays. Plant Physiol. Biochem. 39: 917-926.

Xie D, Sharma SR, Paiva NL, Ferreira D, Dixon RA (2003) Role of anthocyanidin reductase, encoded by *BANYULS* in plant flavonoid biosynthesis. Science 299: 396-399.

Zhong R, Morrison WH, Negrel J, Ye ZH (1998) Dual methylation pathways in lignin biosynthesis. Plant Cell 10: 2033-2045.

Molecular Breeding and Functional Genomics for Tolerance to Biotic Stress

M. Fujimori[1], K. Hayashi[3], M. Hirata[2], S. Ikeda[2], Y.Takahashi[2], Y. Mano[1], H. Sato[1], T. Takamizo[1], K. Mizuno[3], T. Fujiwara[3] and S. Sugita[1]
[1]National Institute of Livestock and Grassland Science, Nishinasuno, Tochigi 329-2793 Japan. [2]Japan Grassland Farming and Forage Seed Association, Nishinasuno, Tochigi 329-2742 Japan. [3]Yamaguchi Agricultural Experiment Station, Oouchi-mihori, Yamaguchi 753-0214, Japan. (Email: masafuji@affrc.go.jp).

Key words: resistance to biotic stress, crown rust, ryegrass, resistance gene, DNA markers, gene isolation

Abstract:

Resistance to biotic stress is one of the most important targets in the improvement of forage and turf grass. Resistance to crown rust in Italian ryegrass is an attractive target of molecular analysis, including linkage analysis and gene isolation, because of its importance in forage and turf grasses. To analyze the major resistance gene in the resistant line 'Yamaiku 130', we performed bulked segregant analysis using amplified fragment-length polymorphism (AFLP) in an F1 population segregated at a 1:1 ratio of resistant to susceptible. We constructed a linkage map of regions flanking the resistance gene locus, designated as Pc1. Three AFLP markers were tightly linked to Pc1 with a map distance of 0.9 cM, and 3 AFLP markers were on the opposite side with a distance of 1.8 cM. ATC-CATG153 co-segregated with Pc1. We performed linkage analysis using DNA markers tightly linked to Pc1 in an F1 population derived from the Italian ryegrass cv. 'Harukaze'. Another resistance gene, designated as Pc2, was identified. Gene isolation of Pc1, using a map-based technique, and the identification of other resistance genes are in progress. Identification of both the DNA markers tightly linked to resistance genes and the plant materials carrying the resistance gene will open new strategies for the development of resistance varieties in Italian ryegrass and related species.

A. Hopkins, Z.Y. Wang, R. Mian, M. Sledge and R.E. Barker (eds.), Molecular Breeding of Forage and Turf, 21-35.

1. INTRODUCTION

Biotic stress is caused by many factors, including infection with fungi, bacteria, viruses and nematodes, and competition with weeds. Biotic stress causes serious losses in the yield and quality of forage and turf grasses, and its control is therefore very important. The use of fungicides and insecticides is limited in forage grass because of their cost and considerations of safety to humans and the environment. One of the most efficient strategies for controlling diseases and pests is the development of resistant varieties. A potential grass variety must contain genes for resistance against biotic stress.

Plants have many mechanisms for protecting themselves against biotic stress. Because plants cannot move to escape from the stress, they have developed defensive strategies, such as hypersensitive reaction and the production of phytoalexins. Much attention has been directed at the isolation of genes for resistance to biotic stress and the analysis of molecular mechanisms of resistance. It has become clear that common genes trigger resistance networks and control resistance to biotic stress (Takken and Joosten 2000).

Resistance against biotic stress is one of the most important targets in the improvement of varieties of forage and turf grass, and many resistant varieties have been bred around the world. On the other hand, our knowledge of the molecular mechanisms of resistance to biotic stress in forage and turf grasses is limited. Here, we summarize the main achievements in biotic stress resistance in plants and also describe our work on crown rust resistance in Italian ryegrass (*Lolium multiflorum* Lam.). We focus on the possibility of using molecular techniques to develop varieties with durable resistance to biotic stress.

2. BIOTIC STRESS IN PLANTS

2.1 Resistance to Biotic Stress

Plants frequently encounter many potential pathogens. However, a limited number of these pathogens can infect plants and cause disease. A large amount of work has been done in this field. The gene-for-gene hypothesis has been demonstrated by molecular evidence (reviewed by Takken and Joosten 2000; Bonas and Lahaye 2002). A resistance gene (R gene) recognizes an elicitor produced by the Avr gene of the pathogen and activates the plant defense network, which includes oxidative burst, ion fluxes, cross-linking and strengthening of the plant cell wall, production of anti-microbial compounds, and induction of pathogenesis-related proteins (Hammond-Kosack and Jones 1996; Sudha and Ravishankar 2002).

2.2 Resistance Genes to Biotic Stress in Plants

More than 30 resistance genes have been isolated in several species, and these have been classified into 8 distinct structural categories by Hulbert et al. (2001). Recently, several novel types of resistance genes were reported, such as tomato Ve gene (Kawchuk et al., 2001), Arabidopsis RRS1-R gene (Lahaye, 2002), and barley Rpg1 gene (Brueggeman et al., 2002). The resistance genes consist of common motifs such as a nucleotide binding site (NBS), leucine-rich repeats (LRRs), kinase, coiled-coil domain (CC), Toll/interleukin-1-receptor (TIR), and transmembrane domain (Hammond-Kosack and Jones 1997). The functions of these motifs are not completely clear. All R genes without Pto contain LRRs. The LRRs are hypervariable and probably act as receptors for Avr factors produced by pathogens. An extensive review of the LRR domain has been conducted by Jones and Jones (1997), and the model of specific recognition of R genes has been reviewed by Robert et al. (1998). The majority of R genes encode an N-terminal NBS and a C-terminal LRR region. This NBS–LRR of R genes confers resistance to bacteria, viruses, fungi, nematodes, and insects (Baker et al. 1997; Rossi et al. 1998), suggesting that a common mechanism in the form of a gene-for-gene relationship is present.

2.3 Resistance Gene Loci in Plants

Genetic and molecular studies have demonstrated that R genes are frequently clustered in the genome (reviewed by Michelmore and Meyers 1998). The flax M locus (Anderson et al. 1997), the lettuce Dm3 locus (Anderson et al. 1996), the rice Xa21 locus (Song et al. 1997), the tomato I2 locus (Simons et al. 1998), and the tomato Cf 4/9 locus (Takken et al. 1999) contain multiple R genes. High rates of both unequal crossing-over and gene conversion occur in the Rp1 loci of maize (Sudupak et al. 1993; Richter et al. 1995). This is thought to facilitate the generation of novel R-gene specificity against pathogens.

R loci consist of genetically separable recognition specificities, such as at least 14 specificities in the maize Rp1 locus (Hulbert 1993). The tomato Cf2 locus contains two R genes conferring resistance to tomato leaf mould (Dixon et al. 1996). The tomato CF4 locus confers resistance to tomato leaf mould, and this resistance is conferred by two distinct R genes, Cf-4 and Hcr9-4E, which recognize Avr4 and Avr4E, respectively (Takken et al. 1999). In these cases the genes at one R locus confer resistance to several races of pathogen. The Mi gene in tomato confers resistance to both aphids and nematodes (Rossi et al. 1998). Two R genes at a single R locus in potato confer resistance to distinct pathogens such as potato virus X and nematodes (Van der Vossen et al. 2000). This indicates that mapping of an R gene to one pathogen is potentially helpful for the mapping of R loci against other pathogens.

Meyers et al. (1999) estimated that Arabidopsis and rice contain approximately 200 and 750–1550 R genes, respectively, and that the number of NBS sequences per cluster in Arabidopsis ranges from 2 to 18, with an average of 4.9. Bai et al. (2002) estimated that there are more than 600 NBS-LRR-type genes in the rice genome. These data suggest that more than 600 R genes are present in the genomes of forage grasses.

3. CROWN RUST RESISTANCE IN ITALIAN RYEGRASS

3.1 Genetic Analysis of Crown Rust Resistance in Ryegrass

Crown rust, which is caused by *Puccinia coronata f. sp. lolii*, is a serious disease of ryegrasses such as Italian ryegrass and perennial ryegrass (*Lolium perenne* L.) worldwide, and can cause severe losses in yield and quality (Potter et al. 1987;, Plummer et al. 1990). Resistance to crown rust is a very important trait for forage and turf grasses, because the broad host range of *P. coronata* includes the genera *Lolium, Festuca, Agropyron, Agrostis, Paspalum, Phleum, Poa* and *Puccinellia* (Smiley et al. 1992). Furthermore, crown rust is a new disease in barley (Jin and Steffenson 2002). Therefore, the molecular analysis of resistance genes to crown rust is an important research target.

Development of a new variety takes more than 10 years. If the new variety is then overcome by a new race of pathogen, this effort and time are wasted. Therefore, pyramiding several R genes in a variety is important in preventing breakdowns in disease resistance. Many varieties with resistance to crown rust have been developed in ryegrass. However, susceptible individuals still exist in these resistant varieties, and higher resistance is hoped for. Genetic analysis of crown rust resistance has been carried out (reviewed by Kimbeng 1999). Resistance to crown rust is conferred by both major and minor resistance genes (Wilkins 1975), and a cytoplasmic effect has been observed by Adams et al. (2000). However, information on the map positions of linkage groups and specificity for disease races is still lacking.

3.2 Molecular Analysis of Crown Rust Resistance in Ryegrass

Molecular analysis, such as linkage analysis, has progressed more slowly in grasses than in cereal crops. However, the recent use of advanced techniques, including amplified fragment-length polymorphisms (AFLP) and simple sequence repeats (SSRs) has facilitated the development of linkage maps in ryegrass. Hayward et al. (1998) constructed a genetic linkage map of Lolium using isozyme, restriction fragment-length polymorphism (RFLP), and random amplified polymorphic DNA (RAPD) markers. AFLP markers have been used to construct linkage maps of perennial ryegrass (Bert et al. 1999) and Italian ryegrass (Fujimori et al. 2000). Reference maps have been constructed using RFLP, AFLP, and SSR markers (Jones et al. 2002a, Jones et al. 2002b). Hirata et al. (2000) constructed an Italian ryegrass linkage map using SSR markers. AFLP and SSR enable the easy construction of linkage maps in ryegrass.

Although linkage analysis has become an easy technique, it is still time consuming and expensive. Therefore, appropriate selection of traits for linkage analysis is important. Crown rust resistance is a suitable trait for linkage analysis because of its importance in variety development, easy and precise phenotyping of F1 population for linkage analysis, and the presence of various resistant varieties. Molecular analysis of crown rust resistance has been carried out in ryegrass. QTLs of crown rust resistance in perennial ryegrass have been identified (Barre et al. 2000; Roderick et al. 2002). Dumsday et al. (2003) mapped major effect loci on linkage group 2 in perennial ryegrass.

We demonstrated the presence of a major resistance gene in the highly resistant breeding line Yamaiku 130 at a molecular level (Fujimori et al. 2003). Linkage analysis for a major gene was applied to a population segregated at a 1:1 ratio of resistant to susceptible. We constructed a high-density linkage map of the regions flanking the resistance gene locus, designated as Pc1, using 34 AFLP markers covering a total distance of 36 cM with an average distance of 1.1 cM. Three AFLP markers were tightly linked to Pc1 with a map distance of 0.9 cM, and 3 AFLP markers were on the opposite side with a distance of 1.8 cM. ATC-CATG153 co-segregated with Pc1. Hirata et al. (2003) demonstrated another major resistance gene, designated as Pc2, in Harukaze. Linkage analysis using DNA markers tightly linked to Pc1 was carried out. The result demonstrated that Pc1 and Pc2 are not linked. Linkage analysis using DNA markers tightly linked to known resistance genes allows the rapid identification of novel R loci. Linkage analysis of other resistance genes is in progress in our laboratory.

4. HOW TO USE DNA MARKERS LINKED TO DISEASE RESISTANCE IN BREEDING PROGRAMS

4.1 Identification of Novel R Loci and R Genes

To pyramid R genes in target variety, identification of several major or minor genes is needed. In the first step, identification of several R loci is essential. Linkage analysis using DNA markers tightly linked to known R loci such as Pc1 allows us to find novel R loci. Markers linked to a novel R locus can be mapped to the reference map of perennial ryegrass described by Jones et al. (2002a) and to that of Italian ryegrass described by Fujimori et al. (2000) to assign the gene to the ryegrass linkage map. The linkage map of Italian ryegrass is one of the most informative linkage maps of forage grasses. This map contains SSR markers developed by Hirata et al. (2000) and anchor probes developed by Inoue et al. (2002). Additionally, mapping of cleaved amplified polymorphic sequences (CAPS) and resistance gene analogue (RGA) to this map is in progress by Miura et al. (Japan Grassland Farming and Forage Seed Association).

Mapping of major resistance genes is useful for increasing our knowledge of resistance in at least 3 ways. First, major genes are more easily scored and mapped precisely on linkage maps than are minor genes, because the resolution of QTL analysis is too low to locate minor genes precisely. Welz and Geiger (2000) found that major and minor genes could be mapped at identical chromosomal positions, indicating that the R locus of major genes is potentially useful for the mapping of minor genes. Second, it has been demonstrated that some resistance genes to different pathogens can be mapped at a single locus or at several loci linked closely with each other (Rossi et al. 1998, van der Vossen et al. 2000). Thus, linkage analysis of major genes may allow us to increase our knowledge of the loci conferring resistance to diverse pathogens. Finally, linkage analysis of major genes is useful in the analysis of minor genes. The effects of minor genes may be masked by those of major genes in the linkage analysis. Therefore, major genes may have to be removed from a population for the linkage analysis of minor genes. In developing populations for linkage analysis, DNA markers tightly linked with major resistance genes are useful for selecting parents carrying only minor resistance genes.

Identification of the specificities of R genes to pathogens is important in the development of varieties with durable resistance. To determine the specificities of R genes, lines carrying single resistance genes in a homozygous state are needed as test lines for race identification. Development of these lines by conventional methods is not easy, because of the out-crossing habit of Italian ryegrass. However, DNA markers tightly linked to major resistance genes facilitates the development of test lines. Lines with known R genes in the homozygous state may be useful not only as test lines, but also as breeding material.

4.2 Introgression of R Genes to Related Species

Introgression of R genes from wild species is a basic technique for development of novel resistant materials in crops such as wheat and rice. A line carrying a single R gene in Italian ryegrass is potentially useful for transferring the R gene to a related species such as perennial ryegrass, meadow fescue (*Festuca pratensis*) and tall fescue (*Festuca arundinacea*). Resistance to crown rust in meadow fescue and tall fescue has been successfully transferred to Italian ryegrass, suggesting that the resistance gene in ryegrass is also useful in the *Festuca–Lolium* complex (Oertel and Matzk 1999). Moreover, the genome of *Lolium* is closely related to that of the *Festuca* species (Jauhar 1993), suggesting that information on Italian ryegrass R loci, including DNA markers and genetic position in the linkage group, is useful in the analysis of the same features in *Festuca* species.

4.3 Marker-Assisted Selection in Forage Grasses

To prevent inbreeding in forage grass breeding programs, many individuals have to be selected from a breeding population, so the population must be large. It might not be as easy to apply marker-assisted selection to the breeding of forage grasses as it is in the case of cereal crops. However, using markers can be helpful in small populations, for example, for introgression of a target gene to another species or for the development of lines carrying one R gene, as mentioned above. In these cases, we can easily use DNA markers to develop lines with target genes.

The presence of DNA markers tightly linked to the R gene locus enables plant breeders to monitor the frequency of R genes and the genetic diversity of R loci in the breeding populations. Information about the percentage of useful genes in the breeding population acts as a good index for breeders. Although uniformity in the target locus is important in the majority of traits such as quality, heading date, and yield, genetic diversity in R loci is potentially useful for stabilizing resistance to diverse diseases. Because pathogens can evolve rapidly, they can overcome single R genes easily. Therefore, high diversity of the R locus may be useful in stabilizing resistance to various pathogens. Wolfe (1985) reported that heterogeneity for disease resistance is useful in disease control. SSR markers may be suitable for analyzing the genetic diversity of the R locus in a breeding population, because of its features of locus specificity and high allele number.

5. ISOLATION OF GENES RELATED TO CROWN RUST RESISTANCE

5.1 Strategy of Gene Isolation in Italian Ryegrass

Several R genes have been isolated from various species by using map-based cloning and transposon-base gene tagging. Transposon tagging can be used to isolate genes in species with relatively large genomes, such as maize (2500 Mbp), whereas species with small genome size, such as in rice (430 Mb), facilitates map-based cloning of agronomically important traits. It is difficult to isolate genes from the majority of forage and turf grasses by using a map-based strategy, because many grasses have large genome sizes, are polyploid, and have out-crossing habits. However, ryegrass is a possible material for gene isolation, because it is diploid and has a smaller genome size (1600 Mbp) than, for example, barley (4800 Mbp). Buschges et al. (1997) cloned the barley Mlo gene against powdery mildew by using map-based cloning; this result suggests that map-based cloning is applicable to the isolation of genes from Italian ryegrass. Map-based gene isolation is one option for gene cloning and requires a high-resolution and reliable map of target traits. In self-pollinated species, to increase the reliability of evaluation of target traits, recombinant inbred lines or F3 lines are used for evaluation. However, in Italian ryegrass we have to use individuals for evaluation because of the plant's out-crossing habit. Nevertheless, Pc1 is an attractive target for gene isolation, because the resistance conferred by Pc1 can be evaluated precisely at an individual level.

Large regions of genomic colinearity have been demonstrated among grass species. Chen et al. (1997) demonstrated microcolinearity in the sh2-homologous regions of the maize, rice, and sorghum genomes. The use of information obtained from cereal crops such as rice and wheat relatives is important in ryegrass. Jones et al. (2002a) demonstrated a synteny relationship between perennial ryegrass and *Poaceae*. Inoue et al. (2002) also demonstrated synteny between Italian ryegrass and *Poaceae*. These results allow us to use information from cereal crops, indicating that it is possible that rice genome sequence data or advanced linkage map information on wheat relatives can be used for gene isolation and the development of markers in ryegrass. Microcolinearity is potentially useful for gene isolation in grass species. However, Leister et al. (1998) showed that RGAs were mapped at nonsyntenic locations among cereal species such as rice, barley, and foxtail millet. Han et al. (1999) tried to isolate the barley Rpg1 gene by using microsynteny with rice. They observed excellent synteny between the barley 7H chromosome short arm and the rice chromosome 6 short arm, and they obtained flanking markers tightly linked with Rpg1 in the syntenous position. However, the Rpg1 gene has not been observed in rice, indicating that microsynteny is not always useful for the isolation of R genes

5.2 Use of Resistance Gene Analogue as a DNA Marker for Isolating the Pc1 Gene

Bacterial artificial chromosomes (BACs) are most broadly used for gene isolation. We constructed a BAC library from genomic DNA isolated from a crown-rust-resistant individual carrying the Pc1 gene. The average insert size of the BAC library was 125 kbp, and it contained 115200 clones. In our estimation, the library would provide eight Italian ryegrass genome equivalents, indicating that it may be helpful in the isolation of the Pc1 gene. To isolate the Pc1 gene using a map-based technique, thousands of individuals may be needed for linkage analysis. Evaluation of phenotypes and genotyping in F1 individuals for linkage analysis requires a lot of work.

It has been demonstrated that R loci frequently contain several R genes and that they are highly polymorphic, indicating that RGAs are useful markers for identifying gene clusters in R loci. To find clusters of R genes at the Pc1 locus, we applied bulked segregant analysis and RGA polymorphisms in F1 individuals. PCR amplification with a degenerate primer permits the amplification of NBS-LRR-type R genes. An RGA marker co-segregated with Pc1 was obtained. This is the starting point of gene isolation for crown rust resistance.

Ikeda et al. (2002) have sequenced 12000 clones amplified from NBS-LRR-type RGAs by using primers designed from sequence motifs conserved among R genes. They obtained 79 unique sequences in Italian ryegrass and primer sets of 79 unique NBS-LRR sequences have been developed. It may be helpful to develop the RGA markers in ryegrass

5.3 Analysis of Gene Expression Using a Microarray System

The microarray technique is a powerful tool for the expression analysis of thousands of genes simultaneously. To elucidate the molecular events and genetic mechanisms involved in resistance against crown rust, we are using this technology to analyze the gene expression profile of the resistant line Yamaiku 130 after crown rust infection. We constructed three cDNA libraries using mRNA from Yamaiku 130 leaves inoculated with spores of crown rust to obtain a comprehensive set of genes associated with disease resistance (Fujimori et al. 2002). We randomly selected 9216 clones and used them for microarray analysis. A redundant 9216-cDNA microarray was used for a time-course gene expression analysis of the defense response following crown rust infection. Our results showed that 106 clones (non-redundant) were induced and 125 clones were repressed at least 1 time point by crown rust infection in Yamaiku 130. Many of the clones (32/106 and 58/125) were differently expressed 8 h after treatment. Some of the up-regulated or down-regulated clones included clones potentially involved in hypersensitivity reactions. Some of the clones included novel genes. This indicates that time-course profiling using microarray techniques provides valuable information on the molecular mechanisms of crown rust resistance in Italian ryegrass. Comparison of the changes in transcript levels among lines homozygous for single resistance genes will provide more useful information.

5.4 Transformation of Genes Related to Disease Resistance

Transformation has been used to confirm the functions of genes isolated from several species, and a good deal of work on transformation has been carried out in forage and turf grasses (Wang et al. 2001). Takahashi et al. (2002) have developed a transformation technique based on the particle gun method using the Italian ryegrass cv. 'Waseaoba', which is susceptible to crown rust, indicating that this system may be useful for confirming the function of the Pc1 gene.

Resistance can be found in other species; barriers to interspecific crosses frequently prevent the transfer of resistance by conventional breeding methods. Transformation systems can efficiently transfer R genes without interspecific barriers. Rommens et al. (1995) demonstrated that the tomato Pto gene confers resistance to *Pseudomonas syringae* in *Nicotiana benthamiana*. Whitham et al. (1996) demonstrated that the tobacco N gene confers resistance against tobacco mosaic virus in transgenic tomato. Tai et al. (1999) demonstrated that the pepper Bs2 is useful against bacterial spot disease in transgenic tomatoes. These data demonstrate that isolated R genes are effective in other species, suggesting that the Pc1 gene would be useful if it were transferred to another grass species that is difficult or impossible to cross with Italian ryegrass. This technique would be particularly useful in forage and turf grasses such as Kentucky bluegrass (*Poa pratensis* L.) because of its apomictic characters.

Transformation of resistance-related genes that control plant defense networks has the potential to provide novel resistance mechanisms to plant species. To develop durable and broad specificity of resistance, several approaches have been tried in the transgenic field (reviewed by Punja et al. 2001). It would be desirable to directly manipulate the hypersensitive response and systemic acquired resistance by engineering the signal transduction pathways that lead to their activation.

6. CONCLUSION AND FUTURE PROSPECTS

Resistance to biotic stress is an essential trait in forage grasses. Although many varieties resistant to disease have been developed, the analysis of biotic stress at a molecular level is still lacking. The use of molecular techniques would enable us to clarify the nature of resistance to biotic stress. The use of information from the genome sequence of Arabidopsis and rice opens up new strategies for developing markers and isolating useful genes in forage grasses. However, to use synteny between advanced plants and forage and turf grasses, we need to map target traits onto linkage maps and to confirm the synteny relationships. We do not yet have enough information about biotic stress in Italian ryegrass, because we have just begun to analyze the mechanism of crown rust resistance. However, in the near future we will be able to increase our knowledge of the number of R loci related to biotic stress, and this information will make it easier for us to analyze the mechanism of resistance to biotic stress.

The development of varieties with high and durable resistance to biotic stress will be possible using advanced technology in forage and turf grass. Clarification of the genetic variation within target species may be the essential process. If there are several R genes in the target species, the use of DNA markers will help us to develop resistant varieties with several resistance genes. If there is no R gene against the target pathogen within the target species, the introgression of resistance genes from a related species will be an efficient way of achieving this aim. If there is no material within a related species, transformation may provide useful breeding materials.

To carry out the research mentioned above, analysis by individual laboratories may be insufficient, and collaborative research, for example by the International Lolium Genome Initiative (ILGI), will be hoped for. Each laboratory will carry out linkage analysis of resistance to important diseases and will develop breeding materials and DNA markers linked to this resistance. In the international collaboration, each resistance gene against important diseases in each country will be mapped and assigned to a reference map by using common SSR markers. The information about mapping position and materials will be held collaboratively by laboratories. This strategy will open up new opportunities for efficient breeding and the development of epoch-making resistant varieties.

ACKNOWLEDGEMENTS

We thank Dr. M. Humphreys (Institute of Grassland and Environmental Research, UK) and Dr. J. Forster (Agriculture Victoria, Australia) for their information about crown rust resistance. We also thank our co-workers for their contribution to the work: Dr. A. Arakawa, Dr. S. Sugita, Dr. K. Sugawara, Dr. H. Ohkubo, Dr. Y. Mikoshiba, Dr. H. Cai, M. Inoue, Y. Miura, F. Akiyama and T. Komatsu. Linkage analysis was funded by a research grant from the Japan Racing Association and microarray analysis was supported by a MAFF rice genome project grant no.2114.

REFERENCES

Adams E, Roldan-Ruiz I, Depicker A, van Bockstaele E, de Loose M (2000) A maternal factor conferring resistance to crown rust in Lolium multiflorum cv. 'Axis'. Plant Breed. 119 (2): 182-184.

Anderson PA, Okubara PA, ArroyoGarcia R, Meyers BC, Michelmore RW (1996) Molecular analysis of irradiation induced and spontaneous deletion mutants at a disease resistance locus in Lactuca sativa. Mol. Gen. Genet. 251 (3): 316-325.

Anderson PA, Lawrence GJ, Morrish BC, Ayliffe MA, Finnegan EJ, Ellis JG (1997) Inactivation of the flax rust resistance gene M associated with loss of a repeated unit within the leucine-rich repeat coding region. Plant Cell 9 (4): 641-651.

Bai JF, Pennill LA, Ning JC, Lee SW, Ramalingam J, Webb CA, Zhao BY, Sun Q, Nelson JC (2002) Diversity in nucleotide binding site-leucine-rich repeat genes in cereals. Genome Research 12 (12): 1871-1884.

Baker B, Zambryski P, Staskawicz B, Dinesh-Kumar SP (1997) Signaling in plant-microbe interactions. Science 276 (5313): 726-733.

Barre P, Mi F, Balfourier F, Ghesquiere M (2000) QTLs for morphogenetic traits and sensitivity to rusts in Lolium perenne. Proceeding of Molecular Breeding of Forage crop 2000 Second International Symposium : 60.

Bert PF, Charmet G, Sourdille P, Hayward MD, Balfourier F (1999) A high-density molecular map for ryegrass (Lolium perenne) using AFLP markers. Theor. Appl. Genet. 99 (3-4): 445-452.

Bonas U, Lahaye T (2002) Plant disease resistance triggered by pathogen-derived molecules: refined models of specific recognition. Curr. Opin. Microbiol. 5 (1): 44-50.

Brueggeman R, Rostoks N, Kudrna D, Kilian A, Han F, Chen J, Druka A, Steffenson B, Kleinhofs A (2002) The barley stem rust-resistance gene Rpgl is a novel disease-resistance gene with homology to receptor kinases. Proc. Natl. Acad. Sci. USA 99 (14): 9328-9333.

Buschges R, Hollricher K, Panstruga R, Simons G, Wolter M, Frijters A, van Daelen R, van der Lee T, Diergaarde P, Groenendijk J, Topsch S, Pieter V, Salamini F, Schulze-Lefert P (1997) The barley mlo gene: A novel control element of plant pathogen resistance. Cell 88 (5): 695-705.

Chen M, SanMiguel P, deOliveira AC, Woo S, Zhang H, Wing RA, Bennetzen JL (1997) Microcolinearity in sh2-homologous regions of the maize, rice, and sorghum genomes. Proc. Natl. Acad. Sci. USA 94 (7): 3431-3435.

Dixon MS, Jones DA, Keddie JS, Thomas CM, Harrison K, Jones JDG (1996) The tomato Cf-2 disease resistance locus comprises two functional genes encoding leucine-rich repeat proteins. Cell 84 (3): 451-459.

Dumsday J, Trigg P, Jones E, Batley J, Smith K, Forster J (2003) SSR-based genetic linkage analysis of resistance to crown rust (Puccinia coronata Corda f.sp. lolii) in perennial ryegrass (Lolium perenne L.). In: Abstracts of Plant & Animal Genome XI Conference, p.252. January 11-15, San Diego, CA, USA.

Fujimori M, Hirata M, Sugita S, Inoue M, Cai H, Akiyama F, Mano Y, Komatsu T (2000) Development of a high density map in Italian ryegrass (Lolium multiflorum Lam) using amplified fragment length polymorphism. In: Abstracts of Molecular Breeding of Forage Crops, Second International Symposium, p.52. November 19-24, Lorne and Hamilton, Victoria, Australia.

Fujimori M, Hirata M, Akiyama F, Mano Y, Komatsu T, Yazaki J, Kishimoto N, Kikuchi S, Takamizo T (2002) A cDNA microarray analysis of crown rust resistance in Italian ryegrass (Lolium multiflorum Lam). In: Abstracts of Plant, Animal & Microbe Genome X Conference. p.258. January 12 – 16, San Diego, CA, USA.

Fujimori M, Hayashi K, Hirata M, Mizuno K, Fujiwara T, Akiyama F, Mano Y, Komatsu T, Takamizo T (2003) Linkage analysis of crown rust resistance gene in Italian ryegrass (Lolium multiflorum Lam.). In: Abstracts of Plant & Animal Genome XI Conference. p.46. January 11-15, San Diego, CA, USA.

Hammond-Kosack KE, Jones JDG (1996) Resistance gene-dependent plant defense responses. Plant Cell 8 (10): 1773-1791.

Hammond-Kosack KE, Jones JDG (1997) Plant disease resistance genes. Annu. Rev. Plant Physiol. Plant Mol. Biol. 48:575-607.

Han F, Kilian A, Chen JP, Kudrna D, Steffenson B, Yamamoto K, Matsumoto T, Sasaki T, Kleinhofs A (1999) Sequence analysis of a rice BAC covering the syntenous barley Rpgl region. Genome 42 (6): 1071-1076.

Hayward MD, Forster JW, Jones JG, Dolstra O, Evans C, McAdam NJ, Hossain KG, Stammers M, Will J, Humphreys MO, Evans GM (1998) Genetic analysis of Lolium. I. Identification of linkage groups and the establishment of a genetic map. Plant Breed. 117 (5): 451-455.

Hirata M, Fujimori M, Komatsu T (2000) Development of simple sequence repeat (SSR) marker in Italian ryegrass. In: Abstracts of Molecular Breeding of Forage Crops, Second International Symposium, p.51. November 19-24, Lorne and Hamilton, Victoria, Australia.

Hirata M., Fujimori M., Inoue M., Miura Y., Cai H., Satoh H., Mano Y. , Takamizo T. (2003) Mapping of a new crown rust resistant gene, Pc2, in Italian ryegrass cultivar 'Harukaze'. In: Abstracts of Molecular Breeding of Forage and Turf, Third International Symposium, p.15. May 18-22, Dallas, Texas and Ardmore, Oklahoma, USA.

Hulbert SH, Sudupak MA, Hong KS (1993) Genetic-relationships between alleles of the RP1 rust resistance locus of maize. Mol. Plant Microbe Interact. 6 (3): 387-392.

Hulbert SH, Webb CA, Smith SM, Sun Q (2001) Resistance gene Complexes: Evolution and Utilization. Ann. Rev. Phytopathol. 39:285-312.

Ikeda S, Miura Y, Sasaki T, Ozaki R, Mizuno K (2002) Isolation of disease resistance gene analogs in Italian ryegrass. In: Abstracts of Plant, Animal & Microbe Genome X Conference. p. 92. January 12 – 16, San Diego, CA, USA.

Inoue M, Gao Z, Hirata M, Fujimori M, Cai H-w (2002) Construction of RFLP linkage maps of Italian ryegrass and comparative mapping between Lolium and Poaceae family. In: Abstracts of Plant, Animal & Microbe Genomes X Conference. p.181. January 12 – 16, San Diego, CA, USA.

Jauhar (1993) Cytogenetics of the Festuca-Lolium complex. Springer-Verlag. Berlin.

Jin Y, Steffenson BJ (2002) Sources and genetics of crown rust resistance in barley. Phytopathology 92 (10): 1064-1067.

Jones DA, Jones JDG (1997) The role of leucine-rich repeat proteins in plant defences. Adv. Bot. Res. 24: 89-167.

Jones ES, Mahoney NL, Hayward MD, Armstead IP, Jones JG, Humphreys MO, King IP, Kishida T., Yamada T. Balfourier F, Charmet G, Forster JW (2002a) An enhanced molecular marker based genetic map of perennial ryegrass (Lolium perenne) reveals comparative relationships with other Poaceae genomes. Genome 45: 282-295

Jones ES, Dupal MP, Dumsday JL, Hughes LJ, Forster JW (2002b) An SSR-based genetic linkage map for perennial ryegrass (Lolium perenne L.). Theor. Appl. Genet. 105 (4): 577-584.

Kawchuk LM, Hachey J, Lynch DR, Kulcsar F, van Rooijen G, Waterer DR, Robertson A, Kokko E, Byers R, Howard RJ, Fischer R, Prufer D (2001) Tomato Ve disease resistance genes encode cell surface-like receptors. Proc. Natl. Acad. Sci. USA 98 (11): 6511-6515.

Kimbeng CA (1999) Genetic basis of crown rust resistance in perennial ryegrass, breeding strategies, and genetic variation among pathogen populations: a review. Aust. J. Exp. Agr. 39 (3): 361-378.

Lahaye T (2002) The Arabidopsis RRS1-R disease resistance gene - uncovering the plant's nucleus as the new battlefield of plant defense? Trends Plant Sci. 7 (10): 425-427.

Leister D, Kurth J, Laurie DA, Yano M, Sasaki T, Devos K, Graner A, Schulze-Lefert P (1998) Rapid reorganization of resistance gene homologues in cereal genomes. Proc. Natl. Acad. Sci. USA 95 (1): 370-375.

Meyers BC, Dickerman AW, Michelmore RW, Sivaramakrishnan S, Sobral BW, Young ND (1999) Plant disease resistance genes encode members of an ancient and diverse protein family within the nucleotide-binding superfamily. Plant J. 20 (3): 317-332.

Michelmore RW, Meyers BC (1998) Clusters of resistance genes in plants evolve by divergent selection and a birth-and-death process. Genome Res. 8 (11): 1113-1130.

Oertel C, Matzk F (1999) Introgression of crown rust resistance from Festuca spp. into Lolium multiflorum. Plant Breed. 118 (6): 491-496.

Plummer RM, Hall RL, Watt TA (1990) The influence of crown rust (Puccinia coronata) on tiller production and survival of perennial ryegrass (Lolium perenne) plants in simulated swards. Grass Forage Sci. 45: 9-16.

Potter LR (1987) Effect of crown rust on regrowth, competitive ability and nutritional quality of perennial and Italian ryegrasses. Plant Pathol. 36: 455-461.

Punja ZK (2001) Genetic engineering of plants to enhance resistance to fungal pathogens - a review of progress and future prospects. Can. J. Plant Pathol. 23 (3): 216-235.

Richter TE, Pryor TJ, Bennetzen JL, Hulbert SH (1995) New rust resistance specificities associated with recombination in the Rp1 complex in maize. Genetics 141 (1): 373-381.

Roderick HW, Humphreys MO, Turner L, Armstead I, Thorogood D (2002) Isolate specific quantitative trait loci for resistance to crown rust in perennial ryegrass. In: Proceedings of 24th EUCARPIA Fodder Crops and Amenity Grasses Section Meeting, p.22-26. Braunschweig, Germany (in press).

Rommens CMT, Salmeron JM, Oldroyd GED, Staskawicz BJ (1995) Intergeneric transfer and functional expression of the tomato disease resistance gene Pto. Plant Cell 7 (10): 1537-1544.

Rossi M, Goggin FL, Milligan SB, Kaloshian I, Ullman DE, Williamson VM (1998) The nematode resistance gene Mi of tomato confers resistance against the potato aphid. Proc. Natl. Acad. Sci. USA 95 (17): 9750-9754.

Simons G, Groenendijk J, Wijbrandi J, Reijans M, Groenen J, Diergaarde P, Van der Lee T, Bleeker M, Onstenk J, de Both M, Haring M, Mes J, Cornelissen B, Zabeau M, Vos P (1998) Dissection of the Fusarium I2 gene cluster in tomato reveals six homologs and one active gene copy. Plant Cell 10 (6): 1055-1068.

Smiley RW, Dernoeden PH, Clarke BB (1992) Compendium of Turfgrass Disease. (second edition). The American Phytopathological Society.

Song WY, Pi L, Wang G, Gardner J, Holsten T, Ronald PC (1997) Evolution of the rice Xa21 disease resistance gene family. Plant Cell 9 (8): 1279-1287.

Sudha G, Ravishankar GA (2002) Involvement and interaction of various signaling compounds on the plant metabolic events during defense response, resistance to stress factors, formation of secondary metabolites and their molecular aspects. Plant Cell Tiss. Org. 71 (3): 181-212.
Sudupak MA, Bennetzen JL, Hulbert SH (1993) Unequal exchange and meiotic instability of disease-resistance genes in the Rp1 region of maize. Genetics 133 (1): 119-125.
Tai TH, Dahlbeck D, Clark ET, Gajiwara P, Pasion R, Whalen MC, Stall RE, Staskawicz BJ (1999) Expression of the Bs2 pepper gene confers resistance to bacterial spot disease in tomato. Proc. Natl. Acad. Sci. USA 96 (24): 14153-14158.
Takahashi W, Oishi H, Ebina M, Takamizo T, Komatsu T (2002) Production of transgenic Italian ryegrass (Lolium multiflorum Lam.) via microprojectile bomberdment of embryogenic calli. Plant Biotechnol. 19 (4): 241-249.
Takken FLW, Thomas CM, Joosten MHAJ, Golstein C, Westerink N, Hille J, Nijkamp HJJ, De Wit PJGM, Jones JDG (1999) A second gene at the tomato Cf-4 locus confers resistance to Cladosporium fulvum through recognition of a novel avirulence determinant. Plant J. 20 (3): 279-288.
Takken FLW, Joosten MHAJ (2000) Plant resistance genes: their structure, function and evolution. Eur. J. Plant Pathol. 106: 699-713.
van der Vossen EAG, van der Voort JNAMR, Kanyuka K, Bendahmane A, Sandbrink H, Baulcombe DC, Bakker J, Stiekema WJ, Klein-Lankhorst RM (2000) Homologues of a single resistance-gene cluster in potato confer resistance to distinct pathogens: a virus and a nematode. Plant J. 23 (5): 567-576.
Wang W, Zhai W, Luo M, Jiang G, Chen X, Li X, Wing RA, Zhu L (2001) Chromosome landing at the bacterial blight resistance gene Xa4 locus using a deep coverage rice BAC library. Mol. Genet. Genomics 265 (1): 118-125.
Wang ZY, Hopkins A, Mian R (2001) Forage and turf grass biotechnology. Crit. Rev. Plant Sci.20 (6): 573-619.
Welz HG, Geiger HH (2000) Genes for resistance to northern corn leaf blight in diverse maize populations. Plant Breed 119 (1): 1-14.
Whitham S, McCormick S, Baker B (1996) The N gene of tobacco confers resistance to tobacco mosaic virus in transgenic tomato. Proc. Natl. Acad. Sci. USA 93 (16): 8776-8781.
Wilkins PW (1975) Inheritance of resistance to puccinia coronata corda and rhynchosporium orthosporum caldwell in Italian ryegrass. Euphytica 24: 191-196.
Wolfe MS (1985) The current status and prospects of multiline cultivars and variety mixtures for disease resistance. Ann. Rev. Phytopathol. 23: 251-273.

QTL Mapping of Gray Leaf Spot Resistance in Ryegrass, and Synteny-based Comparison with Rice Blast Resistance Genes in Rice

J. Curley[1], S. C. Sim[1], G. Jung[1], S. Leong[2], S. Warnke[3] and R. E. Barker[4]
[1]Department of Plant Pathology, University of Wisconsin-Madison, Madison, WI 53706, USA. [2]USDA-ARS, Madison, WI 53706, USA. [3]US National Arboretum, Washington, D.C. 20002, USA. [4]USDA-ARS, Corvallis, OR 97331, USA. (Email: jung@plantpath.wisc.edu).

Keywords: ryegrass, gray leaf spot, resistance, QTL comparisons, synteny

Abstract:

Gray leaf spot (GLS) is a serious fungal disease recently reported on the important turfgrass and forage species, perennial ryegrass (*Lolium perenne*) caused by *Magnaporthe grisea*, which also causes rice blast and many other grass diseases. Rice blast is usually controlled by host resistance, but durability of resistance is a problem. Little GLS resistance has been reported in perennial ryegrass; however, work in our lab suggests resistance is present and segregating in an annual x perennial ryegrass mapping population. Quantitative trait locus (QTL) analysis using GLS reaction data along with the linkage map being constructed in our lab has revealed at least two genomic regions associated with QTLs for GLS resistance, and one of the regions appears syntenic with rice linkage group 7.

A. Hopkins, Z.Y. Wang, R. Mian, M. Sledge and R.E. Barker (eds.), Molecular Breeding of Forage and Turf, 37-46.

1. INTRODUCTION

Perennial ryegrass (*Lolium perenne*) is a valuable cool-season turf and forage grass, extensively used on golf course fairways and roughs, as well as on athletic fields and home lawns. It is a diploid (2n = 14), outcrossing, self-incompatible species. Thus, cultivars are synthetic varieties produced as seed from crosses using many parental clones. Lately many improved cultivars have become available, causing renewed interest in and more widespread use of this species. Its positive attributes include fast establishment and versatility as a turfgrass (Hannaway et al. 1999), as well as excellent forage quality that make it the most important pasture grass species in temperate regions (Jones et al. 2002). As a member of the Festuceae tribe of the Pooidae subfamily of the Gramineae (Yaneshita et al. 1993), it is taxonomically related to many important cereal crops. It is most closely related to oat, barley, and wheat, and somewhat less closely related to rice (Kellogg 2000).

In recent years, a new disease known as gray leaf spot has become a serious problem on perennial ryegrass (Viji et al. 2001; Williams et al. 2001). It is caused by *Magnaporthe grisea*, the fungus that causes rice blast disease on rice, as well as other diseases on a very wide host range among the grass family. For example, it causes foliar disease on cereal crops such as wheat (Viji et al. 2001), barley (Sato et al. 2001), and pearl millet (Morgan et al. 1998). The fungus also infects other turf and forage grasses such as tall fescue (*Festuca arundinacea*), St. Augustinegrass (*Stenotaphrum secundatum*) and Italian ryegrass (*Lolium multiflorum*) (Viji et al. 2001; Williams et al. 2001). Gray leaf spot was first reported on perennial ryegrass in Pennsylvania in 1992 (Landschoot and Hoyland 1992) after a spell of hot, humid weather. It was subsequently reported in other midwestern, eastern, and southeastern states, now reaching as far north and west as central Illinois (Pederson 2000).

In general, perennial ryegrass is susceptible to this disease. For example, under warm, humid conditions mature ryegrass plants can be completely destroyed in a matter of several days (Landschoot and Hoyland 1992). The current control practices involve fungicide application and cultural practices such as reduction of leaf wetness, lowered cutting heights, and reduction of nitrogenous fertilizer applications (Williams et al. 2001). Fungicide use is critically important in managing this disease on turf, however, strains of *M. grisea* resistant to one of the most effective classes of fungicides against this pathogen, the strobilurins, have been reported (Vincelli and Dixon 2002). Furthermore, due to the high genetic variability of the pathogen, resistance to the remaining effective fungicides, thiophanate-methyl and certain DMIs is also a concern (Vincelli and Dixon 2002).

Therefore, there is a pressing need for other ways to manage this destructive disease. Host resistance is a very attractive, environmentally sound control strategy, which has been well-studied and tested in other economically important hosts of *M. grisea*. For example, in rice, blast is largely controlled by host resistance. Many major race-specific genes conferring complete blast resistance have been found (excellently summarized in Sallaud et al. 2003), as well as quantitative trait loci (QTL) that control minor gene or field resistance in rice (Tabien et al. 2002; Fukuoka and Okuno 2001; Wang et al. 1994) as well as in barley (Sato et al. 2001) which is more closely related to perennial ryegrass.

Major gene resistance can often break down, while field resistance is thought to be much more durable. In fact, many blast-resistant rice cultivars soon become susceptible after being put into production, due to high genetic variability of the pathogen and narrow resistance specificity of the host (Wang et al. 1994). In contrast, rice cultivars such as 'Moroberekan' and 'IR36', which have been found to contain both major and minor resistance genes (Wang et al. 1994), have remained resistant under years of disease pressure and inoculation with many *M. grisea* isolates (Tabien et al. 2000; Wang et al. 1994). Similar resistance to gray leaf spot in perennial ryegrass would be very valuable in improving the utility of this species, as well as in reducing the level of dependence on environmentally harmful pesticides.

Although resistance to *M. grisea* is well studied in other hosts, it has received little attention in perennial ryegrass. Little resistance has been reported in available perennial ryegrass cultivars (Williams et al. 2001), and although there appears to be some level of variation in disease reaction between the cultivars (Hoffmann and Hamblin 2001), it is not clear if this reduced susceptibility is sufficient to prevent GLS outbreaks in field plantings of these varieties. Further, there is almost no knowledge of the genetics of resistance to GLS in perennial ryegrass. Such knowledge is important to understand the potential durability of resistance. Therefore, the main objectives of this study are to conduct QTL analysis on the segregating population to determine the number, location, and degree of effect of the genomic regions associated with the resistance trait.

2. MATERIALS AND METHODS

2.1 Plant Materials, Fungal Strains and Inoculation

A ryegrass genetic mapping population consisting of 156 progenies derived from a cross between two highly heterozygous ryegrass clones, MFA and MFB, was originally developed by Dr. R.E. Barker (USDA-ARS,

Corvallis, OR). The MFA and MFB parents, in turn, are derived from crosses between two different clones of the perennial cultivar 'Manhattan' and the annual cultivar 'Floregon' (Figure 1). Also provided by Dr. R.E. Barker were the two perennial parents of MFA and MFB, referred to as Manhattan-1 and Manhattan-3, as well as fourteen other perennial ryegrass genotypes derived from crosses among clones of diverse forage and turf cultivars such as Linn, SR4400, and SR4500.

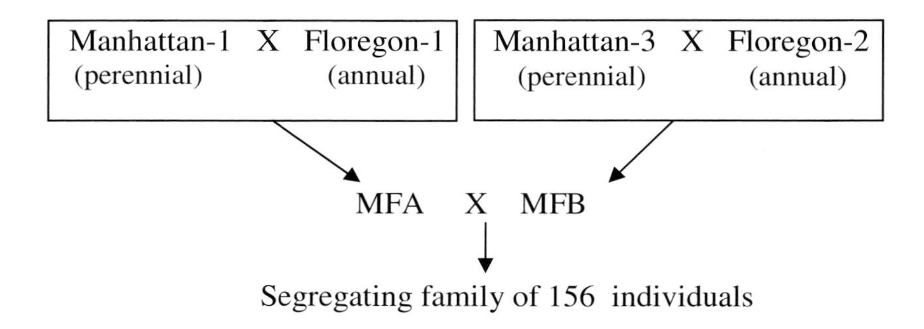

Figure 1. Diagram of crosses used to develop the MFA x MFB mapping population.

Fungal isolates used include several strains isolated from diseased perennial ryegrass fairways, such as GG9, GG11, GG12, GG13, and LP97-1A, provided by Dr. M. Farman, Univ. of Kentucky, and Lin00 and BL00, provided by Dr. A. Hamblin, Univ. of Illinois. Also included is a rice infecting lab strain, 6082, provided by Dr. S. Leong, Univ. of Wisconsin. All isolates were stored as frozen stocks on filter paper at –20 °C, then cultured for spore production on oatmeal agar at room temperature under continuous lighting for two to three weeks.

Ryegrass clones were prepared for inoculation by asexually dividing the plants and transplanting them into small Cone-tainers filled with potting soil, with three to four replicates per clone arranged in a randomized complete block design. Nitrogenous fertilizer was applied once weekly until plants had reached at least 10 cm in height, usually after two to three weeks. Plants were always inoculated when blades were still young.

Gray leaf spot inoculations were carried out using conidial suspensions in 0.2% gelatin solution, with spore concentrations adjusted to 1-3 x 10^5 spores/mL. The spore suspension was misted onto the plants, allowed to dry for about 30 min, and the plants were kept in a mist chamber for three days with continuous leaf wetness to allow symptom development. At the end of three days, the plants were moved to a controlled greenhouse, where lesions were scored five to eight days after inoculation. Lesions were scored following a modification of the rating scale of Smith and Leong (1994), as shown in Table 1. Type 1 and 2 lesions are considered resistant, type 3 lesions are considered intermediate, and type 4 lesions are considered susceptible.

Table 1: Rating scale for gray leaf spot severity

Rating:
0: No visible symptoms
1: Dark brown, non-sporulating, 2-3 mm long lesions
2: Dark brown, non-sporulating lesion with small central necrotic area
3: Circular or small diamond-shaped lesions with prominent dark brown borders and gray or white central sporulating areas
4: Large, expanding, completely unbordered sporulating lesions, often with chlorotic halos

2.2 Linkage Mapping and QTL Analysis

Using the parents and progenies provided to our lab, two partial linkage maps with seven linkage groups were constructed using the "pseudo-testcross" mapping strategy (Grattapaglia and Sederoff 1994). Using this strategy, dominant genetic markers, such as RAPDs or AFLPs, heterozygous in one parent but absent in the other segregate in a 1:1 or testcross ratio, and are used to construct one genetic linkage map for each parent separately. The data for 3:1 segregating markers and codominant (1:1:1:1) segregating markers, such as RFLPs or SSRs, are used to integrate the two linkage maps from each parent. Construction of the maps has been performed using JoinMap version 3.0 (Van Ooijen and Voorrips 2001).

This map includes RAPD and AFLP markers. Also included are two groups of heterologous RFLP probes. The first is a total of 152 "anchor probes" selected from barley, oat, and rice for utility in comparative mapping obtained from Cornell University, Ithaca, NY, USA, described in Van Deynze et al. (1998). The other group consists of a total of approximately 200 oat and barley probes obtained from the USDA probe depository (Albany, CA), many of which have been mapped in other cereals (for example, Causse et al. 1994).

To estimate the number, locations and effects of QTL for GLS resistance, the data were first analyzed using single-factor ANOVA for each pairwise combination of quantitative traits and marker loci. This was done using the Kruskal-Wallis analysis function of MapQTL™ version 4.0 (van Ooijen et al. 2002), to initially indicate genomic regions associated with the trait of interest. The method of interval mapping (Lander and Botstein 1989) using MapQTL™ software was then be used for a more robust location of QTL.

3. RESULTS

Preliminary inoculations of the 14 perennial ryegrass clones together with the mapping parents MFA and MFB showed an interesting variation in reaction. The MFA and MFB clones showed only intermediate lesions upon inoculation with several of the ryegrass isolates, with MFB only slightly more susceptible than MFA, while the perennial clones were almost uniformly susceptible to all isolates tested as measured by the presence and severity of lesions expressed (data not shown). Clone MFA typically expressed type 2 to 3 lesions, and MFB showed type 3 lesions slightly larger than those on MFA (Table 2), although this clone still was not fully susceptible.

Table 2: Gray leaf spot reaction of mapping parents and selected progenies to 3 different isolates

Plant Name	Isolate		
	GG9	BL00	6082
MFA	2-3	2-3	2
MFB	3+	3+	4
8	2	2	2
15	4	4	2-3
16	2-3	2	4
17	2	2	4
19	2	2-3	2
23	4	4	2
25	4	4	2
44	2	2	4
48	4	4	n.d.
54	3-4	3-4	4
69	4	4	2
79	4	4	2
95	3	2	4

Given this small but detectable difference between the two parents, a randomly selected subsample of 14 of the progenies was inoculated with ryegrass isolate GG9, to assess the degree of variation in GLS reaction in the mapping progenies. The disease reaction varied from fully susceptible to very few lesions present, and this was repeated over multiple experiments using several isolates (Table 2). This strongly suggests transgressive segregation is occurring in this population, as the parents MFA and MFB differ only slightly in their reaction to ryegrass isolates, with MFB often being the slightly more susceptible parent. In addition, the ryegrass isolates seem to show similar segregation patterns, while the lab strain 6082 shows a very different segregation pattern, indicating at least some level of race-specificity in the resistance.

The next step was to verify that the whole population shows segregation, so all of the individuals were inoculated with GG9. The phenotypic data comes from two greenhouse inoculations, the first with three replicates and the second with four. Disease reaction was scored using lesion numbers and proportions of resistant lesions, as plants often showed multiple lesions, but varied in lesion number and proportion of susceptible lesions. The same pattern of transgressive segregation was observed as in previous experiments, with the two parents showing less difference than the progenies. In the first experiment, the number of lesions was continuously distributed with a range from 5 to 45 with a mean of 15. The proportion of resistant lesions was continuously distributed and ranged from 0.1 to 1 with a mean of 0.73. The data appeared skewed towards resistant plants in this case. In the second experiment, the number of lesions was continuously distributed and ranged from 5 to 50 with a mean of 22.7. Also, the overall number of lesions appeared higher in the second experiment. The proportion of resistant lesions in the second experiment was continuously distributed and ranged from 0.1 to 1 with a mean of 0.61. The plants seemed to tend towards a higher proportion of susceptible lesions in this experiment.

When interval mapping was conducted using the two parental maps and data from both inoculation experiments separately, two genomic regions were noted. These regions showed elevated LOD scores over both experiments and over two different scoring methods, number of lesions and proportion of type 1 and 2 or resistant lesions. Even though the LOD scores were usually but not always significant (above 3.5), they were always elevated in these two regions, with a range of about 2.0 to 6.0. For both parents, these regions corresponded to linkage group 2, for proportion of resistant lesions, and linkage group 4, for lesion number. Several other regions were noted, on linkage groups 1, 3, and 5, but were not consistent over experiments or parental maps. Further work is needed to determine their importance.

4. DISCUSSION

The pathogen *Magnaporthe grisea* causes foliar disease on many important graminaceous hosts, including both cereal crops and turf and forage species, but by far the most research has been done in rice (Sallaud et al. 2003). Variation in gray leaf spot reaction has been reported for other turf and forage species such as St. Augustinegrass (Holcomb and Shepard 1995), tall fescue (Fraser 1997), and Italian ryegrass (Trevathan 1982). For perennial ryegrass, there is some variation in susceptibility in available cultivars (Hoffmann and Hamblin 2001), as well as possible resistance in exotic plant introductions (Hoffmann and Hamblin 2000), although these have not been evaluated for their adaptability and turf potential under conditions found in the U.S. In addition, as stated above, there is little knowledge of the genetics and potential durability of resistance to GLS, both of which are important to avoid breakdown of resistance, and these points all underscore the importance of this research.

The resistance to GLS present in some of the segregating progeny appears to be stable when tested using several isolates, even including a lab strain which is likely to have very different genetics than the ryegrass field isolates (Table 2). Although field inoculations, and tests with additional isolates are needed, this resistance appears to be sufficient to use in breeding. Additionally, the QTL analysis results suggest that the resistance is controlled by only a few genes, which will simplify the breeding process. Furthermore, as the QTLs are localized to narrower intervals, development of tightly linked markers useful in marker-assisted selection (MAS) will become feasible. This will be helpful in tracking the resistance genes through the breeding process, and will allow incorporation of additional sources of resistance as they are discovered.

A possible genetic interpretation of the putative ryegrass QTL is that the two observed intervals are heterozygous in the parent clones MFA and MFB. This can account for transgressive segregation, with the parents being intermediate (heterozygous) and the most resistant plants having both resistant alleles. This idea will be tested by crossing a resistant progeny individual from the MFA x MFB population with a susceptible perennial parent from a different cultivar background and mapping the resistance, to see if the same marker intervals are still significant.

Another important aspect of this study is the substantial but not complete conservation of molecular marker and gene order between rice and perennial ryegrass, which was observed in this study and elsewhere (Jones et al. 2002). This, as well as the observed conservation between ryegrass, oat, and wheat suggests a substantial degree of synteny between these species. This partial map colinearity at the molecular marker level allows the map location of genes and QTL for *M. grisea* resistance to be compared across these species as well.

In perennial ryegrass, two potentially important genomic regions were found in this study. Interestingly, the interval found in linkage group 4 for lesion number appears to line up with rice linkage group 3; several blast resistance QTL have been found on the opposite end of this linkage group. Similarly, the interval in linkage group 2 for proportion of resistant lesions lines up with rice linkage group 7; QTL, as well as the major blast resistance gene Pi-17(t) (Sallaud et al. 2003), have been detected on this group as well but not on areas that the ryegrass map aligns with.

The addition of more markers mapped in rice will facilitate a more thorough search and comparison of blast QTL between rice and ryegrass. Another approach to strengthen the comparisons of QTL between rice and ryegrass is inoculation of the mapping population with the lab strain 6082. This strain was used in the cloning of the resistance gene Pi-CO39(t) from a rice mapping population (Chauhan et al. 2002). This strain is capable of infecting ryegrass, and produces a different disease reaction on selected progeny than the ryegrass isolates (Table 2). Thus it is likely this isolate can be used to detect additional QTL, and the map locations of these QTL can be compared with the Pi-CO39(t) gene along with the other approximately forty mapped blast resistance genes in rice.

REFERENCES

Causse M, Fulton T, Cho Y, Ahn S (1994) Saturated molecular map of the rice genome based on an interspecific backcross population. Genetics 138: 1251-1274.

Chauhan R, Farman ML, Zhang HB, Leong SA (2002) Genetic and physical mapping of a rice blast resistance locus, *Pi-CO39(t)*, that corresponds to the avirulence gene *AVR1-CO39* of *Magnaporthe grisea*. Mol. Gen. Genomics 267: 603-612.

Fraser ML (1997) Susceptibility of tall fescues to gray leaf spot, 1995. Biol. Cultural Tests 12: 130.

Fukuoka S, Okuno K (2001) QTL analysis and mapping of *pi21*, a recessive gene for field resistance to rice blast in Japanese upland rice. Theor. Appl. Genet. 103: 185-190.

Grattapaglia D, Sederoff R (1994) Genetic linkage maps of *Eucalyptus grandis* and *Eucalyptus urophylla* using a pseudo-test-cross mapping strategy and RAPD markers. Genetics 137: 1121-1137.

Hoffmann NE, Hamblin AM (2000) Reaction of perennial ryegrass to gray leaf spot following inoculation in the greenhouse, 1999. Biol. Cultural Tests 15: 55.

Hoffmann NE, Hamblin AM (2001) Reaction of perennial ryegrass to gray leaf spot following inoculation in the field, 2000. Biol. Cultural Tests 16: T56.

Hannaway D, S Fransen, J Cropper, Teel M, Chaney M (1999) Perennial ryegrass (*Lolium perenne* L.). Oregon State University Extension Publication PNW503.

Holcomb GE, Shepard DP (1995) Reaction of St. Augustinegrass cultivars and selections to gray leaf spot, 1994. Biol. Cultural Tests 10: 41.

Jones E, Mahoney N, Hayward M, Armstead I (2002) An enhanced molecular marker based genetic map of perennial ryegrass (*Lolium perenne*) reveals comparative relationships with other Poaceae genomes. Genome 45: 282–295.

Kellogg E (2000) The grasses: A case study in macroevolution. Ann. Rev. Ecol. Syst. 31: 217-238.

Lander E, Botstein D (1989) Mapping Mendelian factors underlying quantitative traits using RFLP linkage maps. Genetics 121: 185-199.

Landschoot P, Hoyland B (1992) Gray leaf spot of perennial ryegrass turf in Pennsylvania. Plant Dis. 76: 1280-1282.

Morgan R, Wilson JP, Hanna WW, Ozias-Akins P (1998) Molecular markers for rust and *Pyricularia* leaf spot disease resistance in pearl millet. Theor. Appl. Genet. 96: 413-420.

Pederson D (2000) First report of gray leaf spot caused by *Pyricularia grisea* on *Lolium perenne* in Illinois. Plant Dis. 84: 1151.

Sallaud C, Lorieux M, Roumen E (2003) Identification of five new blast resistance genes in the highly blast-resistant rice variety IR64 using a QTL mapping strategy. Theor. Appl. Genet. 106: 794-803.

Sato K, Inukai T, Hayes PM (2001) QTL analysis of resistance to the rice blast pathogen in barley (*Hordeum vulgare*). Theor. Appl. Genet. 102: 916-920.

Smith JR, Leong SA (1994) Mapping of a *Magnaporthe grisea* locus affecting cultivar specificity. Theor. Appl. Genet. 88: 901-908.

Tabien RE, Li Z, Paterson AH, Marchetti MA, Stansel JW, Pinson SRM (2000) Mapping of four major rice blast resistance genes from 'Lemont' and 'Teqing' and evaluation of their combinatorial effect for field resistance. Theor. Appl Genet. 101: 1215-1225.

Tabien R, Li Z, Paterson AH, Marchetti MA, Stansel JW, Pinson SRM (2002) Mapping QTLs for field resistance to the rice blast pathogen and evaluating their individual and combined utility in improved varieties. Theor. Appl Genet. 105: 313-324.

Trevathan LE (1982) Response of ryegrass plant introductions to artificial inoculation with *Pyricularia grisea* under greenhouse consitions. Plant Dis. 66: 696-697.

Van Deynze AE, Sorrells ME, Park WD, Ayres NM, Fu H, Cartinhour SW, Paul E, McCouch SR (1998) Anchor probes for comparative mapping of grass genera. Theor. Appl. Genet. 97: 356-369.

Van Ooijen JW, Bauer MP, Jansen RC, Maliepaard C (2002) MapQTL® version 4.0, Software for the calculation of QTL positions on genetic maps. Plant Research International, Wageningen, the Netherlands.

Van Ooijen JW, Voorrips RE (2001) JoinMap® 3.0, Software for the calculation of genetic linkage maps. Plant Research International, Wageningen, the Netherlands.

Viji G, Wu B, Kang S, Uddin W, Huff DR (2001) *Pyricularia grisea* causing gray leaf spot of perennial ryegrass turf: population structure and host specificity. Plant Dis. 85: 817-826.

Vincelli P, Dixon E (2002) Resistance to QoI (strobilurin-like) fungicides in isolates of *Pyricularia grisea* from perennial ryegrass. Plant Dis. 86: 235-240.

Wang GL, Mackill DJ, Bonman JM, McCouch SR, Champoux MC, Nelson RJ (1994) RFLP mapping of genes conferring complete and partial resistance to blast in a durably resistant rice cultivar. Genetics 136: 1421-1434.

Williams DW, Burrus PB, Vincelli P (2001) Severity of gray leaf spot in perennial ryegrass as influenced by mowing height and nitrogen level. Crop Sci. 41: 1207-1211.

Yaneshita M, Ohmura T, Sasakua T, Ogihara Y (1993) Phylogenetic relationships of turfgrasses as revealed by restriction fragment analysis of chloroplast DNA. Theor. Appl Genet. 87: 129-135.

Differential Gene Expression in Bermudagrass Associated with Resistance to a Fungal Pathogen

Arron C. Guenzi and Yan Zhang
Department of Plant and Soil Sciences, Oklahoma State University, Stillwater, OK 74078, USA.(Email: acg@okstate.edu).

Key words: cDNA microarrays, *Cynodon dactylon*, disease resistance, functional genomics, *Ophiosphaerella herpotricha,* redox status, signal transduction

Abstract:

Bermudagrass, *Cynodon dactylon* (L.) Pers., is extensively used for turf and forage in many warm climatic regions of the world. Spring dead spot (SDS), caused by *Ophiosphaerella herpotricha,* is a serious fungal disease of turf bermudagrass in the southern USA. Suppression subtraction hybridization (SSH), sequencing of cDNA clones from forward and reverse normalized libraries, and cDNA microarrays were used to identify genes associated with resistance or susceptibility to this disease. During the fall and spring seasons, there were 80 and 66 singletons, respectively, that displayed more than a 2-fold differential expression between the resistant and susceptible cultivars. One hundred and seven responsive genes were grouped into six clusters according to their fall and spring expression profiles. The majority of differentially expressed genes had no homology to current accessions in NCBI GenBank. Of those clones with putative identities, the most interesting classes of genes differentially expressed between the resistant and susceptible cultivars were those involved in signaling pathways and the oxidative burst defense mechanism.

A. Hopkins, Z.Y. Wang, R. Mian, M. Sledge and R.E. Barker (eds.), Molecular Breeding of Forage and Turf, 47-52.

1. INTRODUCTION

Over the past decade, major advances have been made in the molecular biology and genomics of host-microbe interactions with model plant systems (e.g. *Arabidopsis*, tobacco and tomato) (Rafalski 2002). Until very recently, relatively little investment has been directed to the grasses, and investment in bermudagrass has lagged behind other grass species of economic importance. The finding of colinearity (synteny) among grass genomes, coupled with the release of the rice physical map (Chen et al. 2002) and genomic sequence (Goff et al. 2002), now raises the possibility of rapid progress in the molecular analysis and manipulation of grass genomes - even previously intractable ones such as bermudagrass (Wang et al. 2001).

Unfortunately, many of the fundamental tools required for bermudagrass to benefit fully from genomics do not exist [large mutant stocks, bacterial artificial chromosome (BAC) libraries]; or are woefully incomplete [high-throughput transformation protocols, saturated molecular maps, expressed sequence tags (ESTs), microarrays, and integrated databases from which to exploit this information]. EST sequences will also serve to anchor bermudagrass to physical maps of other grass species with completed (rice) or on going (maize and sorghum) genome-sequencing projects. ESTs are also valuable as tags to genomic regions from which molecular markers can be developed for marker-assisted selection strategies (Morgante et al. 2001; Dearlove 2002). Filling these critical gaps in bermudagrass genomics will allow geneticists and breeders to take advantage of existing genetic variation, exploit advances in genomics of other grasses, and make bermudagrass genomics information available for improvement of other turf and forage grasses.

2. SPRING DEAD SPOT DISEASE OF BERMUDAGRASS

Spring dead spot (SDS), caused by *Ophiosphaerella herpotricha*, is a serious disease of turf bermudagrass grown in the southern United States (Anderson et al. 2002; Taliaferro, 2003). The objective of our research is to utilize well-characterized genetic resources to dissect the molecular responses of bermudagrass to this soilborne fungal pathogen. Our long-range goals are to identify markers that can be utilized to select for resistance genes and to ultimately engineer increased levels of resistance not obtainable by genetic variation in this species.

Bermudagrass cultivars have been extensively evaluated for SDS resistance (Martin et al. 2001a; 2001b). No immunity has been identified, however, genotypes have been identified with a wide range of phenotypic responses, from highly resistant to highly susceptible. These well-characterized genetic resources have been, and will continue to be, essential to dissect molecular interactions associated with this disease.

3. SUPRESSION SUBTRACTION HYBRIDIZATION

Two to five percent of plant genes are involved in stress defense mechanisms (Cushman and Bohnert 2000; Michelmore 2000). We expected to find a large number of bermudagrass genes induced or repressed in response to fungal infection. To identify a maximal number of differentially expressed genes, with limited financial resources, we constructed suppression subtraction hybridization (SSH) cDNA libraries (Diatchenko et al. 1996) from infected crown tissues from resistant (Yukon) and susceptible (Jackpot) cultivars.

cDNA libraries generated by SSH are very rich sources of sequences which are either unique or share partial homology with known genes (Desai et al. 2000). We have analyzed 834 clones from both forward and reverse normalized-subtraction libraries to insure that genes representing low-abundance transcripts were included. Sixty percent of the clones did not match current accessions in NCBI GenBank, which is nearly two fold greater than the average number (30%) of new sequences usually discovered in SSH libraries.

Although SSH normalizes the cDNA library by suppression PCR, there is still a degree of redundancy from highly expressed genes. The most abundant transcript found shared homology to an unknown function protein from *Arabidopsis thaliana*. This transcript was replicated 14 times and was found in the library representing the resistant cultivar. In addition, these SSH libraries contained clones representing rare transcripts such as an ethylene receptor, auxin binding protein and a signal peptidase. Microarray analysis found that approximately one-half of the SSH clones were differentially expressed. Our results indicate that SSH did enrich these libraries to represent differentially expressed genes for both high and low-abundance transcripts.

4. MICROARRAY ANALYSIS

Parallel analysis of expression for thousands of genes with microarrays has revolutionized genetics (Richmond and Somerville 2000). Microarray analyses are beginning to provide insights into the complex genetic networks coordinating plant responses to biotic stresses (Schenk et al. 2003; Glazebrook et al. 2003; Cooper et al. 2003; Chen et al. 2002). For our research, SSH clones were printed as features on glass slides in duplicate.

Subsequent sequence analysis assembled these clones into 154 contigs (Ayoubi et al. 2002). Features belonging to the same contig were also treated as replications for expression analysis. Differential gene expression was evaluated by labeling mRNA from pooled samples of infected tissue from the resistant and susceptible genotypes with either Cy3 or Cy5 dyes. Expression ratios were generated based on global normalization by GenePix Pro software (Axon Instruments, Inc.). TreeView visualization and k-means clustering were done with Genesis software (Sturn et al. 2002).

5. GENES ASSOCIATED WITH RESISTANCE

As highlighted above, no bermudagrass cultivar has been found which is immune to this disease; resistance is a measure of the degree of susceptibility. The chronic exposure of the host to the pathogen makes this interaction very unique. Disease symptoms were first observed three years after inoculation. In addition, cycles of yearly infection in the late fall and early spring overlap with fall acclimation and spring regrowth. By early summer, dead spots are re-colonized and no symptoms are observed until the turfgrass breaks dormancy in the spring of the following year.

During the fall and spring seasons, there were 80 and 66 singletons, respectively, that displayed more than a 2-fold differential expression between the two cultivars. One hundred and seven responsive genes were grouped into six clusters according to their fall and spring expression profiles. The majority of differentially expressed genes had no homology to current accessions in NCBI GenBank. Of those clones with putative identities, the most interesting classes of genes differentially expressed between the resistant and susceptible cultivars were those involved in signaling pathways and the oxidative burst defense mechanism. Known function genes induced in the resistant cultivar included an ethylene receptor, rac GTPase activating protein, DnaJ protein, voltage-dependent anion channel protein, eukaryotic translation initiation factor, ADP-ribosylation factor-like protein, and LLS1 protein. Among these genes, the ethylene receptor (Ciardi et al. 2000) and DnaJ protein (Futamura et al. 1999) have known roles in pathogen and stress defense.

Genes induced in the susceptible cultivar included an ATP synthase, branched chain alpha-keto acid dehydrogenase, histone H3.3, and formate dehydrogenase. These are all involved in cell maintenance and development processes. In addition, ascorbate peroxidase and a putative cysteine proteinase were highly induced in the susceptible cultivar in both the fall and spring. These genes play an important role in the metabolism of H_2O_2 in higher plants to adjust the redox status of cells in response to abiotic and biotic stresses (Jimenez et al. 1997, Navarre and Wolbert 1999). Ascorbate peroxidase and cysteine proteases have also been implicated as mediators of pathogen-induced program cell death in plants (Tenhaken and Rubel 1997, Solomon et al. 1999).

These differentially expressed genes provide targets for future functional analyses to establish their role, if any, in disease development. However, as highlighted above, this is extremely challenging for a species such as bermudagrass in which many of the tools for functional analyses are missing or in the early stages of development. We have decided to use a genetic approach to validate our microarray results. Expression profiles for eighteen bermudagrass cultivars that were phenotyped for SDS resistance from 1997 to 2002 will be established with funding recently provided by the United States Golf Association. A biomedical research approach of using microarrays to establish gene expression patterns associated with disease development will hopefully allow us to validate which genes are associated with resistance or susceptibility (Desai et al. 2002).

ACKNOWLEDGEMENTS

We gratefully acknowledge our colleagues at Oklahoma State University (M.P. Anderson, P. Ayoubi, D.L. Martin, and C.M. Taliaferro), Kansas State University (N.A. Tisserat), USDA-ARS (J.P. Fellers) and the Samuel Roberts Noble Foundation (R.A. Gonzales, N. Aziz and S. Reddy) who have contributed to the research described above. We also thank the United States Golf Association and Oklahoma Agricultural Experiment Station for financial support.

REFERENCES

Anderson MP, Guenzi AC, Martin DL, Taliaferro CM, Tisserat NA (2002) Spring dead spot a major bermudagrass disease: Now and in the future. USGA Green Section Record: 40: 21-23.

Ayoubi P, Jin X, Leite S, Li X, Martajaja J, Abduraham A, Wan Q, Yan W, Misawa E, Prade RA (2002) PipeOnline 2.0: automated EST processing and functional data sorting. Nucleic Acids Res. 30: 4761-4769.

Chen M et al. (2002) An integrated physical and genetic map of the rice genome. Plant Cell 14: 537–545.

Chen W et al. (2002) Expression profile of *Arabidopsis* transcription factor genes suggests their putative functions in response to environmental stresses. Plant Cell 14: 559-574.

Ciardi JA, Tieman DM, Lund ST, Jones JB, Stall RE, Klee HJ (2000) Response to *Xanthomonas campestris* pv. *vesicatoria* in tomato involves regulation of ethylene receptor gene expression. Plant Physiol. 123: 81-92.

Copper B, Clarke JD, Budworth P, Kreps J, Hutchinson D, Park S, Guimil S, Dunn M, Luginbuhl P, Ellero C, Goff SA, Glazebrook J (2003) A network of rice genes associated with stress response and seed development. Proc. Natl. Acad. Sci. USA 100(8):4945-4950.

Cushman JC, Bohnert HJ (2000) Genomics approaches to plant stress tolerance. Curr. Opin. Plant Biol. 3: 117-124.

Dearlove AM (2002) High throughput genotyping technologies. Briefings in Func. Genomics and Proteomics 1(2): 139-150.

Desai S, Hill J, Trelogan S, Diatchenko L, Siebert P (2000) Identification of differentially expressed genes by suppression subtractive hybridization. In: Functional Genomics: A Practical Approach, Hunt S, Livesey R (eds.), p. 81-112. Oxford Univ. Press, NY.

Desai KV, Xiao N, Wang W, Gangi L, Greene J, Powell JI, Dickson R, Furth P, Hunter K, Kucherlapati R, Simon R, Liu ET, Green JE (2002) Initiating oncogenic event determines gene-expression patterns of human breast cancer models. Proc. Natl. Acad. Sci. USA 99: 6967-6972.

Diatchenko L, Lau YFC, Campbell AP, Chenchik A, Moqadam F, Huang B, Lukyanov S, Lukyanov K, Gurskaya N, Sverdlov E, and Siebert PD (1996) Suppression subtractive hybridization: A method for generating differentially regulated or tissue-specific cDNA probes and libraries. Proc. Natl. Acad. Sci. USA 93: 6025-6030.

Futamura N, Ishiiminami N, Hayashida N, Shinohara K (1999) Expression of DnaJ homologs and Hsp70 in the Japanese willow (*Salix gilgiana*). Plant Cell Physiol. 40: 524-531.

Glazebrook J, Chen W, Estes B, Chang H-S, Nawarth C, Metraux J-P, Zhu T, Katagiri F (2003) Topology of the network integrating salicylate and jasmonate signal transduction derived from global expression phenotyping. Plant J. 34: 217-228.

Goff SA et al. (2002) A Draft Sequence of the Rice Genome (*Oryza sativa* L. *japonica*). Science 296: 92-100.

Jimenez A, Hernandez JA, del-Rio LA, Sevilla F (1997) Evidence for the presence of the ascorbate-glutathione cycle in mitochondria and peroxisomes of pea leaves. Plant Physiol. 114: 275-284.

Martin DL, Bell GE, Taliaferro CM, Tisserat NA, Baird JH, Dobson DD, Kuzmic RM, Anderson JA (2001a) Spring dead spot resistance in inter-specific hybrid bermudagrasses. Intl. Turf. Soc. Res. J. 9: 685-688.

Martin DL, Bell GE, Baird JH, Taliaferro CM, Tisserat NA, Dobson DD, Kuzmic RM, Anderson JA (2001b) Spring dead spot resistance and quality of seeded bermudagrasses under different mowing heights. Crop Sci. 41: 451-456.

Michelmore R (2000) Genomic approaches to plant disease resistance. Curr. Opin. Plant Biol. 3: 125-131.

Morgante M, Hanafey M, Powell W (2001) Microsatellites are preferentially associated with nonrepetitive DNA in plant genomes. Nat. Genet. 30: 197-200.

Navarre, DA, Wolpert, TJ (1999) Victorin induction of an apoptotic/senescence-like response in oats. Plant Cell 11: 237-249.

Rafalski JA (2002) Plant genomics: Present state and a perspective on future developments. Briefings in Func. Genomics and Proteomics 1: 80-94.

Richmond T, Somerville S (2000) Chasing the dream: plant EST microarrays. Curr. Opin. Plant Biol. 3: 108-116.

Schenk P, Kazan K, Manners JM, Anderson JP, Simpson RS, Wilson IW, Somerville SC, Maclean DJ (2003) Systemic gene expression in *Arabidopsis* during an incompatible interaction with *Alternaria brassicicola*. Plant Physiol. 132: 999-1010.

Solomon M, Belenghi B, Delledonne M, Menachem E, Levine A (1999) The involvement of cysteine proteases and protease inhibitor genes in the regulation of programmed cell death in plants. Plant Cell 11: 431-444.

Sturn A, Quackenbush J, Trajanoski Z (2002) Genesis: cluster analysis of microarray data. Bioinformatics 18: 207-208.

Taliaferro CM (2003) Bermudagrass. In: Turfgrass Biology, Genetics, and Breeding, Casler MD, Duncan RR (eds.), Wiley, USA.

Tenhaken R, Rubel C (1997) Salicylic acid is needed in hypersensitive cell death in soybean but does not act as a catalase inhibitor. Plant Physiol. 115: 291-298.

Wang ZY, Hopkins A, Mian R (2001) Forage and turf grass biotechnology. Crit. Rev. Plant Sci. 20(6): 573-619.

Genetic Diversity and Pathogenicity of the Grass Pathogen *Xanthomonas translucens* pv. *graminis*

R. Kölliker, R. Krähenbühl, F. X. Schubiger and F. Widmer
Swiss Federal Research Station for Agroecology and Agriculture, FAL-Reckenholz, 8046 Zurich, Switzerland. (Email: Roland.Koelliker@fal.admin.ch).

Key words: 16s rDNA, AFLP, molecular markers, bacterial wilt, forage grasses, *Lolium multiflorum* Lam.

Abstract:

Bacterial wilt, caused by *Xanthomonas translucens* pv *graminis* (*Xtg*), is one of the most serious diseases of forage grasses throughout Europe, the USA and Australasia. Breeding of resistant cultivars is the only practical measure for controlling the disease in grasslands. Knowledge on genetic diversity and pathogenicity of the prevailing pathogen isolates is indispensable for efficient resistance breeding. Forty-five isolates collected throughout Switzerland were genetically characterised together with reference isolates using 16S ribosomal RNA gene (rDNA) sequencing and AFLP analysis. Pathogenicity of selected isolates was investigated by artificial inoculation of three Italian ryegrass (*Lolium multiflorum*) cultivars with different levels of resistance to bacterial wilt. 16S rDNA sequencing allowed the identification of a DNA signature specific for *Xtg* and closely related isolates. Cluster analysis based on 16S rDNA grouped most of the collected *Xtg* isolates in a single cluster with only minor sequence differences between the individual isolates. AFLP analysis proved highly effective for detecting genetic differences between *Xtg* isolates. However, the observed genetic diversity among the *Xtg* isolates was relatively small and several identical isolates, collected from different locations and various host species were identified. Artificial inoculation revealed significant differences in pathogenicity between some of the isolates tested. Most of the isolates showed medium to high pathogenicity which was congruent with the moderate genetic diversity detected through AFLP analysis. The method presented provides a valuable tool for the selection of *Xtg* isolates particularly suited for resistance breeding based on inoculation with standardised *Xtg* strains.

A. Hopkins, Z.Y. Wang, R. Mian, M. Sledge and R.E. Barker (eds.), Molecular Breeding of Forage and Turf, 53-59.

1. INTRODUCTION

Bacterial wilt of forage grasses was first discovered on *Dactylis, Lolium* and *Festuca* species in breeding nurseries in Switzerland around 1970 (Egli et al. 1975). The pathogen was identified as a *Xanthomonas* species, first named *X. graminis*, later renamed to *X. campestris* pv. *graminis* (Dye et al. 1980) and most recently reclassified to *X. translucens* pv. *graminis* (*Xtg*) (Vauterin et al. 1995).

Xtg infection occurs mainly via wounds and leads to necrosis starting from the infection site and progressing towards the base of the leaf or the plant. Once the bacteria reach vascular tissue, the disease spreads rapidly throughout the plant causing wilt of several leaves and may kill the plant within a few days (Leyns 1993). Since its discovery, bacterial wilt has been recognised as one of the most important diseases of forage grasses in temperate regions. Up to 80% of pastures and meadows have been found to contain infected plants in Scotland (Channon and Hissett 1984) and Belgium (Leyns et al. 1981) and severe yield losses have been observed in experiments involving natural (Schmidt 1988) and artificial (Wang and Sletten 1995) infection. Among the various means for control of *Xanthomonas* diseases, few are applicable to forage grasses. Breeding for resistance is considered the only feasible and efficient measure for controlling bacterial wilt of forage grasses.

In addition to *Xtg* which infects a wide range of species including *Lolium* spp., *Festuca* spp., *Dactylis glomerata*, *Phleum pratense* and *Phalaris arundinacea*, three other *Xanthomonas* pathovars, i.e. pv. *phlei*, pv. *poae* and pv. *arrhenatheri*, with a host range limited to the respective plant genus were identified (Egli and Schmidt 1982). *Xtg* is the most abundant of these four pathovars with a particularly wide geographic distribution and shows very uniform pathogenic behaviour across a range of plant species and genotypes indicating a low strain specificity. Although some studies indicate a high degree of similarity within the *Xtg* also at the genome level (Alizadeh et al. 1997), there is only little information on variability among *Xtg* strains prevalent in pastures and meadows. The aim of this study was to characterise genetic diversity and variation in pathogenicity in a collection of putative *Xtg* isolates collected throughout Switzerland and neighbouring Europe. This information is essential for the selection of bacterial isolates particularly suited for resistance breeding.

2. MATERIALS AND METHODS

The 45 putative *Xtg* isolates used in this study were collected from pastures and meadows throughout Switzerland and neighbouring Europe by isolating single colonies from the exudate of infected plants on GYCA medium (2% Glucose, 1% Yeast Extract, 2% $CaCO_3$, 1.5% agar). In addition, selected *Xanthomonas* type strains were obtained from the Belgian Co-ordinated Collections of Micro-organisms (BCCM, Brussels, Belgium).

2.1 Genetic Analyses

Bacterial cells were lysed in 1% SDS and genomic DNA was obtained using a modified Proteinase K/Phenol extraction protocol.

2.2 16S rDNA Sequencing

Selected isolates were classified by sequence comparison of the 16S RNA gene (De Parasis and Roth 1990). 16S rDNA was amplified using the primers 1627F (5'-AGAGTTTGATCMTGGCTCAG-3') and 1378R (5'-CGGTGTGTACAAGGCCCGGGAACG-3'). Amplification products were cloned and sequenced using an ABI Prism 310 Genetic Analyzer (Applied Biosystems). Sequence alignments and phylogenetic analyses were performed using additional 16S rDNA sequences of *Xanthomonas* spp. type strains obtained from the Ribosomal Database Project II (http://rdp.cme.msu.edu).

2.3 AFLP Analysis

Genetic diversity among all 45 isolates was determined by means of AFLP analysis as described by Vos et al. (1995). Due to the small genome size of *Xtg*, only one PCR amplification was performed using the primer combinations *Eco*RI+0/*Mse*I+C and *Pst*I+C/*Mse*I+0 (Valsangiacomo et al. 1995; Vos et al. 1995). Fragments were analysed on an ABI Prism 3100 Genetic Analyzer (Applied Biosystems) and AFLP markers were scored for presence or absence. Euclidean squared distance was calculated between all pairs of isolates and cluster analysis was performed using the UPGMA method.

2.4 Assessment of Pathogenicity

Pathogenicity of 31 putative *Xtg* isolates and 4 other *Xanthomonas* species was assessed by artificial inoculation of the three *Lolium multiflorum* cultivars Axis (Switzerland), Adret (France) and Ligrande (Germany). Each isolate was tested on all three cultivars using twelve plants per cultivar in four replications. Plants were inoculated 25 days after sowing by cutting using scissors dipped in bacterial suspension containing $5x10^9$ cells/ml.

Disease progress was scored on each plant 12 and 14 days after inoculation using a scale from 1 (no symptoms, not pathogenic) to 9 (complete wilting, highly pathogenic). Means of the two assessments were used for statistical analysis.

3. RESULTS AND DISCUSSION

3.1 Phylogenetic Affiliation of Isolates

Comparison of 1358bp 16S rDNA sequences showed high homology (>95%) among 29 of the selected putative *Xtg* isolates investigated. 28 of the 29 isolates formed a distinct cluster with the *Xtg* type strain (LMG 726, BCCM) and were clearly separated from the 6 outgroup sequences included in the comparison (Figure 1). These isolates were therefore classified as true *Xtg*. Isolate 29.01 was grouped outside the *Xtg* cluster close to the *X.t.* pv. *translucens* reference strain. Since this isolate was initially isolated from *Arrhenatherum elatius*, it was assumed to be *X.t.* pv. *arrhenatheri* (*Xta*). The remaining isolate revealed high sequence homology to *Clavibacter* spp. and was therefore excluded from further analysis.

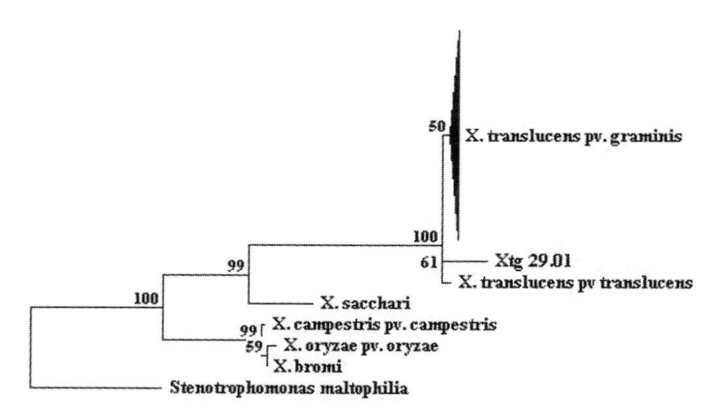

Figure 1. 16S rDNA sequence alignment of selected *Xanthomonas* isolates and reference sequences. Black triangle indicates cluster containing 29 *Xtg* isolates and the *Xtg* reference strain.

3.2 Genetic Diversity among *Xtg* Isolates

AFLP analysis of 44 *Xtg* and one *Xta* isolates using two primer combinations resulted in 137 scorable AFLP markers. 113 of these markers were polymorphic across the entire dataset. However, 79 markers were polymorphic only between *Xtg* and *Xta* but not among *Xtg* isolates, indicating a low genetic diversity among these isolates. Considering the small genome size of *Xanthomonas* of roughly 5Mb (Da Silva et al. 2002), the relatively low number of 34 markers still theoretically accounts for one polymorphism every 150bp. Cluster analysis based on Euclidean squared distance identified three distinct clusters with bootstrap values larger than 50% (1, 3 and 4, Figure 2).

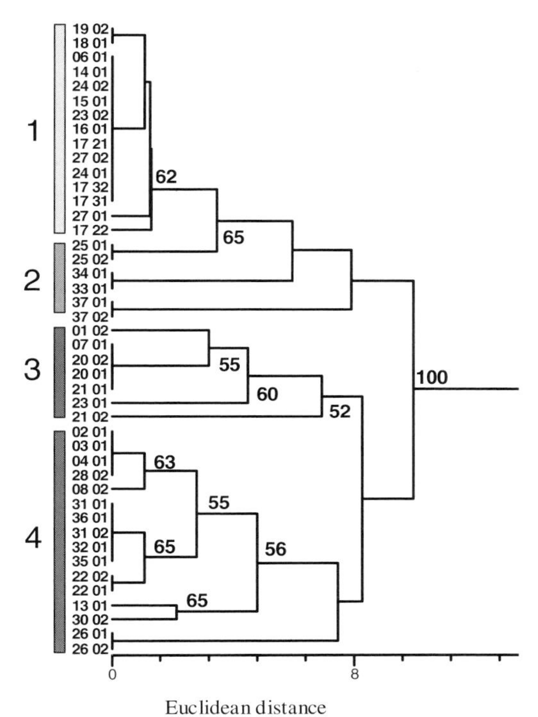

Figure 2. Cluster analysis for 44 *X. translucens* pv. *graminis* isolates based on 137 AFLP markers. Isolates are labeled by four digit numbers and bars indicate the four main clusters.

3.3 Pathogenicity

Pathogenicity screening on three *L. multiflorum* cultivars showed that only the 28 isolates classified as *Xtg* were significantly pathogenic while the *Xta* isolate and other *Xanthomonas* spp. showed no or not typical symptoms. Statistical analysis revealed significant differences in susceptibility to *Xtg* among the three cultivars assessed (Table 1). While some *Xtg* isolates showed significantly higher pathogenicity than others with scores ranging from 3.00 to 8.09, 24 out of 30 pathogenic isolates showed intermediate pathogenicity with scores ranging from 4.5 to 6.5 (data not shown). This variation is comparable to the findings of Michel (2001) who observed significant variation in pathogenicity in a small collection of *Xtg* isolates.

Table 1. Pathogenicity score for 29 *Xtg* isolates on three *Lolium multiflorum* cultivars.

	Pathogenicity[1]		
Cultivar	Minimal score	Maximal score	Mean score[2]
Adret	4.32	8.09	7.00 a
Ligrande	3.30	6.44	5.25 b
Axis	3.00	6.03	4.32 c
Mean	3.54	6.65	5.53

[1] Pathogenicity was assessed by scoring disease symptoms (1 = no symptoms, 9 = complete wilting)
[2] Means with different letters are significantly different according to Duncan's multiple range test ($P<0.05$)

4. CONCLUSIONS

The method presented here provides a valuable tool to classify *Xanthomonas* isolates collected from forage grasses and to assess the variability among *Xtg* isolates. With two exceptions, all putative *Xtg* isolates collected throughout Switzerland were confirmed as *Xanthomonas translucens* pv. *graminis* by means of 16S rDNA sequencing and AFLP analysis. AFLP diversity among *Xtg* isolates was relatively small when compared to *Xta,* a closely related, non-pathogenic isolate. The relatively small variation in pathogenicity and the limited AFLP diversity indicate that a small number of bacterial isolates may be representative for an efficient resistance screening in breeding programs.

REFERENCES

Alizadeh A, Arlat M, Sarrafi A, Boucher CA, Barrault G (1997) Restriction fragment length polymorphism analyses of Iranian strains of *Xanthomonas campestris* from cereals and grasses. Plant Dis. 81: 31-35.

Channon AG, Hissett R (1984) The incidence of bacterial wilt caused by *Xanthomonas campestris* pv *graminis* in pasture grasses in the West of Scotland. Plant Pathol. 33: 113-121.

Da Silva ACR, Ferro JA, Reinach FC, Farah CS, Furlan LR, Quaggio RB et al. (2002) Comparison of the genomes of two *Xanthomonas* pathogens with differing host specificities. Nature 417: 459-463.

De Parasis J, Roth DA (1990) Nucleic acid probes for identification of phytobacteria: identification of genus-specific 16s rRNA sequences. Phytopathology 80: 618-621.

Dye DW, Bradbury JF, Goto M, Hayward AC, Lelliott RA, Schroth MN (1980) International standards for naming pathovars of phytopthogenic bacteria and a list of pathovar names and pathotype strains. Rev. Plant. Pathol. 59: 153-168.

Egli T, Goto M, Schmidt D (1975) Bacterial wilt, a new forage grass disease. Phytopath Zeitschr 82: 111-121.

Egli T, Schmidt D (1982) Pathogenic variation among the causal agents of baterial wilt of forage grasses. Phytopath Zeitschr 104: 138-150.

Leyns F (1993) *Xanthomonas campestris* pv. *graminis:* Cause of bacterial wilt of forage grasses. In: *Xanthomonas,* Swings JG, Civerolo EL (eds.), pp.55-57. Chapman & Hall, London.

Leyns F, Van den Mooter M, Swings J, De Cleene M, De Ley J (1981) Distribution of *Xanthomonas campestris* pv. *graminis* in fields of forage grasses in northern Belgium. Parasitica 37:131-133

Michel VV (2001) Interactions between *Xanthomonas campestris* pv. *graminis* strains and meadow fescue and Italian rye grass cultivars. Plant Dis. 85: 538-542.

Schmidt D (1988) Le flétrissement bactérien des graminées fourragères: essais pour limiter la dispersion de la maladie lors du fauchage. Rev. Suisse Agric. 20: 351-357.

Valsangiacomo C, Baggi F, Gaia V, Balmelli T, Peduzzi R, Piffaretti J (1995) Use of amplified fragment length polymorphism in molecular typing of Legionella pneumophila and application to epidemiological studies. J. Clinical Microbiol. 33: 1716.

Vauterin L, Hoste B, Kersters K, Swings J (1995) Reclassification of *Xanthomonas*. Int. J. Syst. Bact. 45: 472-489.

Vos P, Hogers R, Bleeker M, Reijans M, Vandelee T, Hornes M, Frijters A, Pot J, Peleman J, Kuiper M, Zabeau M (1995) AFLP - a new technique for DNA fingerprinting. Nucleic Acids Res. 23: 4407-4414.

Wang HM, Sletten A (1995) Infection biology of bacterial wilt of forage grasses. J. Phytopathol. 143: 141-145.

Molecular Breeding and Functional Genomics for Tolerance to Abiotic Stress

M. W. Humphreys[1], J. Humphreys[1], I. Donnison[1], I. P. King[1], H. M. Thomas[1], M. Ghesquière[2], J-L. Durand[2], O. A. Rognli[3], Z. Zwierzykowski[4] and M. Rapacz[5]

[1]Institute of Grassland and Environmental Research, Plas Gogerddan, Aberystwyth, SY23 3EB, UK. [2]INRA, Unite de Genetique et d'Amelioration des Plantes Fourrageres, 86600, Lusignan, France. [3]The Agricultural University of Norway, Department of Chemistry and Biotechnology, PO Box 5040, 1432 Aas, Norway. [4]Institute of Plant Genetics, Polish Academy of Sciences, Strzeszynska 34, 60-479 Poznan, Poland. [5]Agricultural University of Cracow, Department of Plant Physiology, Podluzna 3, Krakow, Poland. (Email: mike.humphreys@bbsrc.ac.uk).

Key words: *Lolium/Festuca*, drought resistance, winter hardiness, introgression mapping, androgenesis, precision breeding.

Abstract: Sustainability is a measure of our ability to produce food with the maximium of efficiency combined with the minimum of damage to the environment. Grasslands represent over 40% of all agricultural land in the European Union, and over 70% in the United Kingdom. Whilst *Lolium* in Europe is considered to be the ideal source of profitable and safe high quality animal forage, its general poor persistency limits its use to favourable growing areas. Fortunately, genes for abiotic stress resistance are transferred readily from closely related *Festuca* species by conventional breeding technologies. Introgression mapping allows the assembly of desirable gene combinations and molecular markers to assist with their selection in breeding programmes. Additional new androgenesis techniques have led to novel genotypes rarely observed as outcomes of breeding programmes. *Lolium* x *Festuca* hybrids display promiscuous chromosome recombination enabling genes from one species to be transferred readily to homoeologous chromosome regions where they both function normally and remain stable. Despite the close homology between *Lolium* and *Festuca* species, repetitive DNA sequences differ sufficiently for their genomes to be distinguished, by genomic in situ hybridisation (GISH). This enables the physical mapping of genes for abiotic stress resistance transferred from *Festuca* to *Lolium*. Further chromosome recombination between homoeologous *Lolium* and *Festuca* sequences enables *Festuca* introgressions to be "dissected", and recombination series created. Knowledge of synteny and gene sequences within model species amongst the Poaceae, combined with the development of sequenced molecular markers, and bacterial artificial chromosomes is enabling the isolation of genes for abiotic stress resistance.

A. Hopkins, Z.Y. Wang, R. Mian, M. Sledge and R.E. Barker (eds.), Molecular Breeding of Forage and Turf, 61-80.

1. INTRODUCTION:

The diversity of demands on grassland agriculture has never been higher. Grasslands occupy vast areas of land used in European agriculture (over 40% in the EU). Whilst they are required as sources of good quality and healthy animal fodder, they now have multifunctional requirements which include environmental, social, and cultural aspects. This complexity of objectives places enormous demands on the agricultural industry. These new demands require precision breeding for targeted traits and necessitate better understanding of the complicated underlying mechanisms, their interaction, and their genetic control. Recent developments leading to better understanding of adaptations in grasses to a range of abiotic stresses are reviewed in this article. The developments will in the near-future make "designer breeding" a feasible objective whereby grass varieties will be tailored to specific climatic and edaphic growth conditions. Progress made in the course of the EU-funded Framework V project "Sustainable grasslands withstanding environmental stresses" (SAGES) will be presented.

Whilst technologies such as gene transformation may become effective procedures for improving adaptation of grass varieties to abiotic stresses, conventional breeding through sexual hybridisation is still the principal route for the development of stress resistant varieties. The persistency and productivity of high quality *Lolium* swards is fundamentally affected by abiotic stress, being generally poorly adapted to growth under severe stress conditions. Improvements in the resistance of *Lolium* to both summer and winter stresses involve the use of gene transfer from closely-related and more robust *Festuca* species.

2. BREEDING FOR IMPROVED STRESS RESISTANCE

The *Lolium-Festuca* complex comprises species that occupy different eco-geographical habitats, but some species have sympatric distributions and a few natural hybrids are found in N. W. Europe and the British Isles. Species and ecotypes within the *Lolium-Festuca* complex offer the breeder options to design cultivars with gene combinations that determine valuable and complementary traits. These include good establishment, high productivity, and nutritious forage (*Lolium* traits) and persistency, adaptations to summer or winter stresses, and to growth under low fertiliser applications (*Festuca* traits). The four major species of agricultural importance are *L. perenne*, *L. multiflorum*, *F. pratensis* (all $2n = 2x = 14$) and *F. arundinacea* ($2n = 6x = 42$). Other *Festuca* species, used primarily as sources of improved drought and heat-tolerance for *Lolium* are *F. mairei* and *F. glaucescens* (both $2n = 4x = 28$).

2.1 Amphiploidy

One method of combining desirable traits from related species is amphiploidy in which complete parental genomes are combined and homologous chromosome pairing and disomic inheritance encouraged by chromosome doubling. The disadvantages of this procedure, are the frequent instability of such hybrids (Thomas and Humphreys 1991), and the presence of deleterious gene combinations that may reduce crop yields and quality and negate benefits accrued from improved stress resistance. The two amphidiploid IGER bred varieties "Elmet" (*L. multiflorum* x *F. pratensis*) and "Prior" (*L. perenne* x *F. pratensis*) performed well in Europe and showed particular promise for grazing in Canada and Australia (Suzuki 1972; Breese and Lewis 1984). The two varieties also attracted interest in the USA (Casler 1987) and were used in the development of the variety Spring Green. More recently, greater success at developing amphidiploid varieties between these diploid *Lolium* and *Festuca* species has been made by plant breeders in Central and Eastern Europe (for example, Zwierzykowski et al. 1998).

In amphiploid breeding where complete genomes of the progenitors are combined and maintained as entities, chromosome pairing should, ideally, be restricted to homologous partners with no homoeologous pairing. However, the level of preferential chromosome pairing is insufficient to completely preclude homoeologous pairing causing genetic instability (Jauhar 1975; Lewis 1980). Genomic in situ hybridisation (GISH) can differentially paint *Lolium* and *Festuca* chromosomes and the extent of homoeologous recombination can be visualised in advanced generations of the amphiploids (Thomas et al. 1994; Zwierzykowski et al. 1998; Canter et al. 1999).

Various allopolyploid hybrids involving *Lolium* and polyploid *Festuca* species have also been developed including that between *L. multiflorum* x *F. arundinacea*.The allotetraploid hybrid between *L. multiflorum* and *F. glaucescens*, one of the ancestral species of *F. arundinacea*, is presently being developed in France as a potential cultivar (Ghesquière et al. 1996), and combines good agronomic characters of *Lolium* with the drought resistance of *Festuca*.

2.2 Introgression

The alternative to the amphiploidy approach at combining traits from *Lolium* and *Festuca* is introgression, which allows a limited number of genes to be introduced from a donor species into the reconstituted genome of the recurrent species by recombination and selection. Whilst the amphiploid approach requires homologous chromosome pairing, introgression requires widespread homoeologous chromosome pairing and recombination for gene transfer. This occurs regularly in *Lolium* x *Festuca* hybrids and GISH has proved of particular value in locating sites of intergeneric recombination both in backcross and androgenesis-derived populations (Thomas et al. 1994; Humphreys et al. 1998a).

The first example of the use of GISH to locate genes for abiotic stress resistance introgressed into a crop plant involved the transfer of genes for salt-tolerance from the grass *Thinopyrum bessarabicum* into wheat (King et al. 1993). The use of the technique to locate the presence of *Festuca* genes in *Lolium* either for drought resistance or winter-hardiness traits is described later in this article. The effects of genome composition and homology on preferences for chromosome pairing were key to the development of the introgression breeding approach in *Festulolium* hybrids. A genome combination comprising as part, a diploid set of *Lolium* chromosomes, was found a prerequisite for successful introgression in *Lolium* x *Festuca* hybrid breeding programmes (Humphreys et al. 1998a). Whilst triploid *L. multiflorum* x *F. pratensis* (LmLmFp) hybrids are generally fertile, hybrids with two genomes of *Festuca* (FpFpLm) are sterile. Until recently, the sterility of the FpFpLm triploid hybrids has prevented successful introgression of *Festuca* genes into *Lolium*. However, androgenesis of amphiploid *Festulolium* cultivars has overcome this difficulty (see below). Dihaploid plants produced by androgenesis from amphiploid cultivars have *Lolium-Festuca* recombinant chromosomes and sufficient fertility to be backcrossed onto diploid *F. pratensis* (unpublished data) providing opportunities for the transfer of genes for high quality herbage production to the stress-tolerant *Festuca* species.

F. pratensis has a number of traits of value if introgressed into *Lolium* and it is generally well adapted to harsh winter conditions (Humphreys et al. 1998b). Canter (2000a) reported early indications that ice-tolerance had been transferred from *F. pratensis* to *L. perenne*. One recent outcome of the SAGES project was the observation that freezing-tolerance of *L. perenne* was enhanced by an introgression from chromosome 3 of *F. pratensis* onto a homoeologous *Lolium* chromosome (unpublished). In introgression lines derived from another *L. perenne* x *F. pratensis* hybrid population, plants with good winter hardiness (over three Norwegian winters) and freezing-tolerance were also found that carried with remarkable consistency, *Festuca*-derived genes at the same chromosome location.

In pioneering work on the efficacy of introgression mapping genes transferred from *Festuca* into *Lolium*, the delayed-senescence gene (*sid*) was transferred from *F. pratensis* into *Lolium* species and its physical location mapped by using GISH on the introgression line (Thomas et al. 1994; Thomas et al. 1997). When homozygous, the allele confers a staygreen phenotype in *Lolium,* and the introgressed allele has been tagged with genetic markers, some as close as 1cM (Moore, personal communication). Near isogenic introgression lines of *L. temulentum* have been used in representational difference analysis to identify candidate cDNAs for *sid*, or genes regulated by *sid* (Thomas et al. 1999). "Breeders' toolkits" have been designed to facilitate the selection of *sid* in amenity grass breeding programmes. The first gene in *Lolium* to be mapped in this way to a known chromosome (chromosome 1, formerly known as chromosome 6) was the isozyme locus *pgi/2* (Humphreys *et al.*, 1997). The *pgi/2* locus is now known to be associated closely with QTL for water soluble carbohydrate, and freezing-tolerance (Alm et a.l. 2003, submitted for publication). It is also linked to the *s*-incompatibility locus (Cornish et al. 1980).

The pentaploid hybrids between autotetraploid *L. multiflorum* and *F. arundinacea* (6x) are generally male fertile (Humphreys 1989). After only two backcrosses onto *L .multiflorum* (2x), the diploid genome of the recurrent parent is reconstituted (Humphreys 1989). Using such a pentaploid hybrid Humphreys and Thomas (1993) reported that *L. multiflorum* lines had been produced capable of withstanding 3 months continuous drought and that this drought resistance had been inherited into a further generation. An introgressed chromosome segment from *Festuca* was subsequently found on the long arm of chromosome 3 of *Lolium* (formerly referred to as chromosome 2) (Humphreys and Pašakinskienė 1996).

F. glaucescens and *F. mairei,* two tetraploid (2n=4x=28) species, are potentially useful sources of genes for drought resistance or heat tolerance of *Lolium* species. *F. glaucescens* is very drought resistant and has attracted interest from geneticists and plant breeders alike (Ghesquière et al. 1991 1996; Humphreys J. et al. 2002). *F. mairei,* a North African species is a potential source of both improved drought and heat tolerance for *Lolium* (Chen and Sleper 1999). To introgress traits from these tetraploid *Festuca* species into *Lolium* backgrounds, an ecotype of the tetraploid *Festuca* is crossed to a synthetic tetraploid *Lolium* to produce a fertile tetraploid F_1 that is backcrossed twice onto diploid *Lolium*. The BC_2 are predominantly diploid with introgressed *Festuca* chromosome segments (Morgan et al. 2001). Traits recovered in the BC_2 included a potentially useful rhizomatous character and recently during the SAGES project, drought resistance (Humphreys J. et al. 2002).

3. MARKER ASSISTED SELECTION AND THE DEVELOPMENT OF INTROGRESSION MAPPING

A prerequisite for introgression breeding programmes is access to technologies that facilitate selection of desirable donor genes during recurrent selection. Increasingly, introgression programmes are assisted by the use of genetic markers in order that traits of interest are selected through different plant generations. Markers such as restriction fragment length polymorphisms (RFLPs), amplified fragment length polymorphisms (AFLPs), single nucleotide polymorphisms (SNPs), and simple sequence repeats (SSRs), are now being used regularly to genetically map agronomic traits of interest. Such traits are often under the complex control of a number of genes and map as quantitative trait loci (QTL) on one or often several linkage groups.

An unusual combination of features of *Lolium* x *Festuca* hybrids are that firstly the chromosomes pair freely and recombine in the hybrids and secondly the chromosomes can be differentiated by GISH. This has led to a powerful new approach to genetic mapping. Introgression mapping arises from a fusion of physical and genetic mapping (Thomas et al. 1994; Humphreys et al. 1997; King et al. 1998). In a *Lolium* diploid plant with only one introgressed *Festuca* segment, the gene(s) for any *Festuca* derived trait expressed by the plant must be located within the segment. Therefore, *Lolium* introgression genotypes expressing a *Festuca* derived trait are first analysed by GISH to identify genotypes with only one *Festuca* chromosome segment. That genotype is then screened with molecular markers alongside the parental genotypes and any marker present in the *Festuca* parent but absent in the *Lolium* must be located in the *Festuca* segment (and therefore linked to the *Festuca* derived trait). These markers can be mapped in a further backcross generation and a dense but highly localised map of the *Festuca* segment made in isolation of the *Lolium* genome (Armstead et al. 2001; King et al. 2002). The close homology between *Lolium* and *F. pratensis* allows chromosome substitution lines to be developed where a *Lolium* chromosome is substituted by its *Festuca* homoeologue. Following another backcross onto *Lolium*, a recombination series is created with plants containing different sized segments of the *Festuca* chromosome. The intention is to identify seven genotypes of *L. perenne*, each with a different *Lolium* chromosome substituted by its *F. pratensis* homoeologue. The chromosome can be mapped both genetically and physically in a BC_2 recombinant series (King et al. 2002).

4. ANDROGENESIS

Incidents of intergeneric recombination occur in all regions of the *Lolium* and *Festuca* genome (Canter et al. 1999), and provide a vast source of genetic variation that may be recovered through the gametes. Whilst conventional plant breeding programmes provide many opportunities for combining useful traits, pre- and post-zygotic selections will inevitably limit access to all potentially useful gene combinations. In addition, many useful gene combinations in breeding programmes may remain "hidden" due for example, to the non-expression of recessive alleles, epistasis, or pleiotropy. Humphreys et al. (2001) and Zare et al. (2002) analysed androgenic plants (that is, the product of haploid or polyhaploid gametes) derived from *Lolium/Festuca* hybrids. Certain genotypes displayed extremes of drought resistance and/or freezing tolerance in excess of either their *Lolium* or *Festuca* parent or the F_1 hybrid from which they were derived. An androgenic population in effect "dissects" the genome of its parent plant and reveals an array of phenotypic variation that may be used by the plant breeder.

L. multiflorum x *F. pratensis* amphidiploid (2n = 4x = 28) cultivars have recently received much attention as sources of fertile androgenic lines (Leśniewska et al. 2001). Such cultivars represent the products of several generations of seed multiplication and follow considerable genome restructuring. The outcome is almost perfect mixing of the genomes with no obvious deleterious effects with regard to hybrid fertility and stability. This combination of high levels of intergeneric recombination and balanced gametes, is an ideal starting point for androgenesis, where novel gene combinations and viable culture-responsive pollen microspores are desired. However, the procedure has no value to plant breeding unless it is later possible to produce fertile androgenic genotypes and new populations from selected stress resistant genotypes. Fortunately, genotypes with both male and female fertility were found in the dihaploid androgenic *L. multiflorum* x *F. pratensis* populations, and it was possible to produce progeny following hybridisation with *L. multiflorum* or with *F. pratensis*.

5. DROUGHT RESISTANCE AND WINTER HARDINESS IN GRASSES

5.1 Drought resistance:

Drought is a complex phenomenon, involving not just a lack of available water, but also high transpiration rates, supra-optimal temperatures, photooxidation, mineral deficiency, and hard soil. Their relative importance varies with location and year. Definitions of drought resistance must ultimately take into account the climate and prevailing agriculture. An over-riding requirement is that the crop should survive and re-grow rapidly when autumn rains set in. A secondary requirement is that yields should not be greatly reduced during mild drought (water deficit less than 50 – 100 mm), a response which occurs in many Mediterranean grasses which rapidly become "quiescent" at onset of drought.

Drought resistance (plant survival, rapid growth following re-watering, minimum growth during drought) results from a combination of traits. The traits are not all independent one from the other. The resistant ideotype has to insure some trade off. Plants need to adjust transpiration to absorption. Transpiration is increased by leaf area, while water absorption is increased by root depth. Leaf area cannot be minimised too much for a minimum productivity is expected during drought. Hence, maximum root depth is to be combined with optimum leaf expansion and good control of water loss per unit leaf area (via the cuticule and stomata). Four main groups of traits can be defined and given hierarchy, by order of importance: a) floral phenology, which determines indirectly the amount of growth devoted to roots and the density of vegetative tillers; b) root depth and water status; c) leaf production and extension; and d) regulation of transpiration

Outcomes of the EU funded SAGES programme, include the development of mapping families for drought resistance in *L. multiflorum* resulting from introgression from *F. arundinacea* and *F. glaucescens.* In both mapping families, *Festuca* genes were introgressed onto *Lolium* chromosome 3, but at different chromosome locations (unpublished). A putative association with the presence of *F. glaucescens* genes and osmotic adjustment was found. The principal physiological studies at INRA concerned the depth of water extraction, and on stomatal conductance. Pre-dawn leaf water potential was found to be a sensitive measurement of depth of water extraction. Water extraction was measured at 15cm depth intervals using ^{18}O. Most extraction occurred in the top 50cms of soil, but introgression lines were found to have a deeper extraction than even the *Festuca* control. Stomatal conductance is better controlled under drought in *Lolium* than *Festuca.* Indications were that certain *Lolium* x *Festuca* hybrids displayed rooting traits of the *Festuca* parent and the stomatal conductance of *Lolium.* At IGER, water content in soils was found to be greater under plots of *F. arundinacea* than *L. multiflorum* both under irrigated and drought conditions. A population derived from a *L. multiflorum* x *F. arundinacea* hybrid displayed the greater water-use-efficiency found in *F. arundinacea* and withdrew less water from the soil. Other genotypes with introgressions from *F. arundinacea* and *F.*

glaucescens grew efficiently (better than *Lolium*) under low nitrogen levels. Both findings may be significant for future developments of more sustainable high quality grasslands. Root development was assessed on a *L. perenne* mapping family. Quantitative Trait Loci (QTL) associations with root development were found across the genome but chromosomes 2 and 6 were especially important.

5.2 Winter Hardiness:

To survive the winter, a plant must evolve mechanisms whereby freezing-sensitive tissues can avoid freezing or undergo freezing tolerance compatible with the normal variations of the local climate, co-ordinate the induction of the tolerance at the appropriate time, maintain adequate tolerance during times of risk, and properly time the loss of tolerance and resumption of growth when the risk of freezing has passed (Guy 1990). Winter-hardiness is a complex trait, the factors involved including: a) freezing temperatures; b) fluctuating temperatures; c) wind; d) snow cover; e) ice encasement; f) heaving; g) low-light and h) low-temperature pathogens. Development of winter-hardiness requires exposure of plants to low non-freezing temperatures typically 0°C to 10°C, and shortened photoperiod. Many physiological and biochemical changes occur during cold acclimation (CA), including slowed or arrested growth, reduced tissue water content, altered cell pH, protoplasm viscosity and photosynthetic pigments, reduced ATP levels (Levitt, 1980), transient increases in ABA (Chen et al. 1983), changes in membrane lipids (Uemura and Steponkus 1994), accumulation of compatible solutes including proline, betaine, polyols and soluble sugars, and accumulation of antioxidants (Tao et al. 1998).

The rate and extent of de-hardening is a critical factor in winter survival. Overwintering plants are particularly susceptible to freezing damage in the spring if the deacclimation process occurs prematurely or too rapidly, or if unpredictable temperature fluctuations occur (Levitt 1980; Gay and Eagles 1991). Eagles (1989) suggested that the nature of an adaptive CA process would vary with the stability and predictability of winter conditions in a particular environment. In stable and predictably cold continental climates where the onset of freezing temperatures is rapid, a photoperiod-triggered and rapid acclimation process is desirable, while in the more variable and less severe conditions of a maritime climate, a temperature dependent response might enable plants to exploit a mild autumn or spring by continuing to grow. However, in cultivars adapted to maritime climates, deacclimation may occur in response to fluctuations in winter and spring temperature with a risk of damage by subsequent frosts (Eagles 1994).

Temperate grasses store fructans, a soluble polymer capable of rapid polymerization and depolymerization. The partitioning of solutes may be important as survival depends on survival of apices, particularly the lateral buds rather than mature leaf tissue (Eagles et al. 1993).

Molecular mechanisms involved in CA are largely unknown but information from model species whose genomes have recently been sequenced such as *Arabidopsis* and rice and the development of microarray technologies are giving insight into the complexity of the processes. Differential display can be used to identify acclimatory changes in gene expression. Consistent differences have been found between cDNAs from cold-acclimated and non-acclimated plants. Representational difference analysis was used to selectively amplify cDNA fragments from cold-induced *F. pratensis* seedlings, and to identify cDNAs upregulated during cold acclimation (Canter et al. 2000b). These included homologues of small ribonucleic proteins (SnRNPs) and a homologue of the chloroplast encoded gene *psbA* which codes for the D1 protein of photosystem II (PSII). The importance of PSII in CA will be emphasised later.

The first gene isolated during CA was *cor14b* (Cattivelli and Bartels, 1990) which is homologous to wheat *wsc19* and is induced only by low temperature, and enhanced by exposure to light (Crosatti et al. 1999). Plant exposure to cold also promotes the accumulation of early light inducible proteins (ELIPs) which conregate in the chloroplast stroma in the vicinity of the D1 protein (Shimosaka et al. 1999).

Freezing-tolerant plants withstand extracellular ice formation. During CA proteins accumulate in the apoplast of winter rye and extracts from the apoplast modify ice (Griffith et al. 1992). The accumulation of rye AFPs during CA correlates with the development of freezing-tolerance (Marentes et al. 1993). A heat-stable antifreeze protein has been isolated from *L. perenne* (Sidebottom et al 2000). The *Lolium* AFP causes only slight thermal hysteresis (0.1°C) but is very efficient at inhibiting ice recrystallization. Recently, a homoeologous *F. pratensis* AFP sequence has been isolated at IGER using a bacterial artificial chromosome (BAC) library (J. Humphreys, unpublished). The AFP sequence is encoded by two closely linked gene loci on *F. pratensis* chromosome 2.

5.2.1 The Potential Role of Photoreceptors in Plant Responses to Cold

Cold acclimation and freezing tolerance are the result of a complex interaction between low temperature, light, and photosystem II (PSII) excitation pressure. At low temperatures, plants have two principal difficulties. The first is maintenance of cell membranes in a fluid state. This can be compromised further by ice formation (Thomashow 1999). The second problem relates to thermodependency of photosynthetic electron transport and carbon fixation, which are slowed at low temperature (Guy 1990; Huner et al. 1996). Modulations in light intensity, wavelength or daylength affect a large number of developmental and physiological processes in plants, including germination, greening, phototropism, floral initiation, and development of circadian rhythms (Ahmad 1999). The PSII reaction centre is the key site for regulation of light energy and also the main site of photoinhibitory damage. The D1/D2 protein dimer at the core of PSII appears to be crucial in maintaining the integrity of the complex (Mattoo et al. 1989). Photoinhibition is related to the redox-state of PSII expressed as excitation pressure. The redox-state of PSII reflects fluctuations in the photosynthetic energy balance and so acts as a sensor of any environmental stresses that disturb that balance. Changes to the redox-state of PSII, triggered by a low temperature shift, have been proposed to be one of several potential temperature-sensing mechanisms involved in cold acclimation (Rapacz 2002a, b).

Cessation of elongation growth during cold acclimation and also compact plant morphology (Andersson and Olsson 1961) have long been considered one of the main requirements for proper hardening and overwintering. Elongation growth is considered as competition for cold acclimation because it causes an increase in water content and water potential in cells, facilitating ice formation and cell death during freezing events. Changes related to growth under elevated PSII excitation pressure are detectable in herbaceous plants in early autumn and this phase of plant acclimation was named pre-hardening (Rapacz 1998). Since changes in the redox-state of PSII seem to control changes in plant morphology and elongation growth rate during pre-hardening and cold acclimation, PSII may be a temperature-sensor involved during cold de-acclimation.

Recent work undertaken at the Agricultural University of Cracow as part of the EU-funded SAGES project (Rapacz et al. 2003 - submitted for publication) using dihaploid *Festulolium* androgenic plants reported on the relations of cold tolerance and maximum quantum yield of PSII. The overall aim was to discover if any of the androgenic forms had developed mechanisms of avoidance of photoinhibition, and thus greater cold tolerance, than the parent material since it is known that androgenesis may result in gene expression not present or not revealed in parent lines (Humphreys et al. 1998a). There was a strong correlation over two years (-0.71 and -0.67) between maximum quantum yield of PSII (F_v/F_m) before winter and winter survival. Plants with higher F_v/F_m had lower winter survival, but there were no consistent changes in current quantum yield of PSII, suggesting that a major role was played by cold-induced photoinactivation of PSII. Cold acclimation under controlled conditions demonstrated diverse mechanisms of photosynthetic acclimation and decreased F_v/F_m. In most winter hardy plants, increases in non-photochemical energy quenching of chlorophyll *a* fluorescence (NPQ) and in one case an increase in photochemical quenching was observed during cold acclimation. Exposure of plants to high light at low temperatures showed that winter hardy *Festulolium* plants are more resistant to cold-induced inactivation of PSII. In contrast to previous reports for herbaceous Poaceae that the increased resistance to photoinhibition during cold acclimation was always accompanied by increases in photosynthetic capacity, in these genotypes, the most common reason for higher resistance to cold induced photoinhibition was an increase in NPQ. However, in one of the androgenic genotypes, reduced NPQ was compensated for by increased electron transport, a different mechanism of avoidance of photoinhibition to that observed in the parent material. Thus the process of producing androgenic lines had revealed further sources of variation in mechanisms of avoidance of photoinhibition giving rise to higher cold tolerance.

NDong et al. (2002) reported that expression of the gene *Wcs19* was correlated with PSII excitation pressure measured in vivo as the relative reduction state of PSII. Three different groups of late embryogenesis abundant (LEA) proteins - LEA3-L1; LEA3L-2, & LEA 3-L3 share identities with WCS19. The *Wcs19* gene localises within the chloroplast stroma of wheat and rye. Western analysis demonstrated correlations with LEA3-2 accumulation and freezing tolerance in rye and wheat. When the *Wcs19* gene was introduced into *Arabidopsis*, this resulted in an increase in freezing tolerance.

6. GENETIC ANALYSIS OF ABIOTIC STRESS RESISTANCE

Successful breeding depends on broad understanding of the genetic architecture of relevant traits. Genes with major effects and genes contributing to the expression of quantitative traits (QTL) both have a role in controlling abiotic stress tolerance. The molecular basis of a QTL can be explained either by a gene directly involved in the biochemical pathway leading to a phenotype or by a transcription factor controlling the expression of many genes. Candidate genes are being used in searches for possible co-segregation with known QTL. Two dehydrin loci (*Dhn1/Dhn/2* and *Dhn/9*) are located in the same region of chromosome group 5 in the Triticeae as the vernalisation and frost-resistance loci *Vrn*-1 and *Fr1*, cold and salt-tolerance QTL and ABA QTL have been mapped. Other dehydrins (*Dhn3, Dhn4,*

Dhn5, Dhn7, and *Dhn/8*) on chromosome 6 are associated with drought-tolerance QTL. Unfortunately, only a relatively few studies provide genetic evidence that a QTL for stress-tolerance can be explained by the presence of a co-mapping stress-related gene. This raises the possibility that the molecular basis of a QTL for stress-tolerance could be explained by a regulatory gene able to control the expression of many-stress related genes. Vagujfalvi et al. (2000) demonstrated that the expresion of *cor14b* in wheat is controlled by two loci (*Rcg1* and *Rcg2*) in the *Vrn-1A/Fr1* region of chromosome 5A.

An interesting finding by Ceccarelli et al. (2002) was the genome plasticity of the polyploid grass species, *F. arundinacea* which appeared to change in numbers of interspersed DNA repeats as a response to temperature changes in seedlings exposed to 10ºC or 30ºC. The relevant sequences were dispersed along the length of all chromosomes, but conregated around centromeric regions of certain chromosomes.

Comparisons with the genetic maps of the Triticeae cereals, oats, and rice have revealed significant regions of colinearity, enabling the *Lolium* genetic map to be integrated with those of other Poaceae species (Jones et al. 2002). The map was constructed with RFLP, AFLP, isoenzyme, and EST data from collaborating laboratories within the International *Lolium* Genome Initiative (ILGI). For the *Lolium* markers that could be assigned to corresponding positions in the Triticeae map (which covered 70% of the *Lolium* map), linkage groups 1, 3, and 5 showed complete synteny, whilst linkage groups 2, 4, and 7 each contained small nonsyntenic regions, and linkage group 6, a larger nonsyntenic region. A lower level of synteny was reported between *Lolium* and oat than between *Lolium* and the Triticeae, despite the closer taxanomic affinity between the species.

Drought, low temperature and salinity are the most important abiotic stress factors limiting crop productivity. A genomic map (Cattivelli et al. 2002) of major loci and QTL affecting stress tolerance in the Triticeae identified the importance of group 5 chromosomes for heading date, frost, and salt tolerance. A conserved region with a major role in drought tolerance has also been found on chromosome 7. Some stress-related genes were shown to be linked to stress-tolerance QTL. Multiple-stress QTL and linked markers have also been detected, suggesting the existence of common mechanisms for different stresses, or of clusters of genes controlling different stress tolerance processes. Adaptation mechanisms are likely to be widely conserved among different plant species and knowledge gained from the Triticeae may be transferred directly to *Lolium* and *Festuca.* The LEA (late-embryogenesis-abundant) gene family is a good example, being induced following dehydration, cold treatment, ABA activity, salt, and osmotic stress. LEA genes have been separated into 3 gene classes in the Triticeae: LEA-1 genes encoding glycine-rich hydrophilic proteins with 1-4 copies of a conserved 20 amino acid repeat; LEA-2 genes or dehydrins encoding one or more lysine-rich 15 amino acid sequences, LEA-3 genes containing tandem 11 amino acid repeats. Basic dehydrins are highly expressed during dehydration, but not during cold treatment, whilst acidic dehydrins are mainly cold-responsive (Choi et al. 1999). Danyluk et al. (1998) found the acidic dehydrin WCOR410 accumulated in the vicinity of the plasmalemma in wheat, suggesting a putative role for cryoprotection of the cell membrane.

A number of sequences known as *cor* (or *blt*) genes have been found to be up-regulated by low temperature. Their expression may be affected also by ABA, dehydration or light. These stress-related sequences may have tissue-specific expression. For example, *blt14* in cold-acclimated barley is expressed only in the inner layers of the cortex and in cells surrounding vascular bundles (Pearce et al. 1998).

A QTL map for frost and drought resistance has been produced for *F. pratensis* (Alm *et al.* 2003, submitted for publication). The linkage map involved 501 AFLP and RFLP markers. Two major QTL for freezing-tolerance and 4 QTL for winter-survival were found. Heterologous wheat anchor probes indicated that *Frf4_1* on LG4 of *F. pratensis* was orthologous to the frost-tolerance loci *Fr1* and *Fr2* in wheat. QTL for winter survival were found on chromosomes 1, 2, 5, and 6. The QTL on chromosome 1 for winter survival was orthologous to a water-soluble carbohydrate QTL in *Lolium*. A putative QTL for resistance to moderate drought was found on chromosome 6 and may be orthologous to a mapped ABA locus in wheat. Other reported associations in this work included interestingly that survival against severe drought was associated with genes on chromosome 3. This supported the results described earlier from the GISH analysis made by Humphreys and Pašakinskienė (1996), that a *Festuca* introgression on *Lolium* chromosome 3 had enhanced the drought resistance of the *Lolium* genotype.

7. MICROARRAY TECHNOLOGIES

Increasingly genetic research of complex traits is moving from procedures aimed at simplification by "dissection" towards alternative attempts aimed at identifying the range of genes expressed when a genotype is challenged by a certain event such as the onset of soil water-deficit, or sub- or supraoptimal growth temperatures. Of necessity, the principal research undertaken thus far has concentrated on model species, especially *Arabidopsis* where detailed knowledge exists of the entire genome composition. At this stage in the development of "omic" based technology, it may be questionable how much is gained in understanding the mechanisms that underly such complex traits, and the basis of their genetic control. Cause (an adaptive response) and effect (an injury response) may be difficult to distinguish, but no one can argue that increases, or decreases in expression of very large numbers of genes result follow the onset of an abiotic stress. Having said that, stress-inducible genes have been used to improve the stress-tolerance of crops by genetic manipulation (for example, Bajaj et al. 1999). Hanson and Hitz (1982) argued that when stress is imposed rapidly, a greater number of responses will be injury-induced than under a slower long-term application of the relevant stress factor.

Microarray technologies (Schena et al. 1995; Eisen and Brown 1999) can be used to display cDNA sequences on a glass slide at a density of up to 1000 genes cm^{-2}. The arrayed sequences are hybridised to fluorescently-labelled cDNA probes prepared from RNA samples of different cell or tissue type. The products of stress-inducible genes, involved in the stress response, can be classified into two groups: those that protect directly against environmental stresses, and those that regulate gene expression and signal transduction (Bray 1997; Thomashow 1999). Expression analysis of drought, cold, and salt-inducible genes have shown the existence of several regulatory systems of stress-responsive gene expression. Seki et al. (2002) monitored the expression profiles of 7000 *Arabidopsis* genes under drought, cold, and high-salinity stress. They found in total 227 drought-inducible, 53 cold-inducible, and 194 high-salinity inducible genes by cDNA microarray technologies. Among these genes, 22 were induced by all three stresses including *rd29A/cor78*, *cor15a*, *kin1*, *kin2*, *rd17.cor47*, and *erd10* (Seki et al. (2002) and references therein). Thirty genes were induced by cold and drought stress, and 24 by cold and salt-stress. However, 70% of the salt-induced genes were also induced by drought stress indicating strong relationships between plants' reponses to these two stress factors.

Bray (2002), also in *A. thaliana* has attempted to categorise genes expressed under soil water deficit into those that support plant adaptation to drought and those which are expressed as result to lesions in metabolic and cellular function. As described above, this is a difficult challenge. Eleven percent of the identified genes were involved in detoxification as a consequence of oxidative stress, but it is not known whether they were induced directly by cellular water deficit or merely as result of oxidative stress. Bray (2002) identified approximately 130 genes in *A. thaliana* that were up-regulated by water-deficit, including those involved with signalling, and cellular detoxification. Whilst most work undertaken on gene reponse to abiotic stress has concentrated on up-regulation of gene expression, other genes may be down-regulated. In *Arabidopsis* subjected to drought stress, proline accumulation occurs. This in part occurs by the repression of genes involved in proline breakdown (Kiyosue et al. 1996).

Advances in understanding the effectiveness of stress responses are also being made using transgenic plant analyses (Hasegawa et al. 2000). The imminent availability of genomic sequence and global and cell-specific transcript expression data, combined with determinant identification based on gain- and loss-of-function molecular genetics provides the means of analysing adaptive traits. Genetic activation and suppression screens will lead to an understanding of the interrelationships of the multiple signaling systems that control stress adaptive responses in plants. The increasing evidence that genes responsible for abiotic stress resistance in rice and other cereals have a similar role in *Lolium* and *Festuca* should accelerate our understanding of the genetic basis of stress resistance in these important grasses and aid our methodologies aimed at precision breeding for targeted traits.

REFERENCES

Ahmad M (1999) Seeing the world in red and blue: insight into plant vision and photoreceptors. Curr. Opin. Plant Biol. 2: 230-235.

Alm V, Busso CS, Larsen A, Humphreys MW, Rognli OA (2003). Comparative mapping of quantitative trait loci controlling frost and drought tolerance in meadow fescue (*Festuca pratensis* Huds). Genetics (submitted).

Andersson G, Olsson G (1961) Cruciferous oilseeds. In: Kappert H, Rudolf W (eds) Breeding of special cultivated plants. In: Manual of plant breeding, Römer T, Rudorf W (eds.), Vol.5, pp. 1–66. Paul Barey, Berlin.

Armstead IP, Bollard A, Morgan WG, Harper JA, King IP, Jones RN, Forster JW, Hayward MD, Thomas HM (2001) Genetic and physical analysis of a single *Festuca pratensis* chromosome segment substitution in Lolium perenne. Chromosoma 110: 52-57.

Bajaj S, Targolli J, Liu LF, Ho THD, Wu R (1999) Transgenic approaches to increase dehydration-stress tolerance in plants. Mol. Breed. 5: 493-503.

Bray EA (1997) Plant responses to water deficit. Trends Plant Sci. 2: 48-54.

Bray EA (2002) Classification of genes differentially expressed during water-deficit stress in *Arabidopsis thaliana*: an analysis using microarray and differential expression data. Ann. Bot. 89: 803-811.

Breese EL, Lewis EJ (1984) Breeding versatile hybrid grasses. Span 27: 3-5.

Canter PH, Pašakinskiene I, Jones RN, Humphreys MW (1999) Chromosome substitutions and recombination in the amphiploid *Lolium perenne* x *F. pratensis* cv. Prior (2n=4x=28). Theor. Appl. Genet. 98: 809-814.

Canter PH (2000a) The use of genomic *in situ* hybridization (GISH) to locate introgressed chromosome segments from winter-hardy *Festuca* in a cold-sensitive *Lolium* background. Newsletter Gen. Soc., September 2000: 29-30.

Canter PH, Bettany AJE, Donnison I, Timms E, Humphreys MW, Jones RN (2000b) Expressed sequence tags (ESTs) during cold-acclimation in *Festuca pratensis* include a homologue of the chloroplast gene *psba*. J. Ex. Bot. 51 pp.72.

Casler MD (1987) Wisconsin forage grass performance trial summary. In: Ann. Rep., pp. 6-15. Univ. of Wisconsin, Madison.
Cattivelli L, Bartels D (1990) Molecular cloning and characterization of cold-regulated genes in barley. Plant Physiol. 93: 1504-1510
Cattivelli L, Baldi P, Crosatti C, Di Fonzo N, Faccioli P, Grossi M, Mastrangelo AM, Pecchioni N, Stanca AM (2002) Chromosome regions and stress-related sequences involved in resistance to abiotic stress in *Triticeae*. Plant Mol. Biol. 48: 649-665.
Ceccarelli M, Esposto MC, Roscini C, Sarri V, Frediani M, Gelati MT, Cavallini A, Giordani T, Pellegrino RM, Cionini PG (2002) Genome plasticity in *Festuca arundinacea*: direct response to temperature changes by redundancy modulation of interspersed DNA repeats. Theor. Appl. Genet. 104: 901-907.
Chen H-H, Li PH, Brenner ML (1983) Involvement of abscisic acid in potato cold acclimation. Plant Physiol. 71: 362-365.
Chen C, Sleper DA (1999) FISH and RFLP marker-assisted introgression of *Festuca mairei* chromosomes into *Lolium perenne*. Crop Sci. 39: 1676-1679.
Choi DW, Koag MC, Close TJ (1999) The barley (*Hordeum vulgare* L.) dehydrin multigene family: sequences, allele types, chromosome assignments, and expression characterstics of 11 *Dhn* genes of cv. Dictoo. Theor. Appl. Genet. 98: 1234-1247.
Cornish MA, Hayward MD, Lawrence MJ (1980) Self-incompatibility in ryegrass III. The joint segregation of S and PGI-2 in *Lolium perenne* L. *Heredity* 44: 55-62.
Crosatti C, Polverino de Laureto P, Bassi R, Cativelli L (1999) The interaction between cold and light controls the expression of the cold-regulated barly gene *cor14b* and the accumulation of the corresponding protein. Plant Physiol. 119: 671-680.
Danyluk J, Perron A, Houde M, Limin A, Fopwler B, Benhamou N, Sarhan F (1998) Accumulation of an acidic dehydrin in the vicinity of the plasma membrane during cold acclimation of wheat. Plant Cell 10: 623-638.
Eagles CF (1989) Temperature-induced changes in cold tolerance of *Lolium perenne*. J. Agric. Sci. 113: 339-347
Eagles CF (1994) Temperature, photoperiod and dehardening of forage grasses and legumes. In: Crop adaptation to cool climates, K. Dorffling, B. Brettschneider, H. Tantau, K. Pithan (eds). Report of Cost 814 European Commision workshop, Hamburg 12th – 14th Oct 1994.
Eagles CF, Williams J, Louis DV (1993) Recovery after freezing in *Avena sativa* L., *Lolium perenne* L. and *L. multiflorum* Lam. New Phytol. 123: 477-483.
Eisen MB, Brown PO (1999) DNA arrays for analysis of gene expression. Meth. Enzymol. 303: 179-205.
Gay AP, Eagles CF (1991) Quantitative analysis of cold hardening and dehardening in *Lolium*. Ann. Bot. 67: 339-345.
Ghesquière M, Durand F, Le Quilliec PP, Gaullier F (1991) Use of electrophoretic markers in chromosome manipulations of *Festuca* x *Lolium* hybrids. In: Proc. EUCARPIA Fodder Crops Section Meeting, Algero, Italy.
Ghesquière M, Emile J-C, Jadas-Hercart J, Mousset C, Traineau R, Poisson C. (1996) First in vivo assessment of feeding value of Festulolium hybrids derived from *Festuca arundinacea* var. *glaucescens* and selection for palatability. Plant Breed. 115: 238-244.
Griffith M, Ala P, Yang DSC, Hon W-C, Moffatt BA (1992) Antifreeze protein produced endogenously in winter rye leaves. Plant Physiol. 100: 593-596.
Guy CL (1990) Cold acclimation and freezing stress tolerance: role of protein metabolism. Ann. Rev. Plant Physiol. Plant Mol. Biol. 41: 187-223.
Hanson AD, Hitz WD (1982) Metabolic responses of mesophytes to plant water deficits. Ann. Rev. Plant Physiol. 33: 163-203.
Hasegawa, PM Bressan, RA Zhu, JK Bohnert, HJ (2000). Plant cellular and molecular responses to high salinity. Ann. Rev. Plant Physiol. Plant Mol. Biol. 51: 463-499.

Humphreys J, Thomas HM, Jones RN, Humphreys MW (2002) Sustainable grasslands withstanding environmental stresses (SAGES). Multi-function Grasslands. Quality Forages, Animal Products and Landscapes. In: Proc. European Grassland Federation, 7:310-311. 27-30 May 2002, La Rochelle, France.

Humphreys MW (1989) The controlled introgression of *Festuca arundinacea* genes into *Lolium multiflorum*. Euphytica 42: 105-116.

Humphreys MW, Thomas H (1993) Improved Drought Resistance in Introgression Lines Derived from *Lolium multiflorum* x *Festuca arundinacea* Hybrids. Plant Breeding 111: 155-161.

Humphreys MW, Pašakinskienė I (1996) Chromosome painting to locate genes for drought resistance transferred from *Lolium multiflorum* x *Festuca arundinacea* hybrids. Heredity 77: 530-534.

Humphreys MW, Thomas HM, Harper J, Morgan G, James A, Ghamari-Zare A Thomas H (1997) Dissecting drought- and cold-tolerance traits in the *Lolium-Festuca* complex by introgression mapping. New Phytol. 137: 55-60.

Humphreys MW, Zare A-G, Pašakinskienė I, Thomas H, Rogers W.J, Collin HA (1998a) Interspecific genomic rearrangements in androgenic plants derived from a *Lolium multiflorum* x *Festuca arundinacea* (2n = 5x = 35) hybrid. Heredity 80: 78-82.

Humphreys MW, Pašakinskienė I, James AR, Thomas H (1998b) Physically mapping quantitative traits for stress-resistance in the forage grasses. J. Exp. Bot. 49: 1611-1618.

Humphreys MW, Zwierzykowski Z, Collin HA, Rogers WJ, Zare A-G, Leśniewska A (2001) Androgenesis in grasses - methods and aspects for future breeding. In: B. Bohanec (Ed.) Biotechnological Approaches for Utilization of Gametic Cells, pp. 5-13. Proc. COST 824 final meeting, 1-5 July 2000, Bled, Slovenia.

Huner NPA, Maxwell DP, Gray GR, Savitch LV, Krol M, Ivanov AG, Falk S (1996) Sensing environmental temperature change through imbalances between energy supply and energy consumption: redox state of photosystem II. Physiol. Plant. 98: 358-364.

Jauhar PP (1975) Chromosome relationships between *Lolium* and *Festuca* (Gramineae). Chromosoma 52: 103-121.

Jones ES, Mahoney NL, Hayward MD, Armstead IP, Jones G, Humphreys MO, King IP, Kishida T, Yamada T, Balfourier F, Charmet G, Forster JW (2002). An enhanced molecular marker-based genetic map of perennial ryegrass (*Lolium perenne* L.) reveals comparative relationships with other Poaceae genomes. Genome 45: 282-295.

King IP, Purdie KA, Orford SE, Reader SM, Miller TE (1993). Detection of homoeologous chiasma formation in *Triticum durum* x *Thinipyrum besssarabicum* hybrids using genomic in situ hybridisation. Heredity 71: 369-372.

King IP, Morgan WG, Armstead IP, Harper JA, Hayward MD, Bollard A, Nash JV, Forster JW, Thomas HM (1998) Introgression mapping in the grasses. I. Introgression of *Festuca pratensis* chromosomes and chromosome segments into *Lolium perenne*. Heredity 81: 462-467.

King J, Armstead IP, Donnison IS, Thomas HM, Jones RN, Kearsey MJ, Roberts LA, Jones A, King IP (2002) Physical and genetic mapping in the grasses *Lolium perenne* and *Festuca pratensis*. Genetics 161: 315-324.

Kiyosue T, Yoshiba Y, Yamaguchi-Shinozaki K, Shinozaki K (1996) A nuclear gene encoding mitochondrial proline dehydrogenase, an enzyme involved in proline metabolism, is upregulated by proline but downregulated by dehydration in *Arabidopsis*. Plant Cell 8: 1323-1335.

Leśniewska A, Ponitka A, Slusarkiewicz-Jarzina A, Zwierzykowska E, Zwierzykowski Z, James A, Thomas H, Humphreys MW (2001) Androgenesis from *Festuca pratensis* x *Lolium multiflorum* amphidiploid cultivars in order to select and stabilise rare gene combinations for grass breeding. Heredity 86: 167-176.

Levitt J (1980) Responses of plants to environmental stress. In: Chilling, freezing, and high temperature stresses, 2nd edition, Vol I, Academic Press, New York.

Lewis EJ (1980) Chromosome pairing in tetraploid hybrids between *Lolium perenne* and *L. multiflorum*. Theor. Appl. Genet. 58: 137-143.
Marentes E, Griffith M, Mlynarz A, Brush RA (1993) Proteins accumulate in the apoplast of winter rye leaves during cold acclimation. Physiol. Plant. 87: 499-507.
Mattoo AK, Marder JB, Edelman M (1989) Dynamics of the photosystem II reaction centre. Cell 56: 241-246.
Morgan WG, King IP, Koch S, Harper JA, Thomas HM (2001) Introgression of chromosomes of *Festuca glaucescens* into *Lolium multiflorum* revealed by genomic in situ hybridisation (GISH). Theor. Appl. Genet. 103: 696-701.
Ndong C, Danyluk J, Wilson KE, Pocock T, Huner NPA, Sarhan F (2002) Cold-regulated cereal chloroplasr late embryogenesis abundant-like proteins. Molecular characterization and functional analyses. Plant Physiol. 129: 1368-1381.
Pearce RS, Houlston CE, Atherton KM, Rixon JE, Harrison P, Hughes MA, Dunn MA (1998) Localization of expression of three cold-induced genes *blt101*, *blt49*, and *blt14*, in different tissues of the crown and developing leaves of cold-acclimated cultivated barley. Plant Physiol. 117: 787-795.
Rapacz M (1998) Physiological effects of winter rape (*Brassica napus* var. *oleifera*) prehardening to frost. II. Growth, energy partitioning and water status during cold acclimation. J. Agron. Crop Sci. 181: 81–87.
Rapacz M (2002a) Regulation of frost resistance during cold deacclimation and reacclimation in oilseed rape. A possible role of PSII redox state. Physiol. Plant. 115: 236-243.
Rapacz M (2002b) Cold-deacclimation of oilseed rape (*Brassica napus* var. *oleifera*) in response to fluctuating temperatures and photoperiod. Ann. Bot. 89: 543-549.
Rapacz M, Gasior D, Zwierzykowski Z, Lesniewska-Bocianowska A, Humphreys MW, Gay AP (2003) Changes in cold tolerance and the mechanisms of acclimation of photosystem II to cold hardening generated by anther culture of *Festuca pratensis* × *Lolium multiflorum* cultivars. New Phytologist (submitted).
Schena M, Shalon D, Davis RW, Brown PO (1995) Quantitative monitoring of gene expression patterns with a complimentary DNA microarray. Science 270: 467-470.
Seki M, Narusaka M, Ishida J, Nanjo T, Fujita M, Oono Y, Kamiya A, Nakajima M, Enju A, Sakurai T, Satou M, Akiyama K, Taji T, Yamaguchi-Shinozaki K, Carninci P, Kawai J, Hayashizaki Y, Shinozaki K (2002) Monitoring the expression profiles of 7000 *Arabidopsis* genes under drought, cold and high-salinity stresses using a full-length cDNA microarray. Plant J. 31: 279-292.
Shimosaka E, Sasauma T, Handa H (1999) A wheat cold-regulated cDNA encoding an early light-inducible protein (ELIP): its structure, expression and chromosomal location. Plant Cell Physiol. 40: 319-325.
Sidebottom C, Buckley S, Pudney P, Twigg S, Jarman C, Holt C, Telford J, McArthur A, Worrall D, Hubbard R, Lillford P (2000) Heat stable antifreeze protein from grass. Nature 406: 256-257.
Suzuki M (1972). Winter kill patterns of forage grasses and winter wheat in P.E.I. in 1972. Canadian Plant Disease Survey. 52: 196-159.
Tao DL, Oquist G, Wingsle G (1998) Active oxygen scavengers during cold acclimation of Scots pine seedlings in relation to freezing tolerance. Cryobiology 37: 38-45.
Thomas H, Humphreys MO (1991) Progress and potential of interspecific hybrids of *Lolium* and *Festuca*. J. Agric. Sci. 117: 1-8.
Thomas H, Evans C, Thomas HM, Humphreys MW, Morgan G, Hauck B, Donnison I (1997) Introgression, tagging and expression of a leaf senescence gene in *Festulolium*. New Phytol. 137: 29-34.
Thomas H, Morgan WG, Thomas AM, Ougham H (1999) Expression of the stay-green character introgressed into *Lolium temulentum* Ceres from a senescence mutant of *Festuca pratensis*. Theor. Appl. Genet. 99: 92-99.

Thomas HM, Morgan WG, Meredith MR, Humphreys MW, Thomas H, Leggett JM (1994) Identification of parental and recombined chromosomes of *Lolium multiflorum* x *Festuca pratensis* by genome in situ hybridization. Theor. Appl. Genet. 88: 909-913.

Thomashow MF (1999) Plant cold acclimation: freezing tolerance genes and regulatory mechanisms. Ann. Rev. Plant Physiol. Plant Mol. Biol. 50: 571-599.

Uemura M, Steponkus PL (1994) A contrast of the plasma membrane lipid composition of oat and rye leaves in relation to freezing tolerance. Plant Physiol. 104: 479-496.

Vagujfalvi A, Crosatti C, Galiba G, Dubcovsky J, Cativelli L (2000) Two loci on wheat chromosome 5A regulate the differential cold-dependent expression of the *cor14b* gene in frost-tolerant and frost-sensitive genotypes. Mol. Gen. Genet. 263: 194-200.

Zare A-G, Humphreys MW, Rogers JW, Mortimer AM, Collin HA (2002) Androgenesis in a *Lolium multiflorum* x *F. arundinacea* hybrid to generate genotypic variation for drought resistance. Euphytica 125:1-11.

Zwierzykowski Z, Tayyar R, Brunell M, Lukaszewski AJ (1998) Genome recombination in intergeneric hybrids between tetraploid *Festuca pratensis* and *Lolium multiflorum.* J. Heredity 89: 324-328.

Application of AFLP and GISH Techniques for Identification of *Festuca* Chromosome Segments Conferring Winter Hardiness in a *Lolium perenne* x *Festuca pratensis* Population

Siri Grønnerød[1], Siri Fjellheim[1], Michael W. Humphreys[2], Liv Østrem[3], Peter H. Canter[2], Zanina Grieg[1], Øyvind Jørgensen[1], Arild Larsen[4], and Odd Arne Rognli[1]

[1]Department of Chemistry and Biotechnology, Agricultural University of Norway (AUN), PO Box 5040, N-1432 Aas, Norway. [2]Institute of Grassland and Environmental Research (IGER), Plas Gogerddan, Aberystwyth, Wales SY23 3EB, UK. [3]Norwegian Crop Research Institute (NCRI), Fureneset, N-6967 Hellevik i Fjaler, Norway. [4]Norwegian Crop Research Institute (NCRI), Vågønes Research Station, N-8010 Bodø, Norway. (Email: odd-arne.rognli@ikb.nlh.no).

Key words: *Festuca*, *Lolium*, freezing-tolerance, AFLP, GISH

Abstract:

Diploid BC_1, BC_2 and BC_3 introgression lines derived from the amphiploid *Lolium perenne* x *Festuca pratensis* cv. Prior were scored for freezing tolerance and analysed using AFLP markers and GISH in order to associate molecular markers to introgressed *Festuca* segments. The freezing tests showed highly significant differences between clones of the introgression lines. Two introgression lines were more frost tolerant than the diploid *Lolium*-control cultivars. Thirteen AFLP-primer combinations were tested. It was not possible to identify single markers that were present in the most freezing tolerant plants, and absent in the most susceptible ones. The GISH analyses of 20 diploid PriorBC_1-plants with *Lolium*-morphology showed that 3 genotypes had no visible *Festuca*-segments, while the others had 1-5 segments of different sizes. Analyses of a Prior-BC_3 population showed that introgression in the satellite region of chromosome 3 seem to confer enhanced frost tolerance to *Lolium*. Winter hardiness and freezing tolerance are quantitative traits governed by genes on different chromosomes, and combinations of AFLP markers tagging different chromosomal regions may be used together to select superior genotypes during introgression. AFLP markers most frequently present in plants with the highest freezing tolerance, and absent in plants with inferior freezing tolerance, were therefore chosen and associated with winter hardiness and with the presence of *Festuca* introgressions. These markers may be useful in marker-assisted breeding programmes.

A. Hopkins, Z.Y. Wang, R. Mian, M. Sledge and R.E. Barker (eds.), Molecular Breeding of Forage and Turf, 81-86.

1. INTRODUCTION

The *Festuca-Lolium* species complex is very useful for grass breeding because interspecific hybrids and introgression lines can be made to broaden the gene pool that exists within species in each of the two genera. This makes it possible to transfer specific traits like frost tolerance from the more tolerant *Festuca* to *Lolium* species, and improved quality from *Lolium* to *Festuca* (Thomas and Humphreys 1991). Genomic *in situ* hybridisation (GISH) has been used to identify chromosomes and chromosome segments of the parental species in *Festulolium* hybrids and introgression lines (Thomas et al. 1994; Humphreys et al. 1997; Canter et al. 1999). GISH analysis may identify single *Festuca* chromatin segments in plants with enhanced stress tolerance. By associating molecular markers with these introgressed segments, chromosomal segments or genes responsible for desirable agronomic traits could be tagged and marker assisted selection (MAS) could be utilized.

As a part of the EU-project SAGES (Sustainable Grasslands Withstanding Environmental Stresses (http://www.iger.bbsrc.ac.uk/igerweb/SAGES/)), we have attempted to associate molecular markers (AFLPs) with introgressed chromosomal segments from *Festuca pratensis* into *Lolium perenne* that confer enhanced frost tolerance in the introgression lines.

2. MATERIALS AND METHODS

The plant material used in the present investigation was generated from a population called PriorBC$_1$. The hybrid cultivar Prior is an intergeneric amphiploid created at the Welsh Plant Breeding Station (now IGER) in the early seventies. Prior is the product of crosses between autotetraploids of the parental species *Lolium perenne* ($2n = 4x = 28$) and *Festuca pratensis* (Canter et al. 1999). Prior was then backcrossed at IGER to diploid *Lolium* ($2n = 2x = 14$) to producetriploid hybrids ($2n = 3x = 21$). These triploid hybrids were backcrossed at IGER to diploid *Lolium* varieties ('Aurora', 'Corbiere', 'Frances', 'Liprior' and 'Peramo'), to produce the BC$_1$-generation. During the years 1992-1995, 136 plants from 14 different combinations of crosses of BC$_1$ were tested in field trials at NCRI-Fureneset. In 1995, 70 surviving plants were polycrossed, and 20 genotypes with *Lolium perenne*-like phenotypes were selected from this PriorBC$_1$-population.

A BC$_1$ plant from the initial *Lolium* backcrosses (P168/131(5)) was backcrossed at IGER onto a freezing- and ice-sensitive genotype from *L. perenne* (2x) cv. Liprior, and the most tolerant BC$_2$-plants were identified using simulated ice-encasement tests (Canter 2000). Ice-tolerant BC$_2$-plants were then backcrossed at NCRI onto another *L. perenne* cv. Liprior genotype to produce BC_3-populations. The initial BC_1-plant (P168/131(5)), 3 BC_2-plants (P182/170(2), P182/170(17), and P182/170(18)) and 20 BC_3 plants derived from P182/170(2) were included in the test.

The plant materials were screened at IGER for *Festuca* introgressions using GISH, genotyped at AUN with AFLP markers, and freezing tolerance was determined in artificial freezing tests at NCRI-Vågønes and/or at IGER. The GISH analyses were carried out as described in Thomas et al. (1994). AFLP genotyping was performed using conventional procedures (Vos et al. 1995), using the enzyme combinations *Eco*RI/*Mse*I and *Pst*I/*Mse*I and visualised using silver-stained polyacrylamide gels. AFLP-analyses were conducted using the same 13 primer combinations (*Pst*I/*Mse*I) that have been used to map QTLs for frost tolerance and winter survival in the 'B14/16 x HF2/7' full-sib mapping population of *Festuca pratensis* (Alm 2001; Alm et al. 2003). In order to generate additional independent markers, five *Eco*RI/*Mse*I-primer combinations were also applied.

3. RESULTS AND DISCUSSION

Association of AFLP markers with frost tolerance was done by relating presence and absence of markers to the frost tolerance level of the introgression lines. The goal was to detect AFLP markers that were present in all frost tolerant individuals but absent in individuals with low frost tolerance. However, our results show that it was not possible to identify a single marker that was consistently present in the most freezing tolerant plants only, and absent in the most susceptible ones. Therefore, markers that were most frequently present in plants with the highest freezing tolerance are presented in Table 1. Some markers that earlier had been associated with QTL for frost tolerance in *F. pratensis* (Alm 2001) were not polymorphic in the PriorBC_1-population, and others did not amplify. Therefore, only 4 of the *Pst*I/*Mse*I markers in Table 1 have been mapped and linked to QTL for frost tolerance in *F. pratensis*. It is clearly a disadvantage that the identified markers may be specific for a particular cross, and therefore not generally applicable (Stam 1998).

Clones of the genotypes in Table 1 and control genotypes were established and freeze-tested in three series using 5 complete replicates during early winter, late winter and spring/early summer at NCRI-Vågønes. The acclimation conditions (pre-hardening and hardening) were the same for all three series. However, the growth conditions before acclimation, i.e. photoperiod and temperature, were different for the three series because the plants were grown in a conventional glasshouse. The results show highly significant differences between clones, but also significant interactions between clones and series. The explanation may be that the growth climate before hardening had a differential influence on the development of freezing tolerance among the clones. However, the mean frost tolerance values over series are the best ranking we can obtain for general freezing tolerance of the plants. Control plants were the *Lp* cultivars 'Arka' (2x) and 'Maja' (4x) and *Fp*2x 'Fure' and 'Salten'. The BC_1-plants *L*-335 and *L*-284 had the highest frost tolerance and they were also better than the *Lolium*-controls.

Table 1. AFLP screening of Prior-BC_1 and BC_2 plants with different freezing tolerance. *Lolium* chromosomes (Triticeae numbering) with introgressed segments identified with GISH are indicated. *Pst*I/*Mse*I-markers associated with frost tolerance in the 'HF2/7 x B14/16' mapping family of *Festuca pratensis* (Alm 2001) and *Eco*RI/*Mse*I-markers have been used. The genotypes are sorted in decreasing order of freezing tolerance (scale 0 = completely dead, 9 = no visible injury).

	HF2/7	B14/16	*L*-284	*L*-335	*L*-486	168/131-5	*L*-294	*L*-466	*L*-287	*L*-337	*L*-145	*L*-383	*L*-42	*L*-221	*L*-441	18	*L*-468	*L*-452	182/170-2	*L*-285	*L*-463	*L*-393	*L*-152	*L*-357	17	*L*-354
Chromosome						3	4	3				3?									3?		2?			
(GISH)			3	-		4	?	4	3		-	4		4	3	4	2	-	3	4	4?	2	3		4	
Freezing score			6.6	6.6	5.6	5.5	5.3	5.3	5.1	5.0	4.9	4.5	4.4	4.4	4.4	4.3	4.1	4.1	3.9	3.9	3.8	3.7	3.6	3.3	3.0	2.8
P77M72_350	+	-	-	+	+	-	-	+	-	-	-	-	-	+	-	-	+	-	-	+	-	-	-	-	-	-
P77M72_293	-	+	+	-	-	-	-	-	-	-	+	-	-	-	-	-	-	-	-	-	-	-	-	-	-	-
P64M17_G	-	-	-	+	+	+	-	+	-	-	-	-	-	-	-	+	+	-	-	+	-	-	-	-	-	-
P77M66_150	+	-	+	-	-	-	-	?	-	-	-	+	-	-	-	-	-	-	-	-	-	-	-	-	-	-
P77M66_B	+	-	+	+	+	-	-	+	-	-	?	+	+	+	+	-	-	-	-	+	-	-	?	-	-	-
P65M73_100	-	+	+	-	-	-	-	-	-	-	-	-	-	-	-	-	-	-	-	-	-	-	-	-	-	-
P65M73_A	+	-	-	-	-	-	-	+	-	-	-	-	+	-	-	-	-	-	-	-	-	-	-	-	-	-
P66M68_O	+	+	+	+	-	+	-	-	-	-	-	+	+	-	-	-	-	-	+	-	-	-	-	-	-	-
P66M68_Y	+	+	+	+	-	+	+	+	-	-	-	+	-	-	-	+	-	-	-	-	-	-	-	-	-	-
P65M78_M	-	-	+	-	-	+	+	+	-	-	-	+	-	-	-	+	-	-	+	-	-	-	-	-	-	-
E35M17_265	-	+	+	+	-	-	+	-	-	-	+	+	-	-	-	-	-	-	+	-	-	-	-	+	+	-
E35M17_252	+	+	+	+	+	-	-	+	-	-	-	+	+	+	+	-	-	+	-	-	+	-	-	-	-	-
E40M47_3	-	-	-	+	+	-	-	-	+	-	-	-	+	+	-	-	-	-	-	-	-	-	-	-	-	+
E40M47_329	-	-	+	-	-	-	-	+	+	-	+	+	+	-	-	-	-	+	+	-	-	-	-	-	+	-
E40M47_260	+	+	+	+	-	+	+	+	-	+	-	+	+	+	+	+	+	-	+	-	-	+	-	+	-	-
E40M47_245	-	-	+	+	-	-	-	+	+	+	+	+	-	+	-	-	-	-	-	-	+	-	-	+	-	-
Number visible + bands shown	8	7	11	10	5	5	4	9	3	2	4	10	7	6	3	4	3	2	5	3	2	1	0	3	2	1

+ = a visible band on an polyacrylamide gel - = no visible band in the gel

All the plants analysed cytologically were diploids. GISH analysis of the plants classified as *Lolium*-types showed that 3 genotypes (*L*-335, *L*-145, and *L*-452) had no visible *Festuca*-segments while the others had 1-5 *Festuca* segments of different sizes and location. As far as possible the individual *Lolium* chromosomes that carry *Festuca* segments were karyotyped.

However, only the presence of *Festuca* introgressions on *Lolium* chromosomes 2, 3, and 4 (identified most easily), are described in Table 1.

Four genotypes, i.e. *L*-284, *L*-335, *L*-466 and *L*-383, have the highest number of "positive" (+) markers (Table 1). Five AFLP markers are consistently present for the four genotypes. *L*-284 and *L*-466 have a *Festuca*-segment on the satellite of chromosome 3, while it is uncertain whether this is the case for genotype *L*-383. *L*-335 had no visible segments. Seven markers are consistently present in genotypes *L*-284, *L*-466 and *L*-383 (Table 1). *L*-466 and *L*-383 have common segments on chromosome 4. *L*-383 was also the most freezing-tolerant genotype in a freezing test conducted at IGER (data not shown). GISH-analysis showed that the BC_1 plant P168/131(5) carried 3 *Festuca* segments including the entire satellite of chromosome 3 and an entire arm of chromosome 4. Amongst the BC_2 progeny from P168/131(5), P182/170(2) carries a single *F. pratensis* chromosome segment – the translocation representing the entire satellite region of chromosome 3. P182/170(2) was previouslyshown to be ice-tolerant (Canter 2000) and freezing-tolerant intests at IGER (data not shown). P182/170(17) (another ice-tolerant plant), also carries a single *Festuca* segment - replacing one entire arm of chromosome 4 of *Lolium* (Canter (2000). Plant P182/170(18) carries two *F. pratensis* segments. The BC_3 population derived from P182/170(2) were assessed for freezing-tolerance at IGER, and LT_{50} (lethal temperature for 50% of tillers taken at different growth stages) was determined for each genotype (Table 2).

Table 2. AFLP markers identified in 20 genotypes of BC_3-plants from P182/170(2), in the BC_1-plant P168/131(5), and the BC_2-plants P182/170(2), P182/170(17) and P182/170(18). The plants are sorted in decreasing order of freezing tolerance conducted at IGER. Sizes of *Festuca* segments present of chromosome 3 are indicated.

BC3-P182/170(2)	78	115	71	33	75	94	79	57	90	112	51	110	88	120	82	36	31	53	105	86	168/131(5)	182/170(17)	182/170(18)	182/170(2)
LT_{50}	19.0	12.7	11.4	11.3	10.7	10.5	10.5	10.4	10.3	10.1	12.0	11.8	9.1	8.1	8.1	7.9	7.6	6.5	5.3	0.0				
Chromosome size	100	75	100	0	100	100	75	30	100	100	0	0	0	0	0	0	0	0	0	0				
p66m43_E	+	+	+	+	+	+	+	+	+	+	?	-	-	-	-	-	-	-	-	+	+	-	-	+
p64m17_M	-	-	-	-	-	-	-	-	-	-	-	+	+	+	+	+	+	+	+	+	-	+	-	+
p64m17_R	+	+	+	+	+	+	+	-	+	+	-	-	+	-	+	-	-	-	+	-	+	-	-	+
p64m17_S	+	+	+	+	+	+	+	-	+	+	-	-	-	-	-	-	-	-	-	-	+	-	-	+
p64m17_X	+	+	+	+	+	+	+	+	+	+	-	-	-	-	-	-	-	-	-	-	-	-	-	-
e42m50_270	+	+	+	+	+	+	+	+	+	+	-	-	-	-	-	-	-	-	-	-	-	-	+	+
e42m50_157	+	+	+	+	+	+	+	+	+	+	-	-	-	-	-	-	-	-	-	-	-	-	+	+

The GISH results suggest that if introgressed chromosomal segments from *F. pratensis* have conferred freezing tolerance, it derives from different chromosomes. The general dominant nature of AFLPs makes it difficult to monitor introgression, especially when the recipient species (*Lolium*) apparently was much more variable than the donor species (*Festuca*), and when the initial parents of the BC_1-population were not available. This complicates screens for markers linked to genes for freezing-tolerance. As demonstrated in the BC_3 derived from P182/170(2), the creation of a further generation can simplify MAS. In this case the presence of *Festuca* genes on the satellite of chromosome 3 are clearly associated with enhanced freezing-tolerance. The screening of molecular markers that co-segregate with the alien segment should allow the development of MAS systems for freezing-tolerance. Once other regions of the *Festuca* genome bearing genes for freezing-tolerance are located, itshould be possible to select a combination of markers specific for a range of relevant *Festuca* genes and further enhance the freezing-tolerance of *Lolium*.

REFERENCES

Alm V (2001) Comparative genome analyses of meadow fescue (*Festuca pratensis* Huds.): Genetic linkage mapping and QTL analyses of frost and drought tolerance. Doctor Scientiarum Theses 2001:20 Agricultural University of Norway (ISBN 82-575-0469-6).

Alm V, Fang C, Busso CS, Devos KM, Vollan K, Grieg Z, Rognli OA (2003) A linkage map of meadow fescue (*Festuca pratensis* Huds.), and comparative mapping with the Triticeae species, *Lolium*, oat, rice, maize, and sorghum. Theor. Appl. Genet. (submitted).

Canter PH (2000) The use of genomic in situ hybridisation (GISH) to locate introgressed chromosome segments from winter-hardy *Festuca* in a cold-sensitive *Lolium* background. Nature Genetics Newsletter 43: 30-31.

Canter PH, Pasakinskiene I, Jones RN, Humphreys MW (1999) Chromosome substitutions and recombination in the amphiploid *Lolium perenne* x *Festuca pratensis* cv Prior (2n=4x=28). Theor. Appl. Genet. 98: 809-814.

Humphreys M, Thomas HM, Harper J, Morgan G, James A, Ghamari-Zare A, Thomas H (1997) Dissecting drought- and cold-tolerance traits in the *Lolium-Festuca* complex by introgression mapping. New Phytol. 137: 55-60.

Stam P (1998) Crop physiology, QTL analysis and plant breeding. In: Inherent Variation in Plant Growth. Physiological Mechanisms and Ecological Consequences, Lambers H, Poorter H, Van Vuuren MMI (eds), pp. 429-440. Backhuys Publishers, Leiden, the Netherlands.

Thomas H, Humphreys MO (1991) Review: Progress and potential of interspecific hybrids of *Lolium* and *Festuca*. J. Agric. Sci. 117: 1-8.

Thomas HM, Morgan WG, Meredith MR, Humphreys MW, Thomas H, Leggett JM (1994) Identification of parental and recombined chromosomes in hybrid derivatives of *Lolium multiflorum* x *Festuca pratensis* by genomic in situ hybridization. Theor. Appl. Genet. 88: 909-913.

Vos P, Hogers R, Bleeker M, Reijans M, van de Lee T, Hornes M, Frijters A, Pot J, Peleman J, Kuiper M, Zabeau M (1995) AFLP: a new technique for DNA fingerprinting. Nucl. Acids Res. 23: 4407-4414.

A Functional Genomics Approach for Identification of Heat Tolerance Genes in Tall Fescue

Yan Zhang[1], John Zwonitzer[1], Konstantin Chekhovskiy[1], Gregory D. May[2] and M.A. Rouf Mian[1]

[1]Forage Improvement Division, [2]Plant Biology Division, The Samuel Roberts Noble Foundation, Ardmore, OK 73401, USA. (Email: rmmian@noble.org).

Key words: heat stress tolerance, tall fescue, cDNA library, microscopy, suppression subtractive hybridization

Abstract:

Tall fescue (*Festuca arundinacea* Schreb.) is a major cool season forage and turf grass species around the world. Heat stress is one of the limiting factors for forage production and turf management of this grass species. This research uses functional genomics approaches for discovering the molecular basis of heat tolerance in tall fescue genotypes. To identify genes expressed during heat stress, a heat stressed shoot cDNA library was constructed from popular tall fescue cultivar Kentucky 31 with shoot tissues collected at five time points (12 – 78 hrs) after exposure to high heat (42 – 44°C) in a growth chamber. High-throughput sequencing of the cDNA clones is underway for generating 5,000 ESTs. The ESTs related to heat-stress will be identified from this cDNA library. Heat-tolerance related genes are cloned by using suppression subtractive hybridization (SSH) between heat tolerant and heat sensitive cultivars. Heat tolerance-related genes isolated by SSH will be combined with genes cloned from shoot cDNA library to generate a cDNA microarray to study the expression profiles of tall fescue cultivars with different responses to the heat stress. Heat tolerance genes identified from this project will be useful for development of tall fescue cultivars adapted to high heat environments.

A. Hopkins, Z.Y. Wang, R. Mian, M. Sledge and R.E. Barker (eds.), Molecular Breeding of Forage and Turf, 87-96.

1. INTRODUCTION

Higher plants exposed to excess heat exhibit a characteristic set of cellular and metabolic responses, including a decrease in the synthesis of normal proteins and an accelerated transcription and translation of heat shock proteins (HSPs). This response is observed when plants are exposed to temperatures at least 5°C above their optimal growing conditions (Morimoto et al. 1994). In addition to altering patterns of gene expression, heat also damages cellular structures, including organelles and the cytoskeleton, and impairs membrane function. Studies have demonstrated the cellular and metabolic changes required to survive high temperatures (Guy 1999).

Tall fescue (*Festuca arundinacea*) is a major cool season forage and turf grass. With proper management it can be grown in a wide range of soil and climatic conditions. Tall fescue is a hexaploid outcrossing species that belongs to the grass family *Poaceae*, subfamily *Pooideae*, and tribe *Triticeae*. The genome size is approximately 5.27-5.83 x 10^6 kb (Seal 1983). It contains three genomes (PG1G2). The P (2x) genome is from *F. pratensis* while the G1G2 (4x) genome is from *F. arundinacea* var 'glaucescens' (Sleper 1985).

Heat stress is one of the limiting factors for forage production and turf management of tall fescue. Understanding the molecular mechanisms of heat-tolerance will be helpful in development of heat-tolerant tall fescue cultivars. Heat-tolerance of plants is a complex trait, which is controlled by multiple genes. Therefore, a global observation of gene expression patterns is required to evaluate the association between conditions of gene expression and gene function. Genome-wide expression profiling at the transcript level using microarray is one of the most exciting tools to study the cell and its integrative processes. A broad picture of genes coordinately expressed in a cell might provide a dynamic molecular view and help to understand the operative biochemical and regulatory networks. A number of recent studies have demonstrated that microarrays can be successfully used for studying gene expression of plant responses to abiotic stresses (Oztur et al. 2002; Seki et al. 2001; Wang et al. 2000).

In this study, a heat stressed shoot cDNA library was constructed from tall fescue to identify genes expressed during heat stress and acclimation. Differentially expressed gene transcripts were identified by suppression subtractive hybridization (SSH) from two tall fescue cultivars with different levels of heat tolerance. The cDNA microarrays containing the differentially expressed genes will be generated to examine the abundance and expression of tall fescue transcripts in both heat tolerant and sensitive cultivars under various temperatures and durations to dissect the molecular basis of heat tolerance. The information generated from this study will be useful in developing heat tolerant tall fescue cultivars.

2. MATERIALS AND METHODS

2.1 Plant Materials and Heat Treatments

Four tall fescue genotypes with different levels of heat tolerance were used in this study (Table 1). A previous study (Zwonitzer and Mian 2002) demonstrated that two tall fescue genotypes (PI423078 and PI297901) were able to survive well through the heat stress at a temperature up to 42°C and continued to grow well after decreasing the temperature in the growth chamber. However, cultivar Kentucky 31 barely survived the stress and a heat sensitive genotype (PI283316) died under the same condition.

Table 1. Tall fescue genotypes used in the study

Genotypes	Response to Heat Stress
PI423078	High Tolerance
PI297901	High Tolerance
Kentucky 31	Low tolerance
PI283316	Sensitive

Four tall fescue genotypes were grown in growth chambers in a controlled environment (light intensity at 375 μmol $m^{-1}s^{-2}$, 16 hours day length, and 60% relative humidity). Plants were exposed to high heat (up to 44°C) and the tissue samples were collected at five time points as listed in Table 2. At each sampling time, the entire shoot mass from three plants of each genotype (under both heat stressed and unstressed conditions) was collected.

Table 2. Heat stress treatments to tall fescues used in heat-tolerance and expression study

Temperature (°C)		Duration (Hours)
Day Time	Night Time	
24	16	72
29	21	72
34	26	72
39	31	48 (sampled at 12 and 36 hrs)
42	36	48 (sampled at 12 and 36 hrs)
44	36	12 (sampled at 12 hrs)

2.2 Construction of Heat Stressed cDNA Library

After collection of samples from all five time points, the frozen shoots and leaves were ground to fine powder with mortar and pestle. Total RNA was extracted from ground leaf tissues from each sampling time by Tri-reagent (MRC, Cincinnati, OH) and the total RNA was quantified and was checked for quality. Based on concentration, an equal amount of each RNA sample was pooled together to form a composite sample of total RNA for mRNA isolation to construct the cDNA library. The mRNA was purified by Oligotex mRNA midi kit (Qiagen, Valencia, CA).

The cDNA library was constructed by using the ZAP-cDNA Gigapack III Gold Cloning Kit (Stratagene, La Jolla, CA). The cDNA was size fractionated following Stratagene's drip column procedure. The fractions containing cDNA sizes around and larger than 1000 bp were selected for library construction.

2.3 EST Sequencing and Data Processing

High-throughput sequence analyses of these libraries are underway at the Noble Foundation with a target of generating 5,000 ESTs. The sequence data generated from this project are processed at the Virginia Bioinformatics Institute (VBI). The partial nucleotide sequence data obtained for the clones have been stored in our in-house database. The data will be made available to the public through GenBank at a later date.

The EST Analysis Pipeline (ESTAP) is used for processing the EST sequences. The ESTAP software is a series of automated procedures to verify, store and analyze EST (Expressed Sequence Tag) data generated using high-throughput platforms. ESTAP automatically cleanses raw sequence data by removing vector, low quality, and contaminating sequences. The cleansed sequences are compared to a set of DNA or protein databases using the BLASTx algorithms with expected value at 1e-4. ESTAP also clusters and assembles EST collections into singlets and contigs. The repeat, ribosomal and mitochondrial sequences are masked. The raw and cleansed data and analysis results are stored in a relational database.

2.4 Examination of Gene Expression by RT-PCR

The expression level of ascorbate peroxidase in heat stressed tall fescue (KY 31) was examined by RT-PCR analysis. Four μg of total RNA from each time point sample was used to synthesize cDNA by Ready-To-Go RT-PCR kit (Amersham, Piscataway, NJ). By following the two-step protocol suggested by manufacturer, first-strand cDNA generated with lyophilized pd $(T)_{12-18}$ was used as a template for polymerization. Two µl of cDNA template was added into each PCR reaction with Arabidopsis Actin primers and tall fescue ascorbate peroxidase gene specific primers (5': AAGTGCTACCCCAAGGTCA, and 3': CACCAGAAAGGCCAACAAT). PCR was conducted with the following program: inactivated the reverse transcriptase at 95°C for 5 min, then followed with 94°C for 1 min, 55°C for 1 min, 72°C for 2 min with 25 cycles.

2.5 Histology Study of Tall Fescue Cell Structure and Lignin Content

To study the cell structures of four tall fescue genotypes, second leaf from the bottom of one tiller was collected for each genotype plant and fixed in 2.5% glutaraldehyde in 25 mM potassium phosphate buffer (pH 7.1) for 24 h. All fixed tissues were washed with the buffer, dehydrated through a graded ethanol series, and embedded in JB-4 methacrylate plastic (Polysciences, Warrinton, PA). Serial 3-µm sections were cut and stained with 1% aqueous acid fuchsin and counterstained with 0.05% toluidine blue. Sections were photographed on a Leitz Laborlux microscope.

A modified Maule staining method (Lin and Dence 1992) was used for characterizing the lignin content of two tall fescue genotype plants. Sections of tall fescue leaf were hand-cut and immersed in 1% neutral potassium permanganate solution for 1 min at room temperature. The sections were rinsed in distilled water, decolorized with 12% HCl for 5 min, washed thoroughly in water and treated with a few drops of 1.5% sodium bicarbonate solution.

2.6 Suppression Subtractive Hybridization (SSH)

The total RNA was isolated from tall fescue shoot and leaf tissues of PI297901 and PI283316 heat stressed (39°C, 12 h) plants by using Tri-reagent (MRC, Cincinnati, OH), and the mRNA was purified by Oligotex mRNA midi kit (Qiagen, Valencia, CA). Differentially expressed genes were isolated by using PCR-Select cDNA Subtraction Kit (Clontech, Palo Alto, CA) starting with 4 µg of poly A^+ RNA. In the last step of secondary PCR, a seven-minute 72°C extension was applied to ensure that all PCR products were full length and 3' adenylated. Three sets of subtractions were performed, including both experimental forward and reverse as well as control subtractions following the manufacturer's instructions. The subtracted cDNA population was cloned into a TOPO TA cloning vector and transformed with One Shot TOP10F´ chemically competent cells (Invitrogen, Carlsbad, CA).

3. PRELIMINARY RESULTS

3.1 Heat Stressed Shoot cDNA Library

A cDNA library was constructed by using low heat-tolerant cultivar Kentucky 31 exposed to five different time or temperature (12 and 24 hrs at 42°C; 12, 48, 78 hrs at 44°C). The primary library has a titer of 2 x 10^5. The average size of cDNA inserts was ~1200 bp.

High-throughput sequence analyses of these libraries are underway at the Noble Foundation with a target of generating 5,000 ESTs. Preliminary analysis of the first set of 2584 ESTs showed the average readable sequence length was ~800 bp. The sequence data generated from this project are processed at the Virginia Bioinformatics Institute (VBI). From the ESTs we found that genes encoding proteins involved in stress responses (ascorbate peroxidase, heat shock protein) as well as signal transduction process (small GTP-binding protein, transcription factor, RING finger protein) were expressed under heat stress.

3.2 Expression of Stress Response Gene

Ascorbate peroxidase (APX) exists as isoenzymes and plays an important role in the metabolism of H_2O_2 in higher plants. A high level of endogenous ascorbate is essential to effectively maintain the antioxidant system that protects plants from oxidative damage due to biotic and abiotic stresses. Studies indicated that APX isoenzymes are critical components that prevent oxidative stress in photosynthetic organisms (Jimenez et al. 1997). The expression level of ascorbate peroxidase in heat stressed tall fescue (KY 31) was examined by RT-PCR analysis (Fig. 1 A). The expression of the gene varied for the temperatures and durations. APX was induced shortly after the exposure to the heat conditions (12 and 36 hrs at 39°C). The level of APX expression was reduced with the increased temperatures and exposure time. Expression of APX was not detectable under normal temperature (Fig. 1 B).

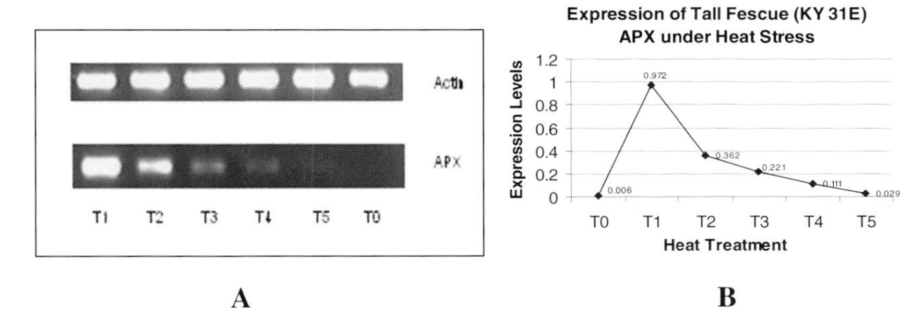

Figure 1. **A** and **B:** Expression of tall fescue ascorbate peroxidase under heat stress. The PCR products were checked with 1% agarose gel in 1X TAE with EtBr. T1: 42°C, 12 hrs; T2: 42°C, 24 hrs; T3: 44°C, 12 hrs; T4: 44°C, 48 hrs; T5: 44°C, 78hrs. T0: control plant. Twelve μl of PCR product was loaded in each lane.

3.3 Cell Structure of Tall Fescue Leaf and Lignin Content

The morphology of the four genotypes of tall fescue plants is very different even under normal growing conditions. Therefore, the histological study of tall fescue leaf tissue was conducted to illustrate the cell structure of the plants (Fig. 2).

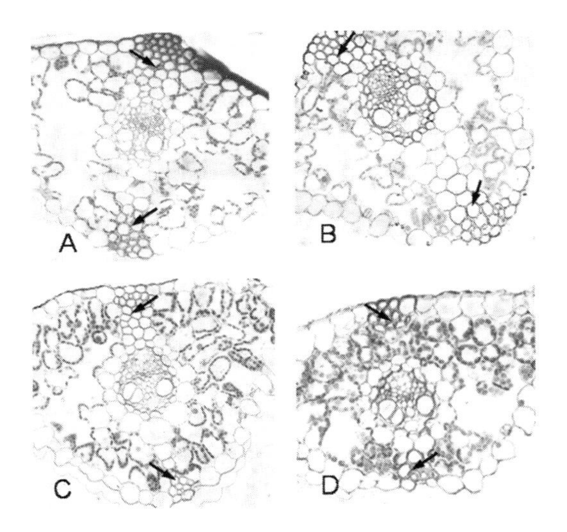

Figure 2. TS of the leaf of tall fescue. A: PI297901, B: PI423078, C: Kentucky 31, D: PI283316. Arrows indicate the sclerenchyma and collenchyma structure in each genotype (LM x 40).

Except for the size of the cell is smaller in genotype Kentucky 31 and PI 283316, the major difference are the sclerenchyma and collenchyma (supporting tissue) in between the vascular bundle and epidermal cells. Therefore, heat tolerant genotypes (PI423078 and PI297901) have more supporting tissues embed compared to heat sensitive genotypes (Kentucky 31 and PI283316), especially in the collenchymas.

Lignin content was examined by light microscopy (Fig.3). S-lignin was stained to red, which is not digestible. G-lignin was stained to yellow. Comparing lignin contents of two genotypes with free hand sectioned leaf tissue, no significant difference for the S-lignin deposit was detected between two genotypes.

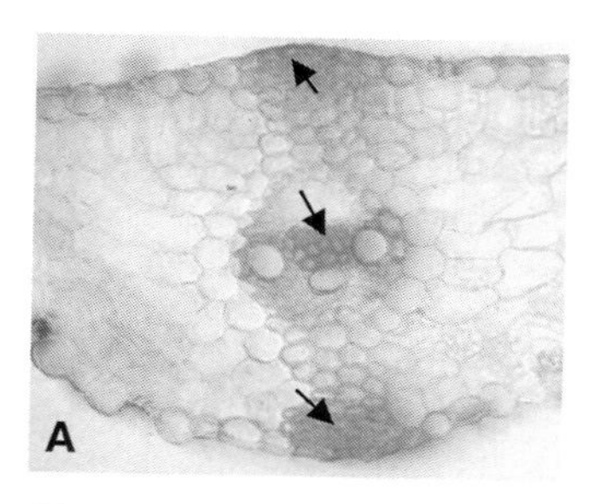

PI297901 minor vein (LMx20)

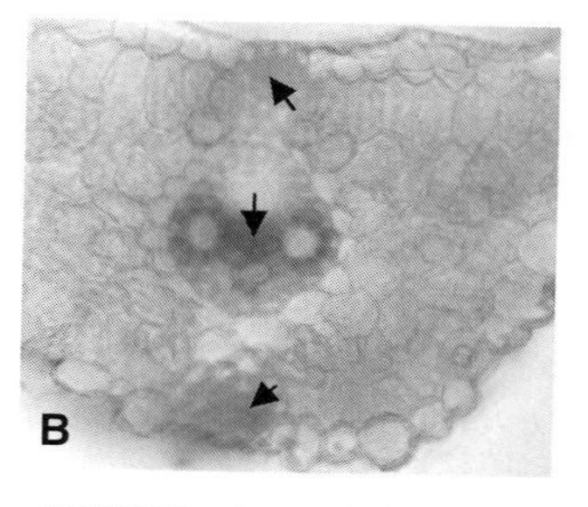

PI283316 minor vein (LMx20)

Figure 3. Free-hand sections of tall fescue leave tissue. A: PI297901 (heat tolerant). B: PI283316 (heat sensitive). S-Lignin stained to red color (indicated by arrows) and G-Lignin stained to yellow color.

3.4 Identify Heat Tolerance Genes by Suppression Subtractive Hybridization (SSH)

One subtraction was conducted with tall fescue plants treated in 39°C for 12 h to identify genotype specific genes that are differentially expressed for heat tolerance or sensitivity. A set of 288 clones was generated by SSH in which mRNA from heat stressed shoot of PI297901 (heat tolerant) was used as 'tester' and the mRNA from shoot of PI283316 (heat sensitive) was used as 'driver'. This set of cDNA clones was enriched for genes over-expressed in heat tolerant genotype when compared to the heat sensitive one. To obtain clones that were over expressed in the heat sensitive genotype, a reverse subtraction was performed. In this case, mRNA from heat stressed PI283316 was used as 'tester' and mRNA from PI297901 was used as 'driver'. This set of cDNA clones was enriched for genes over-expressed in the heat sensitive genotype. About 390 clones were selected from this reverse subtraction. All clones will be sequenced to identify heat tolerance-related genes.

4. DISCUSSION

In order to mimic the natural conditions of tall fescue fields, the growth chamber experiment was designed with slow step-wise temperature increases. Therefore, all the tall fescue plants were acclimatized to the heat conditions rather than a direct heat shock. Key et al. (1983) indicated that plants can acquire thermotolerance if subjected to a non-lethal high temperature for a few hours before encountering heat shock conditions. An acclimated plant can survive exposure to a temperature that would otherwise be lethal. The acclimation process is thought to involve new proteins, synthesized in response to high temperature, that confer thermotolerance to the organism. From the EST data, several genes were identified that encode protein homologs that are involved in the general stress responses including cysteine proteinase, ascorbate peroxidase, glutathione S-transferase, and late embryogenesis abundant (LEA) protein. Therefore, tall fescue genotypes PI297901 and PI423078 have higher acclimation ability than Kentucky 31 and PI283316. Interestingly, genes that regulate other stress conditions, such as senescence, ripening, cell death, freeze, cold acclimation, aluminum, wound, and pathogen (stem rust) stresses, were also present in this heat stressed tall fescue cDNA library. This result suggested plants may apply some common molecular mechanisms (e.g. signal transductions and transcriptional regulations) as defenses in response to all stress conditions.

The effect of high temperature on higher plants is primarily on photosynthetic functions. The heat tolerance limit of leaves of higher plants is determined by the thermal sensitivity of primary photochemical reactions occurring in the thylakoid membrane system. Studies have shown that long-term acclimations can be superimposed upon fast adaptive adjustment of the thermal stability (Weis and Berry 1988). Light could also cause an increase in tolerance to heat, and this stabilization is related to the light-induced proton gradients. In addition to irreversible effects, high temperature may also cause large, reversible effects on the rate of photosynthesis (Weis and Berry 1988). Several genes involved in photosynthesis were identified from our heat stressed cDNA library. Although the function of these genes is unknown under heat stressed condition, they could play important roles in heat acclimation for tolerant genotypes PI297901 and PI423078.

Lignin is a complex substance containing various phenolics which is deposited in the cellulose walls of the sclerenchyma and tracheary elements; it increases their strength and renders the walls impermeable to water (Goodwin and Mercer 1983). High lignin content dramatically reduces the digestibility of tall fescue. The ideal tall fescue genotype should have high heat tolerance and average/low level of lignin. Comparing lignin contents of two genotypes with free hand sectioned leaf tissue, no significant difference for the S-lignin deposit was detected between two genotypes. Therefore the digestibility of these two tall fescue genotypes is similar and suitable for the heat stress study.

ACKNOWLEDGEMENTS

We thank The Virginia Bioinformatics Institute (VBI) for processing the sequence data. Special thanks go to Angela Scott in The Samuel Roberts Noble Foundation for tall fescue EST sequencing.

REFERENCES

Guy C (1999) The influence of temperature extremes on gene expression, genomic structure, and the evolution of induced tolerance in plants. In: Plant Responses to Environmental Stresses, Lerner HR (ed.), pp.497-548. Marcel Dekker, New York.

Key JL, Lin CY, Ceglarz E, Schoffl F (1983) The heat shock response in soybean seedlings. In: Structure and Function of Plant Genomes, Ciferri O and Dure L (eds.), Ser. A 63 pp.25-36. NATO ASI ser.

Jimenez A, Hernandez JA, del-Rio LA, and Sevilla F (1997) Evidence for the presence of the ascorbate-glutathione cycle in mitochondria and peroxisomes of pea leaves. Plant Physiol. 114: 275-284.

Lin SY and Dence CW (1992) Methods in lignin chemistry. Springer-Verlag, Berlin; New York.

Morimoto RI, Jurivich DA, Kroeger PE, Mathur SK, Murphy SP, Nakai A, Sarge K, Abravaya K, Sistonen LT (1994) Regulation of heat shock gene transcription by a family of heat shock factors. In: The Biology of Heat Shock Proteins and Molecular Chaperones, Morimoto, RI, Tissieres A, and Georgopoulos C (eds.), pp. 417-455. Cold Spring Harbor Laboratory Press, Cold Spring Harbor.

Oztur ZN, Talame V, Deyholos M, Michalowski CB, Galbraith DW, Gozukirmizi N, Tuberosa R, Bohnert HJ (2002) Monitoring large-scale changes in transcript abundance in drought- and salt-stressed barley. Plant Mol Biol. 48: 551-573.

Seal AG (1983) DNA variation in *Festuca*. Heredity 50: 225-236.

Seki M, Narusaka M, Abe H, Kasuga M, Yamaguchi- Shinozaki K, Carninci P, Hayashizaki Y and Shinozaki K (2001) Monitoring the expression pattern of 1300 *Arabidopsis* genes under drought and cold stresses by using a full-length cDNA microarray. Plant Cell 13: 61–72.

Sleper DA (1985) Breeding tall fescue. J. Plant Breed. Rev. 3: 313-342.

Weis E and Berry JA (1988) Plants and high temperature stress. Symp. Soc. Exp. Biol. 42: 329-346.

Wang R, Guegler K, LaBrie ST, and Crawford NM (2000) Genomic analysis of a nutrient response in Arabidopsis reveals diverse expression patterns and novel metabolic and potential regulatory genes induced by nitrate, Plant Cell 12: 1491–1510.

Zwonitzer JC and Mian MAR (2002) Heat tolerance of tall fescue genotypes differing in persistence in the Southern Great Plains. 94th Annual Meetings of ASA abstracts, Nov. 10-14, 2002, Indianapolis, IN.

Quantitative Trait Locus Mapping of Winter Hardiness Metabolites in Autotetraploid Alfalfa (*M. sativa*)

B. Alarcón Zúñiga[1,2], P. Scott[3], K. J. Moore[2], D. Luth[2] and E. C. Brummer[2]
[1]*Animal Science Department, Universidad Autónoma Chapingo, Chapingo, México.* [2]*Raymond F. Baker Center for Plant Breeding, Department of Agronomy, Iowa State University, Ames, IA 50011,USA.* [3]*USDA-ARS, Corn Insects and Crop Genetics Research Unit. Ames, IA, USA. (Email: brummer@iastate.edu).*

Key words: alfalfa, winter hardiness, metabolites, QTL mapping

Abstract:

In winter hardy alfalfa cultivars, cold acclimation occurs prior to the onset of freezing temperatures and normally is accompanied with a series of metabolic and morphological adjustments. We are studying the accumulation pattern of metabolites throughout the autumn previous to freezing and relating them to winter survival in an F1 segregating population between the cross of *M. sativa* subsp. *sativa* and subsp. *falcata*. Morphological components and soluble carbohydrates, protein, amino-N groups, and free fatty acids were measured in 2001 and 2002 in the field. Broad sense heritability was intermediate for shoot and root mass and height, and for metabolites, ranged from low (TNC=0.04) to high (starch=0.80). The genetic correlation between winter injury was not significant for most of the metabolites, except for soluble protein and amino-N group concentrations. The presence of allele *a1* of *MSAIC B*, a cold-related gene, was positively associated with autumn plant height but negatively associated with root mass in the WISFAL-6 parent. Numerous QTL were detected for concentrations of metabolites. Our results suggest that winter injury and autumn biomass are controlled by different loci in this population.

A. Hopkins, Z.Y. Wang, R. Mian, M. Sledge and R.E. Barker (eds.), Molecular Breeding of Forage and Turf, 97-104.

1. INTRODUCTION

Alfalfa originated in the Caucasus Mountains and surrounding regions from which it spread throughout much of Eurasia and North Africa, resulting in alfalfa germplasm with a wide diversity of winter hardiness. The most winter hardy alfalfa germplasm is *M. sativa* subsp. *falcata*, which grows primarily in colder and drier areas of Europe and Asia. Winter hardy alfalfa becomes dormant in response to shortening photoperiod and cooling temperatures typical of autumn, thus acclimating for the harsh environmental conditions of winter (McKenzie et al. 1988).

The accumulation of various compounds, including regulators of osmotic potential like sucrose and proline and of membrane fluidity like linoleic acid, takes place during hardening, helping cells to tolerate dehydration and maintain membrane integrity. The soluble carbohydrate concentration in alfalfa taproots during early autumn is not associated with cold tolerance, but winter hardy germplasm accumulates higher concentrations of soluble sugars in late autumn than do non-hardy cultivars (Cunningham et al. 1998; Alarcón-Zúñiga et al. 2001). Taproot starch concentrations may also be partially involved with winter hardiness as they can be converted to sugars when needed for acclimation (Boyce and Volenec, 1992). The accumulation of oligosaccharides, including raffinose and stachyose, has been linked to decreased winter injury (Cunningham et al. 2003). The concentrations of other compounds and the expression levels of a suite of genes has been shown to change under cold or winter stresses in alfalfa and many other plants (Thomashow, 1999).

Alfalfa cold acclimation is a complex phenomenon comprising a multitude of physiological and biochemical processes (Volenec et al. 2002). The objective of this paper is to provide a preliminary examination of the genetic relationships among concentrations of carbohydrates, soluble protein, amino-N groups, and fatty acids during late autumn, autumn plant growth, and winter injury using both quantitative genetic and molecular marker methods to detect quantitative trait loci (QTL).

2. RESEARCH METHODOLOGY

2.1 Experimental Population

Two genotypes, ABI408 (*Medicago sativa* subsp. *sativa*) and WISFAL-6 (*M. sativa* subsp. *falcata*) were crossed to form an F1 population of 200 individuals segregating for yield, winter injury, and autumn growth (Brummer et al. 2000). The 200 F1 individuals, two parents, and eight checks were clonally propagated by stem cuttings in the greenhouse and transplanted at Ames, IA on June 1, 2001 and on May 15, 2002. The plot design in both years was a 14 by 15 quadruple alpha lattice design with 3 replications and 3 destructive harvest dates (August, November and April). Five plants per genotype were space planted at 16 cm apart within a plot in each replication with rows separated by 80 cm. The plots received no fertilization (P and K were above recommended levels for alfalfa cultivation) and were hand-weeded; insects were controlled chemically.

2.2 Growth Components, Winter Injury and Metabolite Analysis

Height was measured in late August and early November each year, after which the plants were dug, washed in water, and taproots, crowns, and shoots separated. The 3 tissues were frozen with liquid nitrogen and stored in dry ice for transportation to the lab, where they were kept at -80°C until freeze-drying. The freeze-dried samples were weighed and finely ground with a 1 mm mesh screen (UDY Cyclone, UDY Manufacturing, Fort Collins, Co). The following April, the 3rd block was scored for winter injury by the method of McCaslin and Woodward (1995), which included the digging of the complete plant and visual scoring on a scale of 1 = no injury, all plants symmetrical with equal shoot length to 5 = all dead plants.

The ground taproot samples were scanned by Near Infrared Reflectance Spectroscopy and reflectance measurements (log 1/R) between 1100 to 2500 nm, recorded at 4-nm intervals obtained with a scanning monochromator (NIRS Systems, Silver Springs, MD 20910). Fifty calibration samples, representing the range of H-values for the entire sample set, were selected for all the metabolites to analyze by wet chemistry.

Starch, sucrose, glucose and fructose were extracted using 80% ethanol and quantified by enzymatic assay (Sigma Chemical Co., St. Louis, MO; SCA20, sucrose; FA20, fructose; GAG020, glucose). Total non-structural carbohydrates (TNC) was estimated by means of the anthrone assay and scored using a glucose standard (Koehler, 1952). Residual extracted samples were dried at room temperature and the starch content determined by the alpha-amylase-amyloglucosidase-glucose and oxidase-peroxidase enzymatic assays (Sigma Chemical Co., St. Louis, MO; STA20). Soluble protein and amino-N groups were extracted with sodium phosphate and quantified by the BCA procedure (Smith et al., 1985). Total N concentration was obtained by dry combustion, and crude protein was estimated as total N x 6.25. Fatty acids were extracted with methanolic HCl and a C:17 internal standard was added to each sample. Fatty acids were esterified using 0.5% BHT in hexane (Sukhija and Palmquist, 1988) and quantified by gas chromatography. GC-

MS was used to verify the correct elution time of the free fatty acid methyl esters found by the GC (Roessner et al., 2001).

2.3 Statistical Analysis for Quantitative Traits and QTL Qapping

The analysis of variance of each trait was done in each harvest date and combining across years with all effects in the model considered random. Statistical analysis was performed by MIXED, CORR, and GLM procedures of Statistical Analysis System (SAS Institute, 1990; Little et al., 1996). Orthogonal contrasts were developed to determine differences among offspring and parents for each trait (Lynch and Walsh, 1998). The heritability on an entry mean-basis was estimated according to Holland et al. (2003):

$$h^2 = \frac{\sigma_g^2}{\sigma_g^2 + \frac{\sigma_{gl}^2}{l} + \frac{\sigma_e^2}{rl}}$$

where σ^2_g, σ^2_{gl}, and σ^2_e represent variance components due entry, entry x year, and experimental error, respectively; l is the number of years and r the number of replications. Standard error of heritability estimates was determined from variance components of the MIXED and IML procedures with entries as fixed effects (Falconer and Mackay, 1996; Lynch and Walsh, 1998). The genetic correlation for all the pairwise traits was determined from least square means for each entry in each environment. MANOVA was used to obtain SS of the cross product and determine genetic correlations (SAS Institute, 1990).

Single marker analysis was performed by the MIXED procedure to estimate association between QTL traits and marker loci with a probability level of $P<0.01$. The mean for each trait for individuals in the population for which the given marker was present (+) or abscent (-) were calculated to estimate the phenotypic effect of the marker allele. Markers were assigned to tentative linkage groups based on Robins et al. (2003).

3. RESULTS AND DISCUSSION

3.1 Quantitative Genetic Analysis

The metabolite concentration in the progeny was intermediate to the parents (Table 1), with no difference between the mean progeny and mean parental values (i.e., no mid-parent heterosis was present). Concentration of TNC, sucrose, glucose, soluble protein and amino-N groups were higher in ABI408 than WISFAL-6; for starch, linolenic acid, and total fatty acid concentrations, WISFAL-6 was higher. The progeny showed transgressive segregation for all metabolites. The broad sense heritability on an entry mean basis ranged from low for TNC (0.07) to high for starch (0.80). Genetic correlations with winter injury in November across years were only significant for amino-N groups and soluble protein (P<0.01). The same low correlation was found between winter injury and metabolite content for the August harvest (data not shown); however, average metabolite content increased up to twofold from August to November, suggesting their importance for winter survival. Similar results have been found by Cunningham et al. (2003) and Dhont et al. (2003).

Table 1. Parental and progeny mean values, range of progeny values, broad sense heritability on a entry mean basis, and genetic correlations (WIr_A) with winter injury of metabolites measured on taproots in November 2001 and 2002 in Ames, IA and averaged across years

Trait	F1 Progeny vs parents		Progeny	Range		
	ABI408	Wisfal6	mean	Progeny	H^2	$WIr_{A\pm}$ SE
TNC, mg g^{-1}DM	128**	104 **	111±14	70-155	0.04±0.1	-0.15±0.1
Starch, mg g^{-1}DM	255**	339	317±61	86-502	0.8±0.03	-0.19±0.1
Sucrose, mg g^{-1}DM	78*	62	67±23	18-141	0.5±0.07	0.07±0.1
Glucose, μg g^{-1}DM	1093	980	938±28	150-900	0.4±0.09	0.01±0.01
FAME, mg g^{-1}DM	13 **	16 *	15±1	12-18	0.5±0.08	0.20±0.1
Linoleic, mg g^{-1}DM	7.5	8.9 **	8±0.7	6-10	0.6±0.07	0.18±0.04
Sol. Protein, mg g^{-1}DM	60 **	54	55±4	26-68	0.5±0.07	0.26±0.1**
Amino-N, μmoles g^{-1}DM	122 **	92 *	104±31	40-191	0.5±0.08	0.44±0.1**

3.2 QTL Mapping for Fall Growth Components

We identified several QTL from each parent for all traits (Table 2). Marker locus *UGA769a1* had the strongest association with shoot mass and was also associated with root mass; its presence decreased both traits in the progeny. The presence of allele *a1* of *MSAIC B*, a cold-related gene (Laberge et al., 1993), was positively associated with autumn plant height but negatively associated with root mass in the WISFAL-6 parent. Importantly, little overlap between loci controlling autumn plant height or shoot mass was observed with those controlling winter injury. This is not surprising considering the low genetic correlation between these traits in this population (Brummer et al., 2000). These results raise the hope that biomass yield and winter injury can be manipulated independently, at least in some populations.

3.3 QTL Mapping of Metabolites

Numerous markers located in same linkage group (G) were associated with the concentration of the metabolites related to alfalfa winter hardiness that we measured (Table 3). Some loci were associated with multiple metabolites. ABI408 carries alleles at two loci on G that have opposite effects, both of which explain about 11% of the phenotypic variation observed for this trait. Two markers associated with both linoleic and total fatty acid concentration (*ACG/CTG325* and *UGA328*) were also located in the linkage group G. Marker locus *UGA577* was strongly associated with soluble protein (R^2 = 11.1), although the presence of the marker allele decreased the soluble protein content. Few of these QTL correlate with those controlling the overall agronomically important phenotype of winter injury.

Table 2. Molecular markers associated with QTL for shoot and root biomass, plant height, and winter injury based on mean values across two years at Ames, IA. The parent carrying the marker allele is designated. The probability level, R^2, and phenotypic value for individuals with (+) or without (-) the marker allele are given.

LG	Closest marker	Parent	p-value	R^2	Mean (+)	Mean(-)
			Shoot mass, g			
B	*UGA161b2*	ABI408	0.0009	6.0	23.4	19.8
K	*AGC/CAC211*	ABI408	0.0083	3.7	23.7	21.5
L	*AGC/CAC352*	ABI408	0.0045	4.3	23.9	21.6
B	*UGA189a2*	Wisfal 6	0.0026	5.1	22.1	25.0
D	*UGA769a1*	Wisfal 6	0.0002	7.3	20.9	23.9
D	*bC3C-25aV2a1*	Wisfal 6	0.0015	5.3	23.8	21.2
G	*AGC/CAT452*	Wisfal 6	0.0048	4.9	23.4	20.6
I	*ACG/CTC168*	Wisfal 6	0.0073	3.8	23.7	20.6
M	*MS58a2*	Wisfal 6	0.0077	3.9	23.1	19.7
			Root mass, g			
B	*AGC/CTG247*	ABI408	0.0099	3.6	5.9	5.6
G	*UGA1208b*	ABI408	0.0064	4.1	5.5	5.9
J	*bC2A-9AV28b*	ABI408	0.0008	6.1	5.8	5.7
B	*ACG/CTA142*	Wisfal 6	0.0014	5.4	6.1	5.6
C	*MSAICAa1*	Wisfal 6	0.0077	3.8	5.9	5.6
D	*UGA769a1*	Wisfal 6	0.0005	6.4	5.5	5.9
J	*MSAIC Ba1*	Wisfal 6	0.0005	6.5	5.5	6.1
M	*MS58a2*	Wisfal 6	0.0004	7.0	5.9	5.1
			Autumn Plant Height, cm			
E	*ARC1H11b*	ABI408	0.0033	4.5	13.5	14.5
G	*UGA328b*	ABI408	0.0064	3.9	13.5	14.4
J	*UGA452b2*	ABI408	0.0040	4.7	14.5	13.5
A	*UGA36a4*	Wisfal 6	0.0011	6.1	13.3	14.5
B	*UGA85a3*	Wisfal 6	0.0095	3.8	14.5	13.6
G	*ACG/CTA301*	Wisfal 6	0.0022	4.9	13.7	14.8
J	*MSAIC Ba1*	Wisfal 6	0.000003	11.4	14.8	13.2
			Winter Injury, 1 = none to 5=dead			
B	*V25b1*	ABI408	0.00021	7.8	2.3	2.8
H	*AGC/CTT279*	ABI408	0.00361	4.6	2.4	2.2
J	*UGA191b2*	ABI408	0.00233	4.9	2.4	2.2
K	*UGA246b*	ABI408	0.00575	4.2	2.3	2.5
A	*UGA1208a1*	Wisfal 6	0.00989	3.6	2.2	2.4
D	*UGA452a2*	Wisfal 6	0.00008	8.5	2.2	2.4
D	*UGA769a2*	Wisfal 6	0.00845	3.9	2.4	2.2

Table 3. Molecular markers associated with QTL for soluble carbohydrates, soluble protein, amino-N groups and fatty acids based on mean values across two years at Ames, IA. The parent carrying the marker allele is designated. The probability level, R^2, and phenotypic value for individuals with (+) or without (-) the marker allele are given

LG	Closest marker	Parent	p-value	R^2	Mean (+)	Mean (-)
			Total Nonstructural Carbohydrates, mg g^{-1}DM			
M	*UGA122*	ABI408	0.0054	4.4	112	109
B	*ACG/CTG277*	Wisfal6	0.0010	5.7	112	112
D	*bC3C-25aV2*	Wisfal6	0.0013	5.4	111	112
K	*UGA328-2*	Wisfal6	0.0026	5.2	111	113
			Starch, mg g^{-1}DM			
E	*bN2-20aV55*	ABI408	0.0026	4.9	311	325
G	*UGA1208*	ABI408	0.000005	11.2	305	327
G	*UGA5*	ABI408	0.000007	11.2	331	309
N	*UGA906*	ABI408	0.0029	4.8	314	332
A	*Vg1H6*	Wisfal6	0.0003	7.8	326	308
J	*RC-1-51dT23V20*	Wisfal6	0.0039	4.5	309	323
			Sucrose, mg g^{-1}DM			
G	*UGA1208*	ABI408	0.0013	5.7	69	66
G	*Vg2D11-1/2*	ABI408	0.0032	4.8	67	66
D	*bC3C-25aV2*	Wisfal6	0.0034	4.5	66	69
D	*Hg2G1*	Wisfal6	0.0045	4.5	67	68
			Glucose, µg g^{-1}DM			
E	*UGA28*	ABI408	0.0004	6.9	1015	975
F	*AGC/CTC141*	ABI408	0.0031	4.7	975	1008
C	*V25-1*	Wisfal6	0.0042	5.1	1002	961
G	*AGC/CTT192*	Wisfal6	0.0007	6.1	986	1031
			Linoleic Acid, mg g^{-1}DM			
G	*ACG/CTG325*	ABI408	0.0011	5.6	8.2	7.9
G	*UGA328*	ABI408	0.0009	5.7	7.9	8.1
G	*UGA5*	ABI408	0.0022	5.4	7.9	8.1
A	*Vg1H6*	Wisfal6	0.0029	5.4	7.9	8.2
H	*UGA553*	Wisfal6	0.0092	3.8	8.1	7.8
			Total Fatty Acid Methyl Esters, mg g^{-1}DM			
G	*UGA328*	ABI408	0.0011	5.6	14.4	14.7
B	*RC2B-63BV8*	Wisfal6	0.0086	3.4	14.5	14.8
			Soluble Protein, mg g^{-1}DM			
E	*UGA577*	ABI408	0.000005	11.1	54	56
C	*AGC/CAT159*	Wisfal6	0.00063	6.2	56	54
			Amino-N groups, µmoles g^{-1}DM			
F	*UGA109-1*	ABI408	0.0042	4.8	106	102
D	*Hg2G1*	Wisfal6	0.0066	4.3	104	105
G	*AGC/CAA452*	Wisfal6	0.0033	4.8	104	104
J	*ACG/CAC324*	Wisfal6	0.0056	4.3	103	106

4. CONCLUSION

We have demonstrated that we can identify loci involved in important physiological and biochemical pathways underlying complex, agronomically important traits like winter injury. Further analysis will focus on identifying the precise regions of the genomes in which QTL for metabolite concentration, biomass production, and winter injury reside, and on attempting to link the QTL to potential candidate genes.

ACKNOWLEDGEMENTS

Funding from the USDA-IFAFS competitive grants program (Grant #00-52100-9611) is gratefully acknowledged.

REFERENCES

Alarcón-Zúñiga B, Brummer EC, Scott MP, Luth D, Moore KJ, and Volenec JJ (2001) Metabolic profiling analysis of diploid and tetraploid alfalfa genotypes contrasting in winter hardines. ASA annual meeting, Oct 21-25. Charlotte, NC, USA.

Boyce PJ, Volenec JJ (1992a) Taproot carbohydrate concentrations and stress tolerance of contrasting alfalfa genotype. Crop Sci. 32: 757-761.

Brummer EC, Shah M, Luth D (2000) Reexamining the relationship between fall dormancy and winter hardiness in alfalfa. Crop Sci. 40: 971-977.

Cunningham SM, Volenec JJ, Teuber LR (1998) Plant survival and root and bud composition of alfalfa populations selected for contrasting fall dormancy. Crop Sci. 38: 962-969.

Cunningham SM, Nadeau P, Castonguay Y, Laberge S, Volenec JJ (2003) Raffinose and stachyose accumulation, galactinol synthase expression, and winter injury of contrasting alfalfa germplasm. Crop Sci. 43: 562-570.

Dhont C, Castonguay Y, Nadeau P, Bélanger G, Chalifour F (2003) Alfalfa roots nitrogen reserves and regrowth potential in response to fall harvests. Crop Sci. 43:181-194.

Falconer DS, Mackay TFC (1996) Introduction to quantitative genetics, Longman, UK.

Holland JB, Nyquist WE, Cervantes-Martinez CT (2003) Estimating and interpreting heritability for plant breeding: an update, *In*: Plant Breeding Reviews, Volume 22, Janick J (ed.), pp.9-112. John Wiley & Sons.

Laberge S, Castonguay Y, Vezina LP (1993) New cold- and drought-regulated gene from *Medicago sativa*. Plant Physiol. 101: 1411-1412.

Littell RC, Milliken GA, Stroup WW, Wolfinger RD (1996) SAS system for mixed models. SAS Institute, Cary, NC.

Lynch M, Walsh B (1998) Genetics and analysis of quantitative traits, pp. 980. Sinauer Ass.

McCaslin M, Woodward T (1995) Winter survival. In: Standard test to characterize alfalfa cultivars, Fox CC et al. (eds.), pp. A-7. 3rd ed. NAAC, Beltsville, MD.

McKenzie JS, Paquin R, Duke SH (1988) Cold and heat tolerance. *In:*). In: Alfalfa and alfalfa improvement, Agron. Monogr. 29, Hanson AA et al. (eds.), pp. 259-302. ASA-CSSA-SSSA, Madison, WI.

Robins JG, Viands DR, Campbell TA, Luth D, Brummer EC (2003) Mapping biomass yield in an intersubspecific cross of alfalfa. In: Abstracts of Molecular Breeding of Forage and Turf, Third International Symposium, p.15. May 18-22, Dallas, Texas and Ardmore, Oklahoma, USA.

SAS Institute. 1990. SAS/STAT user's guide. Version 6. SAS Institute, Cary, NC.

Smith PK, Krohn RI, Hermanson GT, Mallia AK, Gartner FH, Provenzano MD, Fujimoto EK, Goeke NM, Olson BJ, Klenk DC (1985) Measurement of protein using bicinchoninic acid. Anal. Biochem. 150: 76-85.

Tomashow MF (1999) Plant cold acclimation: freezing tolerance genes and regulatory mechanism. Ann. Rev. Physiol. Mol. Biol. 50: 571-99.

Volenec JJ, Cunningham SM, Haagenson DM, Berg WK, Joern BC, Wiersma DW (2002) Physiological genetics of alfalfa improvement: past failures, future prospects. Field Crops Res. 75: 97-110.

Molecular Genetics and Modification of Flowering and Reproductive Development

Daniele Rosellini
Dipartimento di Biologia Vegetale e Biotecnologie Agroambientali, Università degli Studi di Perugia, Borgo XX giugno 74, 06121 Perugia, Italy.(Email: roselli@unipg.it).

Key words: abscission, apomixis, fertility, flowering, MADS box, meristem identity, self-incompatibility, sterility

Abstract:

Reproductive traits of forage and turf plants have not been significantly improved because the end products of both forage and turf are not the seed; consequently research on reproduction has not received much attention. However, no variety can be successful if it is not a good seed producer. New interest for reproduction research may be brought about by the need for gene containment of transgenic forage cultivars. From flower induction to seed maturation, the scope of molecular reproduction research is wide even for forage crops, because it can help improve biomass yield and quality, modify the growth habit, exploit heterosis, improve fertility, seed yield and seed quality, and control gene flow from transgenic crops. Apomixis research, which employs some forage species as model organisms, holds promise to provide tools for increased progress in agriculture. This review summarizes the current knowledge on the molecular aspects of reproduction in forage and turf plants. Studies on model plants that may be applicable to forage and turf species will also be referred to.

A. Hopkins, Z.Y. Wang, R. Mian, M. Sledge and R.E. Barker (eds.), Molecular Breeding of Forage and Turf, 105-126.

1. INTRODUCTION

Sexual reproduction is considered of secondary importance in forage and turf research because the seed is not the end product. However, almost all the species are propagated through seed, with the exception of some rhizomatous turf species that can be established by stolonizing and sprigging (Beard 1973). Therefore, if seed is not made available at a reasonable cost, no seed-propagated cultivar can be successful.

Table 1 schematically presents some aspects of plant sexual reproduction and their potential impact on forage and turf species. A better understanding of the molecular basis of reproductive development and a better control of the reproductive process may have a big impact on forage crops; however, relatively little effort has been devoted to research in this field.

Table 1. Possible applications of knowledge on reproduction in forage and turf species

Knowledge on	Can improve
Flower induction and flowering time	Biomass and quality Growth habit Seed yield and quality
Male fertility and sterility Microsporogenesis and Microgametogenesis	Exploitation of heterosis Breeding practice Gene containment
Female fertility and sterility Megasporogenesis and Megagametogenesis	Seed yield Exploitation of heterosis Gene containment
Incompatibility	Breeding practice Exploitation of heterosis
Apomixis	Exploitation of heterosis Seed yield and quality
Fruit abscission	Seed yield and quality

A notable exception is the research on apomixis, only because it is found in several grasses that are used as forages. In fact, the principal objective of the investigations on apomixis in these species is to transfer the trait to seed crops.

An indicator of the amount of molecular data available for forage crop reproduction can be derived from the sequence information available in nucleotide databases. A quick search of NCBI nucleotide database made on 10 February, 2003, yielded 123,443 entries using the keyword ‘flower’. Of these, 6,728 were from *Medicago truncatula* and 1,929 from *Lotus japonicus*, two forage legume model species. For forage crops, 90 *M. sativa* and one *Lolium perenne* sequence were found, while no flower-related sequences were retrieved for *Dactylis, Festuca, Phalaris, Trifolium* or other important forage species.

The reproductive process starts with floral induction and continues with floral meristem formation and the differentiation of female and male flower parts; mega- and microsporogenesis and mega- and microgametogenesis lead to the formation of the female and male gametes; double fertilization, embryo and endosperm development, and seed maturation then follow; eventually, fruit abscission or pod dehiscence often occur in forage crops. This review summarizes the current knowledge on the molecular aspects of reproduction in forage and turf plants. Reference is made to the studies conducted on model plants and non-forage crops, for findings that may be applicable to forage and turf species.

2. FLOWER INDUCTION AND FLOWERING TIME

It has been reported, and is generally accepted, that selection for seed yield in forage grasses may have adverse effects on forage yield by changing the plant architecture toward a high proportion of 'reproductive tillers', whereas forage yield is maximized when tillers remain in the vegetative state (White 1990). Control of flower induction and flowering time could help avoid the possible conflict between the forage and seed yielding attitudes of cultivars. In fact, the ability of keeping flowering repressed when producing forage, and allowing it in seed production fields, can be a way to improve both forage and seed yield and quality.

The most useful application of biotechnological flowering control is perhaps that of gene containment of transgenic forage and turf crops. At a workshop titled 'Biotechnology-derived, Perennial Turf and Forage Grasses: Criteria for Evaluation' (http://www.isb.vt.edu/news/2003/artspdf/feb0301.pdf), organized by the Council for Agricultural Science and Technology (CAST), it was pointed out that manipulation of flowering can be advantageous for productivity. There were suggestions that male sterility or both male and female sterility should be guaranteed for safe release of transgenic varieties. In fact, the release of transgenic forage and turf varieties is particularly difficult under the present regulatory system due to potential weediness, long distance pollen dispersal, and widespread occurrence of interfertile wild plants. When expressing biologically active, high value molecules, such as vaccines, in transgenic forage crops, the necessity of strict control of gene flow would be even higher. Therefore, it is not surprising that flowering control is one of the main objectives of several research groups (Jensen et al., 2003; Forester et al., this volume).

The availability of dependable inducible gene expression systems (reviewed in Zuo and Chua 2000) may facilitate the practical realization of flowering control. A gene repressing flowering driven by a constitutive promoter would maintain the crop in the vegetative state, and an inhibitor of the repressor could be expressed through an inducible system. When seed production is needed, treatment with the inducing chemical would switch on the inhibitor and flowering would be temporarily de-repressed. Recently, a chemically inducible system has been proposed for control of plant gene expression that, for the first time, employs a commercially available chemical that has no toxicity for humans or plants (Padidam et al. 2003). This system is based on a chimeric insect ecdysone receptor and may be suitable for field use.

Alternatively, flowering could be permanently repressed in the marketed varieties. The *cre-lox* site-specific recombination system, that has been used for precise and reliable excision of target sequences in plants (Srivastava and Ow 2001 and references therein; Corneille et al. 2001), may be exploited for this purpose. A commercial, non-flowering hybrid crop could be obtained by crossing a parent engineered to contain the *cre* recombinase with another parent engineered with a repressor of flowering, whose expression is blocked by a sequence flanked by the *lox* recognition sites. The commercial hybrid would express the *cre* enzyme, the blocking sequence would be excised, and the flower inhibitor would be expressed in the released variety. This system would also require pollination control methods capable of assuring 100% crossing in the seed production field. Obviously, any sophisticated flowering control system could only be economically justified for very high-value varieties. Some potential targets are discussed below.

It remains to be assessed whether a non-flowering variety is actually capable of superior performances. In particular, attention should be paid to make sure that persistence of perennial species that rely on root or crown reserves for winter survival is not negatively affected by such a profound alteration of plant development. If the completion of the reproductive cycle is necessary for the accumulation of reserves, persistence of a non-flowering variety could be reduced. However, naturally sterile bermudagrass and zoysia varieties have normal persistence.

Plenty of molecular data have recently accumulated on flowering control, especially in *Arabidopsis thaliana*, with more than 80 genes involved (Simpson et al. 1999; Simpson and Dean 2002), and a complete review of the current knowledge is not possible here. On the contrary, very little is known about floral induction or meristem identity genes in forage plants, but some of the key genes characterized in model plants appear to have conserved functions and could be tested for flowering manipulation in forages. The vast literature on floral organ identity homeotic genes, acting downstream of the meristem identity genes, will not be reviewed here because their usefulness for forage and turf breeding appears to be minor.

It has been shown that the vegetative state is likely to be the result of repression of genes that promote reproductive development (Sung et al. 2003). Of the many flowering repressor genes, the Arabidopsis *Flowering Locus C* (*FLC*) gene (Michaels and Amasino 1999), encoding a MADS-box transcription factor (Alvarez-Buylla 2001), plays a central role. *FLC* is a dominant inhibitor of flowering that negatively regulates the so-called floral pathway integrator genes, responsible for the activation of the downstream meristem identity genes (reviewed in Simpson and Dean 2002). When *FLC* was over-expressed in Arabidopsis by the *35S* promoter, a marked delay of flowering was obtained, with plants forming 10 to >45 rosette leaves vs 6 in non-transformed plants, and thus significantly increasing the biomass produced (Michaels and Amasino 1999).

A gene that shows conserved flowering control functions in diverse plants is Arabidopsis *TERMINAL FLOWER 1* (*TFL1*). This gene encodes a phosphatidyl ethanolamine binding protein that controls floral transition by activating *FLC,* and is the homologue to *Anthirrinum majus CENTRORADIALIS* (*CEN*). *TFL* mutation results in the inflorescence meristem being converted into a terminal flower, that is, from inderminate to determinate, while its over-expression extends all growth phases, resulting in late flowering and more highly branched plants in Arabidopsis (Ratcliffe et al. 1998). In tomato, the mutation of a *TFL1/CEN* homologue, *SELF PRUNING* (*SP*), similarly caused an early termination of the reproductive phase, with the transformation of the apical meristem into a flower, and over-expression of *SP* or *CEN* restored the indeterminate growth phenotype (Pnueli et al. 1998). When *CEN* was expressed under the control of the *35S* promoter in tobacco, a determinate species, the vegetative phase was markedly prolonged, with some transformants flowering after more than ten months of vegetative growth (Amaya et al. 1999). However, over-expression of Arabidopsis *TFL1* did not produce the same phenotype. The rice genes *RCN1* and *RCN2*, isolated by homology with *TFL1*, also induce late flowering: transgenic rice plants over-expressing them formed a minimum of 4 leaves more than control plants before heading, and some of them did not flower at all (Nakagawa et al. 2002).

A *TFL1* homologue, *LpTFL1*, was isolated form *L. perenne* and expressed in Arabidopsis using the *35S* promoter. The transgenic plants formed 34 rosette leaves vs 16 in control plants, flowered 10 days to 3 months later or did not flower (Jensen et al. 2001), producing a much higher biomass. Over-expression of this gene in *Festuca rubra* lead to complete flower inhibition over two flowering seasons in some of the transgenic plants (Jensen et al., 2003). This technology may be applicable to most forage grasses. Forester et al. (2003) report the isolation and characterization of putative *L. perenne* orthologues of several Arabidopsis genes involved in flowering control, with the final aim of achieving controlled delay, acceleration, inhibition or induction of flowering.

Lolium temulentum has been studied as a model for vernalization because flowering is promoted by a single long day (Ormrod and Bernier 1990). One *L. temulentum* gene homologous to Arabidopsis *LFY*, *LtLFY*, and two genes homologous to Arabidopsis meristem identity gene *Apetala 1* (*AP1*), *LtMADS1* and *LtMADS2*, were isolated, and their expression studied in situ (Gocal et al. 2001). The conservation of function between *AP1* and *LTMADS2* was indicated by partial complementation of an Arabidopsis *AP1* mutant by *LtMADS2* under the control of the *AP1* promoter. Nevertheless, the expression pattern of *LtLFY* differed from that of the Arabidopsis homologue, in that *LFY* expression induces *AP1* in Arabidopsis, whereas expression of *LtMADS2* was shown to begin before that of *LtLFY*. In the same species, the expression of *LtGAMYB*, a homologue of the barley gibberellic acid (GA)-regulated *MYB* gene was associated with the transition of the shoot apices to floral development. Its expression is enhanced following the increased levels of GA induced by the long day exposure (Gocal et al. 1999).

CONSTANS (*CO*) is a key gene of the photoperiodic floral induction pathway in Arabidopsis. *Lt-COL*, a *L. temulentum* gene isolated by homology with *CO*, shows the same, circadian clock regulated, expression pattern as *CO*, and is under study using antisense and RNAi strategies (Cheng et al. 2003).

In *Poa annua,* a MADS-box gene was isolated from flowers during floral meristem formation but nothing is known about its function (Li and Griffin 2002).

QTL mapping studies toward identifying the number of factors controlling vernalization requirements in *Festuca pratensis* (Ergon et al., 2003) and first-year flowering in *Leymus cinereus* x *L. triticoides* (Larson et al. 2003) are underway; *Festuca* chromosome 4 seems to host a number of vernalization QTL.

The available knowledge on floral induction and meristem identity genes enables testing some of the genes controlling flower development as tools for forage improvement, capitalizing from the conservation of their functions in different species, even between monocotyledons and eudicotyledons. In particular, both the key flowering repressors and the so-called floral pathway integrator genes (reviewed in Simpson and Dean 2002) are potential targets to attain the inhibition of flowering.

3. FERTILIY/STERILITY

Sterility is the inability of producing viable gametes. It could be attained by inhibiting flowering as discussed above, or by just preventing the formation of viable gametes, using natural mutations or genetic engineering.

Dominant transgenic sterility could be useful for breeding hybrid varieties, especially if coupled with tools for selecting the sterile plants in the seed multiplication fields. Transgenic herbicide resistance linked to male sterility would make it possible to eliminate segregating fertile plants in hybrid seed production fields by spraying with the herbicide. This pollination control and gene containment strategy is actively pursued in forage and turf plants (Luo et al, this volume; Jensen et al. 2003). Female sterility of the pollen parent could also be useful because the seed could be harvested from the entire field, eliminating the need for row-planting of the parental lines. If the seed and the pollen parent were engineered for resistance to a different herbicide (as proposed by Brummer 1999), the hybrid progeny from such a cross would be both male and female sterile and herbicide resistant. Spraying both herbicides on the farmers' fields would leave only male and female sterile hybrid plants, thus providing gene containment. Environmentally friendly herbicides are available for the practical applicability of this strategy, but a very high level of heterosis is necessary.

Conditional expression of the sterility genes using inducible promoters (see above) appears to be desirable. In fact, the management of sterility during breeding may be too cumbersome and costly, especially in polyploid species, unless means for switching sterility on and off are available.

3.1 Male Fertility and Sterility

The *Barnase* male sterility system (Mariani et al. 1990) uses a microbial *RNAse* gene under the control of a tobacco anther tapetum-specific promoter (*pTA29*) to effectively destroy the anther tapetum before pollen maturation. This technology has been applied to tobacco, canola (Mariani et al. 1990), corn (Williams 1993), wheat (DeBlock et al. 1997), chicory and cauliflower

(Reynaerts et al. 1993), demonstrating its flexibility. Alfalfa expressing the *pTA29-Barnase* construct, engineered at Plant Genetic Systems, Gent, Belgium (now CropDesign), is practically male sterile, because very few viable pollen grains are produced; segregation of the transgene is Mendelian and progenies expressing male sterility to a higher level with respect to the primary transformants can be selected (Rosellini et al. 2001).

A new *barnase* system has been recently proposed that can be very useful for forage and turf species. The *barnase* enzyme was engineered by splitting it into two subunits that spontaneously reassociate restoring enzyme activity when tomato plants expressing the single subunits are crossed (Burgess et al. 2002). The obvious advantage of the system is that the two genes encoding the subunits can be bred to homozygosity, each in one of two breeding populations that, when crossed, will produce a sterile hybrid. Control of these genes by male and female organ-specific promoters could allow the engineering of both male and female sterility in a hybrid crop, thus preventing seed and pollen transgene dispersal.

Microsporogenesis mutations leading to multinucleate microspores and male sterility are common in plants, and can be found by quick screening for oversize pollen grains. The *jumbo pollen* (*jp*) mutation leading to the formation of sterile, tetranucleate pollen grains has been mapped in diploid alfalfa by Tavoletti et al. (2000) using two mapping populations. The *jp* gene was mapped to linkage group 6, close to the Vgl1b RFLP marker. In a third mapping population, bi- tri- and tetranucleate microspores were found, with a prevalence of binucleate microspores, and three quantitative trait loci (QTL) contributing to the control of the multinucleated pollen trait were located, one of which was close to Vgl1b (Tavoletti et al. 2000). The authors hypothesize that linkage group 6 contains a cluster of genes involved in multinucleate microspore formation, a particular combination of which results in the *jp* trait.

3.2 Female Fertility and Sterility

Female sterility, that is, the inability to produce viable female gametes, is widespread in plants, and has been found in several forage plants such as *Lotus corniculatus* (Dobrowsky and Grant 1980), *Medicago sativa* (Vyshniakova 1991; Rosellini et al. 1998) and *Trifolium repens* (Pasumarty et al. 1993). However, no molecular data are available.

An ovule sterility trait is being studied in alfalfa using molecular tools. Sterile ovule are recognized at floral maturity by the presence of callose within the nucellus (Rosellini et al. 1998). A cytological study demonstrated that megasporogenesis is blocked at the beginning of meiosis, and that callose is deposited in the nucellar cell walls at the onset of sterility. Microsporogenesis and pollen formation are normal (Rosellini et al. 2003). The trait was mapped in a tetraploid backcross population, and one chromosome region was found to explain 73% of the variation (Brouwer and Rosellini, unpublished), indicating simple genetic control. This is confirmed by 85% narrow-sense heritability estimated by parent-offspring regression (Rosellini et al 1998). Differential display has yielded a number of ESTs differentially expressed between fertile and sterile full-sib plants, some of which are homologous to genes whose function can be related to callose metabolism and cell division. We are studying some of these ESTs further, because of the paucity of information on mutations specifically affecting female megasporogenesis (Bhatt et al. 2001).

The production of 2n eggs is common in plants and, together with the formation of 2n pollen, is at the base of the large occurrence of polyploidy (Bretagnolle and Thompson 1995). Gametes with the somatic chromosome numbers may be useful for scaling the ploidy levels during breeding programs to maximize heterozygosity (reviewed in Bingham et al. 1994) and to introgress favorable genes from the wild diploid to the cultivated tetraploid counterparts. Moreover, 2n gamete production may be a component of apomixis, and if coupled with parthenogenesis (Barcaccia et al. 1997c), could permit apomixis in a legume, which has not been reported to date.

In *Medicago*, the formation of 2n gametes by first or second division restitution has been studied in a *M. falcata* diploid mutant that produces 55-70% 2n eggs (Barcaccia et al. 1997b). This meiotic abnormality allowed mapping of the centromeres using half-tetrad analysis for RFLP mapping (Tavoletti et al. 1996) . In another mapping study employing RAPD, ISSR, and AFLP markers and bulked segregant analysis, a paternal (wt) ISSR marker mapped 9.8 cM from the putative 2n egg locus. Four maternal and three paternal independent QTLs were also mapped, explaining a large proportion of the variation of 2n egg frequency in the 2n-egg-producing plants. One of the paternal QTL co-mapped with the putative 2n egg locus and alone explained 43% of the variation. At least five genes appear to be involved in 2n egg production in alfalfa (Barcaccia et al. 2000).

A differential display study led to the isolation of a gene, *Mob1*-like, that is likely to be involved in the meiotic abnormalities leading to 2n egg formation. In fact, the product of the *Mob1* gene controls spindle pole body duplication, separation of chromosomes completion of cytokinesis (Barcaccia et al. 2001). Northern and *in situ* hybidization analyses revealed that *Mob1*-like is expressed early during ovule development in apomeiotic megaspore mother cells and 2n embryo sacs of the mutant plant, and also at the end of sporogenesis in the three micropilar megaspores undergoing programmed cell death. No expression was observed in functional megaspores. Nine distinct *Mob1*-like members were isolated from genomic DNA amplified with specific primers. Sequencing revealed that all the members from the apomeiotic plants carry two extra stop codons and a 66-bp insertion relative to the wild-type members. A putative motif related to animal cell death gene regulator, *NB-ARC*, was identified. Immunolocalization studies are showing that the Mob1 proteins are present on spindle pole bodies and tubules during nuclear division and on the cytoplasm division septum during cytokinesis (Barcaccia G., personal communication).

4. SELF-INCOMPATIBILITY

Self-incompatibility is a widespread genetic device by which plants are able to recognize and reject self pollen and prevent self-fertilization. The best studied SI systems are those of the *Solanaceae*, *Brassicaceae* and *Papaveraceae* (de Nettancourt 2000). In each of these systems, peculiar molecular mechanisms are involved in the determination of the SI reaction, demonstrating that different ways of preventing self-fertilization evolved independently in flowering plants.

Among forage plants, SI is found both in the grass and legume families. In *Trifolium*, a gametophytic, one-locus system is present (de Nettancourt 2000) but nothing is known at the molecular level. In grasses, SI is based on a two-locus, complementary system, *SZ*, whereby both the pollen alleles at the S and Z loci must be matched in the style for an incompatible reaction to occur. Mapping studies of the S and Z genes in forage species are limited to *Phalaris coerulescens* and *L. perenne*. In *Phalaris coerulescens*, the isolation of a gene, *Bm2*, initially believed to be a pollen S gene, and later identified as an S-linked gene (Langridge et al. 1999) was reported. This gene was isolated by screening a mature pollen-derived cDNA library with cDNA probes from plants with different S and Z genotypes. It encodes a novel thioredoxin, maps about 2 cm from the S locus, is expressed in mature pollen, and may be involved in some aspects of the incompatibility reaction.

In *L. perenne*, a genotyping and mapping study of the S and Z loci was recently accomplished using an F_1 family obtained by crossing a doubled haploid ($S_{11}Z_{11}$) with an unrelated heterozygous pollen parent ($S_{23}Z_{23}$) (Thorogood et al. 2002). The resulting 139-plant progeny had only four expected incompatibility genotypes and the single plants were genotyped by *in vitro* pollination: the compatibility of the crosses was assessed by evaluating pollen germination and tube growth by means of callose staining with aniline blue. RFLP and isozyme markers were employed for mapping the incompatibility loci (Jones et al. 2002). The S and Z loci segregated independently; the S locus was assigned to linkage group (LG) 1 and the Z locus to LG 2. However, some of the markers of LG 3 also showed association with the S locus, indicating that LG 3 interacts with the S locus or with a gene closely linked to it. The genomic region of LG 1 containing the S gene showed synteny with that of rye. The mapping of self-incompatibility loci could be useful to facilitate monitoring S and Z allele diversity in breeding populations, avoiding the risk of reducing it as a consequence of increasing homozygosity for useful alleles located on the same genomic regions. Allele-specific probes for both the S and Z loci would be useful for this purpose (Thorogood et al. 2002). Self fertility of *L. perenne* is also studied by the same group (D. Thorogood, personal communication). Molecular marker mapping showed that it is influenced by the S and Z loci, plus another locus on LG 5. This locus is syntenic with the S5 locus of rye and appears to be a non functional SI locus. A male and female sterility locus was mapped to LG 4 in this population.

Another attempt of isolating genes related to self-incompatibility in *L. perenne* is underway, based on differential display (Van Daele et al., this volume). The c-DNA-AFLP technique is used to isolate fragments of genes differentially expressed between non-pollinated, self-pollinated and cross-pollinated pistils. Mapping of these ESTs in an already available linkage map will also help to confirm the chromosome region hosting the self-incompatibility locus.

5. APOMIXIS

Apomixis, or agamospermy, is the asexual formation of seed. In *gametophytic apomixis*, an unfertilized diploid cell originating from restitutional meiosis (*diplospory*) or a somatic nucellar cell (*apospory*), forms an embryo sac with maternal genes only. Then the egg cell of this embryo sac develops without fertilization (*parthenogenesis*) forming an embryo. Alternatively, an embryo can originate directly from a somatic cell without the formation of an embryo sac (*adventitious embryony*). In most cases, pollination and fertilization of the secondary nucleus of the embryo sac is necessary for the formation of a viable endosperm; this is known as *pseudogamous* apomixis; in *autonomous* apomixis, in contrast, pollination and fertilization are not necessary for seed formation (Savidan 2000).

Research on apomixis is very active for its potential to bring about an 'asexual revolution' in agriculture (Vielle-Calzada et al. 1996) when implemented in the most important seed crops. In fact, it can permit seed-cloning hybrid genotypes, thus eliminating the need for crossing parental lines every year to produce the hybrid seed for the market.

Apomixis has been investigated at the molecular level essentially by genome mapping in several grasses belonging to the genera *Brachiaria, Panicum, Paspalum, Pennisetum, Poa, Tripsacum* (Pessino et al. 1999), some of which are used as forage or turf. Most of these species have aposporous, pseudogamous apomixis, and the confirmed apomictic plants are all polyploid. Several recent reviews on apomixis are available (Koltunow et al. 1995; Pessino et al. 1999; Savidan 2000; Grimanelli et al. 2001). Here, the most relevant findings of molecular studies will be summarized.

- A single genomic region controls apospory in *Brachiaria brizantha* (Pessino et al. 1998), *Panicum maximum* (Nakagawa et al. 2003), *Paspalum simplex* (Pupilli et al. 2001), *Pennisetum ciliare* (*Cenchrus ciliaris*, Jessup et al. 2002), *Pennisetum squamulatum* (Ozias-Akins et al. 1998), and diplospory in *Tripsacum dactyloides* (Kindiger et al. 1996). The term ASGR (Apospory Specific Genomic Region) has been introduced for the chromosome segment containing the apospory locus. Physical fine mapping and fluorescence *in situ* hybridization in *Pennisetum* recently showed that the ASGR is characterized by the abundance of retrotransposon-like sequences and a low gene content, and that it occupies a significant portion of a chromosome arm (Akiyama et al. 2003; Ozias-Akins et al. 2003).

- Recombination is suppressed in the apomixis genomic region in *Brachiaria* (Pessino et al. 1998), *P. simplex* (Pupilli et al. 2001), *P. maximum* (Nakagawa et al. 2003), *P. ciliare* (Jessup et al. 2002), *P. squamulatum* (Ozias-Akins et al. 1998; Roche et al. 1999), and *T. dactyloides* (Grimanelli et al. 1998a). This indicates the possible presence of a complex locus. However, evidence for recombination, even though rare, between apospory and parthenogenesis, the two components of apomixis, has been found in *Poa pratensis* (Matzk et al. 2000; Albertini et al. 2001b). This can be interpreted as an indication of low-frequency recombination within a complex apomixis locus. Hemizygosity, that is, the presence of sequences belonging to the ASGR in single copy in the genome, has been proved in *P. squamulatum* (Ozias-Akins et al. 1998) and in *P. simplex* (Labombarda et al. 2002) and is a good enough explanation for the lack of recombination in these cases.

- Segregation distortion is often present for apomixis and markers linked to it (Grimanelli et al. 1998b; Jessup et al. 2002; Ozias-Akins et al. 1998). The association of apomixis with a gametic lethal factor causing abortion of homozygous gametes has been proposed to explain distorted segregation (discussed in Ozias-Akins et al. 1998).

- Tetrasomic inheritance is the rule in grasses (Grimanelli et al. 1998a; Pupilli et al. 1997; Ozias-Akins et al. 1998). However, one recent report supports for the first time disomic inheritance of markers linked to apospory (Jessup et al. 2002).

- Synteny of apospory/diplospory-associated chromosome regions with regions of cereal species has been observed for *B. brizantha* with the short arm of maize chromosome 5 (Pessino et al. 1997) and linkage group C of sorghum, for *T. dactyloides* with the long arm of maize chromosome 6 (Leblanc et al. 1995), for *P. simplex* with the long arm of rice chromosome 12 (Labombarda et al. 2002), for *P. ciliare* with the linkage group D of sorghum (Jessup et al. 2002). Based on the apparent synteny of different maize and sorghum chromosomes with the apomixis regions of different species, it cannot be excluded that apomixis evolved independently more than once in grasses (Jessup et al. 2002).

Another approach to the isolation of apomixis genes is the cloning of differentially expressed transcripts from developing flowers of apomictic and sexual genotypes. Attempts have been made or are underway in several species (Vielle-Calzada et al. 1996; Leblanc et al. 1997; Chen et al. 1999; Albertini E., personal communication), but no gene confirmed to play a role in apomixis has been as yet isolated.

The identification of apomictic plants in progenies of sexual by apomictic crosses is of fundamental importance for breeding as well as for basic research. Molecular markers are useful for confirming the reproductive behavior of putatively apomictic plants (Mazzucato et al. 1995; Barcaccia et al. 1997a; Ortiz et al. 1997). A useful tool would be the availability of PCR-based markers capable of diagnosing the apomictic *vs* sexual mode of reproduction in segregating progenies or natural population. SCAR markers with these characteristics have been obtained in *P. pratensis* (Albertini et al. 2001a).

The fact that a large genomic region containing several genes likely controls apomixis, probably implies that the manipulation of this trait by genetic engineering and its transfer to non-apomictic species is not yet feasible. Large-insert cloning programs have been udertaken by several labs with the aim of physical mapping the genomic region controlling apomixis (Roche et al. 2002; Nakagawa et al. 2003; Pupilli F., personal communication).

In the model plant *A. thaliana*, mutations responsible for fertilization-independent endosperm and seed development have been described (Spillane et al. 2000), and a possible alternative way for transforming a sexual into an apomictic plant might become available in the future using such mutations. However, natural apomixis does not seem to result from the failure of a single gene of the reproductive pathway, but rather from the epistatic, possibly silencing action exerted on the normal sexual reproduction pathway by a set of genes inherited as a unit and evolved in polyploid plants (Ozias-Akins et al. 1998). Therefore, studying apomixis in polyploids species with a relatively large genome may prove to be an obligate strategy.

6. SEED ABSCISSION AND POD DEHISCENCE

Seed dispersal is a key adaptive factor in wild plants, and forage plants can be considered wild for seed traits because of lack of selective pressure for seed yield (Coolbear et al. 1997). Fruit abscission and pod dehiscence compounded with lack of ripening uniformity, results in seed shattering and loss if harvesting is too late, and in poor seed quality if harvest is too early (McWilliam 1980). Some variation for seed retention is present in natural populations and has been exploited for breeding varieties resistant to shattering (reviewed in Elgersma and van Wijk 1997; Falcinelli 1991), but the problem is still serious.

Fruit abscission and pod dehiscence have not been studied at the molecular level in forages, but studies in other species provide potentially useful information. Evidence is accumulating that dehiscence (causing pod shattering) and abscission (causing seed shedding), as well as other cell separation events, may involve similar mechanism (Roberts et al. 2002).

At the cell separation site, a specialized tissue differentiates, usually made of a few layers of cells that are smaller than adjacent non-separating cells and more densely cytoplasmic. An abscission zone (AZ) develops at the base of the organs to be shed, such as sepals, petals, fruits, and leaves, whereas a dehiscence zone (DZ) is formed along the valve margins of the pods. Some mutations affecting the differentiation of these specialized tissues have been characterized. A tomato mutant (*jointless*), does not develop an AZ in the flower pedicel (Mao et al. 2000); JOINTLESS encodes a protein of the MADS-box family.

Recently, a number of genes contributing to the formation of the dehiscence zone in *A. thaliana* have been identified. Two of these, *SHATTERPROOF 1* and *2* (*SHP1* and *SHP2*), are MADS-box proteins that control the development of the silique dehiscence zone cells and the lignification of the valve margins; when both genes are mutated, dehiscence does not occur (Liljegren et al. 2000). Another MADS-box gene involved in DZ formation is *FRUITFUL* (*FUL*). In *ful* mutants DZs evolve ectopically, whereas in plants over-expressing *FUL*, a complete lack of a DZ is observed. It has been shown that *FUL* is a negative regulator of the *SHP* genes (Ferrandiz et al. 2000). *A. thaliana* plants mutated for *INDEHISCENT 1* (*IND1*), a transcription factor, are also unable to shatter due to the lack of the lignified patches at the valve margins (Liljegren et al. 2000).

Cell separation occurs when cell wall hydrolysis takes place at the cell separation layers. A cocktail of hydrolytic enzymes probably brings about abscission, and β-1,4 glucanase (cellulase) and polygalacturonase (PG) are basic ingredients (reviewed in Roberts et al. 2002). In tomato, antisense inhibition of two cellulases, *Cel1* and *Cel2*, resulted in the need for a greater force to bring about pedicel abscission (Lashbrook et al. 1998; Brummell et al. 1999).

PG has been associated with fruit shedding in the monocotyledon *Elaeis guineensis*, the oil palm (Henderson et al. 2001) and tomato; in the latter species, three fruit abscission PGs have been identified, that are different from the PGs associated with fruit softening (Kalaitzis et al. 1997). In *Arabidopsis*, the expression of an abscission-related PG (*PGAZAT*) has been studied using promoter-GUS or -GFP fusions, and found to be specific to abscission zones at the base of the anther filament, petals and sepals (Gonzalez-Carranza et al. 2002). SAC66 and SAC70 are two homologous PGs from *B. napus* and *A. thaliana*, respectively, that are expressed at the dehiscence zone of the silique and between the seed and the funiculus. A *SAC70::Barnase* construct induced indehiscent anthers and male sterility in transgenic *B. napus*. These plants also showed reduced pod shattering (Jenkins et al. 1999). Similar results were obtained in transgenic *A. thaliana* (Roberts et al. 2000).

The balance between ethylene, accelerating the abscission process, and IAA, delaying it, is at the base of the timing of abscission (Taylor and Whitelaw 2001). Abscissic acid may also be involved in seed abscission (Sargent et al. 1984).

By silencing or controlling the expression of genes involved in the differentiation of the abscission zone, ethylene/IAA balance or cell wall breakdown it may be possible to reduce or prevent abscission or dehiscence (Roberts et al. 2002). The fact that at least some of the molecular mechanisms involved in abscission and dehiscence appear to be conserved, the transgenic approach may prove to be more efficient than conventional breeding.

Map based isolation of grass seed shattering genes is a possible strategy for improving this trait, and QTL studies can provide a starting point (Kennard et al. 2002).

7. CONCLUSIONS

This review shows that relatively little is known on the molecular basis of the reproductive process in forage and turf plants. However, research in this field is now receiving some attention, especially for allowing the full exploitation of biotechnology while ensuring gene containment.

One aspect of the possible contribution of the DNA technologies is that of QTL mapping of seed traits. The necessity of maintaining or improving seed yield during the variety development process is a serious complication for forage and turf breeders, because it requires testing in specialized seed crops and in the seed production locations, that are often different from the forage production areas. The availability of molecular markers linked to QTL for reproductive traits and seed yield components (Fang et al. 2003) may prove to be useful for simplifying the testing of seed potential of the breeding lines. Very little effort has been made toward this objective, but it may be worth pursuing it.

Our understanding and ability to manipulate reproduction will profit from molecular knowledge gained in model species, but even the attempt of just transferring that knowledge to forage and turf plants would require considerable effort. The efficiency of the genetic transformation techniques applied to forage and turf plants is rapidly progressing (Spangenberg 2001; Wang et al. 2001), and many genes involved in reproduction control are available for testing in these species.

REFERENCES

Akiyama Y, Goel S, Hanna WW and Ozias-Akins P (2003) Architecture and dynamics of the chromosome associated with apoaporous apomixis in *Pennisetum squanulatum.* In: Abstracts of Plant & Animal Genome XI Conference, P284. January 11-15, San Diego, CA, USA.

Albertini E, Barcaccia G, Porceddu A, Sorbolini S, Falcinelli M (2001a) Mode of reproduction is detected by Parth1 and Sex1 SCAR markers in a wide range of facultative apomictic Kentucky bluegrass varieties. Mol. Breed. 7: 293-300.

Albertini E, Porceddu A, Ferranti F, Reale L, Barcaccia G, Romano B, Falcinelli M (2001b) Apospory and parthenogenesis may be uncoupled in Poa pratensis: a cytological investigation. Sex. Plant Reprod. 14: 213-217.

Alvarez-Buylla ER (2001) MADS-box gene evolution beyond flower: expression in pollen, endosperm, guard cells, roots and trichomes (vol 24, pg 457, 2000). Plant Journal 25: 593

Amaya I, Ratcliffe OJ, Bradley DJ (1999) Expression of CENTRORADIALIS (CEN) and CEN-like genes in tobacco reveals a conserved mechanism controlling phase change in diverse species. Plant Cell 11: 1405-1417.

Barcaccia G, Albertini E, Rosellini D, Tavoletti S, Veronesi F (2000) Inheritance and mapping of 2n-egg production in diploid alfalfa. Genome 43: 528-537.

Barcaccia G, Mazzucato A, Belardinelli A, Pezzotti M, Lucretti S, Falcinelli M (1997a) Inheritance of parental genomes in progenies of Poa pratensis L. from sexual and apomictic genotypes as assessed by RAPD markers and flow cytometry. Theor. Appl. Genet. 95: 516-524.

Barcaccia G, Tavoletti S, Falcinelli M, Veronesi F (1997b) Environmental influences on the frequency and viability of meiotic and apomeiotic cells of a diploid mutant of alfalfa. Crop Sci. 37: 70-76.

Barcaccia G, Tavoletti S, Falcinelli M, Veronesi F (1997c) Verification of the parthenogenetic capability of unreduced eggs in an alfalfa mutant by a progeny test based on morphological and molecular markers. Plant Breed. 116: 475-479.

Barcaccia G, Varotto S, Meneghetti S, Albertini E, Porceddu A, Parrini P, Lucchin M (2001) Analysis of gene expression during flowering in apomeiotic mutants of *Medicago* spp.: cloning of ESTs and candidate genes for 2n eggs. Sex. Plant Reprod. 14: 233-238.

Beard JB (1973) Turfagrass - Science and Culture, pp657. Prentice Hall, Inc., Englewood Cliffs, NJ, USA

Bhatt AM, Canales C, Dickinson HG (2001) Plant meiosis: the means to 1N. Trends Plant Sci. 6: 114-121.

Bingham ET, Groose RW, Woodfield DR, Kidwell KK (1994) Complementary gene interactions in alfalfa are greater in autotetraploids than diploids. Crop Sci. 34: 823-829.

Bretagnolle F, and.Thompson JD (1995) Gametes with the somatic chromosome number: mechanisms of their formation and role in the evolution of autopolyploid plants. New Phytol. 129:1-22.

Brummell DA, Hall BD, Bennett AB (1999) Antisense suppression of tomato endo-1,4-beta-glucanase Cel2 mRNA accumulation increases the force required to break fruit abscission zones but does not affect fruit softening. Plant Mol. Biol. 40: 615-622.

Brummer EC (1999) Capturing heterosis in forage crop cultivar development. Crop Sci. 39: 943-954.

Burgess DG, Ralston EJ, Hanson WG, Heckert M, Ho M, Jenq T, Palys JM, Tang KL, Gutterson N (2002) A novel, two-component system for cell lethality and its use in engineering nuclear male-sterility in plants. Plant J. 31: 113-125.

Chen LZ, Miyazaki C, Kojima A, Saito A, Adachi T (1999) Isolation and characterization of a gene expressed during early embryo sac development in apomictic guinea grass (*Panicum maximum*). J. Plant Physiol. 154: 55-62.

Cheng XF, Ge YX, Wang ZY (2003) Isolation and characterization of a CONSTANS-like gene in *Lolium temulentum.* In: Abstracts of Molecular Breeding of Forage and Turf, Third International Symposium, p.47. May 18-22, Dallas, Texas and Ardmore, Oklahoma, USA.

Coolbear P, Hill MJ and Win Pe (1997) Maturation of grass and legume seed. In: Forage seed production. 1. Temperate species, Fairey DT, Hampton JG (eds.), pp 71-103. CAB International, Walligford, Oxon, UK.

Corneille S, Lutz K, Svab Z, Maliga P (2001) Efficient elimination of selectable marker genes from the plastid genome by the CRE-lox site-specific recombination system. Plant J. 27: 171-178.

DeBlock M, Debrouwer D, Moens T (1997) The development of a nuclear male sterility system in wheat. Expression of the barnase gene under the control of tapetum specific promoters. Theor. Appl. Genet. 95: 125-131.

De Nettancourt D (2000) Incompatibility and incongruity in wild and cultivated plants, pp 344. Springer-Verlag, New York.

Dobrowsky S and Grant WF (1980) An investigation into the mechanism for reduced seed yield in Lotus corniculatus. Theor. Appl. Genet. 57:157-160.

Elgersma A and van Wijk AJP (1997) Breeding for higher seed yield in grasses and forage legumes. In: Forage seed production. 1. Temperate species. Fairey DT, Hampton JG (eds.), pp 243-270. CAB International, Walligford, Oxon, UK.

Ergon A, Cheng F, Jørgensen O, Aamlid T, Rognli O (2003) Genes controlling vernalization requirement in Festuca pratensis Huds. are present on *Festuca* chromosome 4. In: Abstracts of Molecular Breeding of Forage and Turf, Third International Symposium, p.46. May 18-22, Dallas, Texas and Ardmore, Oklahoma, USA.

Falcinelli M (1991) Backcross breeding to increase seed retention in cocksfoot (*Dactylis glomerata* L. Euphytica 56: 133-135.

Fang C, Alm V, Aamlid T, Rognli O (2003) Comparative mapping of QTLs for seed production and related traits in meadow fescue (*Festuca pratensis* Huds.). In: Abstracts of Molecular Breeding of Forage and Turf, Third International Symposium, p.45. May 18-22, Dallas, Texas and Ardmore, Oklahoma, USA.

Ferrandiz C, Liljegren SJ, Yanofsky MF (2000) Negative regulation of the SHATTERPROOF genes by FRUITFULL during Arabidopsis fruit development. Science 289: 436-438.

Forester N, Gagic M, Veit B, Bryan G, Kardailasky I (2003) Controlled flowering in perennial ryegrass. In: Abstracts of Molecular Breeding of Forage and Turf, Third International Symposium, p.42. May 18-22, Dallas, Texas and Ardmore, Oklahoma, USA.

Gocal GFW, King RW, Blundell CA, Schwartz OM, Andersen CH, Weigel D (2001) Evolution of floral meristem identity genes. Analysis of Lolium temulentum genes related to APETALA1 and LEAFY of Arabidopsis. Plant Physiol. 125: 1788-1801.

Gocal GFW, Poole AT, Gubler F, Watts RJ, Blundell C, King RW (1999) Long-day up-regulation of a GAMYB gene during Lolium temulentum inflorescence formation. Plant Physiol. 119: 1271-1278.

Gonzalez-Carranza ZH, Whitelaw CA, Swarup R, Roberts JA (2002) Temporal and Spatial Expression of a Polygalacturonase during Leaf and Flower Abscission in Oilseed Rape and Arabidopsis. Plant Physiol. 128: 534-543.

Grimanelli D, Leblanc O, Espinosa E, Perotti E, De Leon DG, Savidan Y (1998a) Mapping diplosporous apomixis in tetraploid Tripsacum: one gene or several genes? Heredity 80: 33-39.

Grimanelli D, Leblanc O, Espinosa E, Perotti E, De Leon DG, Savidan Y (1998b) Non-Mendelian transmission of apomixis in maize-Tripsacum hybrids caused by a transmission ratio distortion. Heredity 80: 40-47.

Grimanelli D, Leblanc O, Perotti E, Grossniklaus U (2001) Developmental genetics of gametophytic apomixis. Trends Genet. 17: 597-604.

Henderson J, Davies HA, Heyes SJ, Osborne DJ (2001) The study of a monocotyledon abscission zone using microscopic, chemical, enzymatic and solid state C-13 CP/MAS NMR analyses. Phytochemistry 56: 131-139.

Jenkins ES, Paul W, Craze M, Whitelaw CA, Weigand A, Roberts JA (1999) Dehiscence-related expression of an Arabidopsis thaliana gene encoding a polygalacturonase in transgenic plants of Brassica napus. Plant Cell Environ. 22: 159-167.

Jensen C, Gao C, Andersen C, Didion T, Nielsen K (2003) The potential of the floral repressor LpTFL1 from perennial ryegrass to control the floral transition. In: Abstracts of Molecular Breeding of Forage and Turf, Third International Symposium, p.50. May 18-22, Dallas, Texas and Ardmore, Oklahoma, USA.
Jensen CS, Salchert K, Nielsen KK (2001) A TERMINAL FLOMER1-Like gene from perennial ryegrass involved in floral transition and axillary meristem identity. Plant Physiol. 125: 1517-1528.
Jessup RW, Burson BL, Burow GB, Wang YW, Chang C, Li Z, Paterson AH, HUSSEY MA (2002) Disomic inheritance, suppressed recombination, and allelic interactions govern apospory in buffelgrass as revealed by genome mapping. Crop Sci. 42: 1688-1694.
Jones ES, Mahoney NL, Hayward MD, Armstead IP, Jones JG, Humphreys MO, King IP, Kishida T, Yamada T, Balfourier F, Charmet G, Forster JW (2002) An enhanced molecular marker based genetic map of perennial ryegrass (Lolium perenne) reveals comparative relationships with other Poaceae genomes. Genome 45: 282-295.
Kalaitzis P, Solomos T, Tucker ML (1997) Three different polygalacturonases are expressed in tomato leaf and flower abscission, each with a different temporal expression pattern. Plant Physiol. 113: 1303-1308.
Kang JS, Tiwana MS (2003) Sowing and closing time effect on flowering and seed production of Persian clover (*Trifolium resupinatum* L.) in India. In: Abstracts of Molecular Breeding of Forage and Turf, Third International Symposium, p.49. May 18-22, Dallas, Texas and Ardmore, Oklahoma, USA.
Kennard WC, Phillips RL, Porter RA (2002) Genetic dissection of seed shattering, agronomic, and color traits in American wildrice (Zizania palustris var. interior L.) with a comparative map. Theor. Appl. Genet. 105: 1075-1086.
Kindiger B, Bai D, Sokolov V (1996) Assignment of a gene(s) conferring apomixis in Tripsacum to a chromosome arm: Cytological and molecular evidence. Genome 39: 1133-1141
Koltunow AM, Bicknell RA, Chaudhury AM (1995) Apomixis - molecular strategies for the generation of genetically identical seeds without fertilization. Plant Physiol 108: 1345-1352.
Labombarda P, Busti A, Caceres ME, Pupilli F, Arcioni S (2002) An AFLP marker tightly linked to apomixis reveals hemizygosity in a portion of the apomixis-controlling locus in Paspalum simplex. Genome 45: 513-519.
Langridge P, Baumann U, Juttner J (1999) Revisiting and revising the self-incompatibility genetics of *Phalaris coerulescens*. Plant Cell 11: 1826.
Larson S, Wu X, Palazzo A, Jensen K, Jones T, Wang R, Chatterton J (2003) QTL analysis of plant development and salt tolerance in *Leymus* wildryes. In: Abstracts of Molecular Breeding of Forage and Turf, Third International Symposium, p.48. May 18-22, Dallas, Texas and Ardmore, Oklahoma, USA.
Lashbrook CC, Giovannoni JJ, Hall BD, Fischer RL, Bennett AB (1998) Transgenic analysis of tomato endo-beta-1,4-glucanase gene function. Role of cell in floral abscission. Plant J. 13: 303-310.
Leblanc O, Armstead I, Pessino S, Ortiz JP, Evans C, doValle C, Hayward MD (1997) Non-radioactive mRNA fingerprinting to visualise gene expression in mature ovaries of Brachiaria hybrids derived from B-brizantha, an apomictic tropical forage. Plant Sci. 126: 49-58.
Li AG, Griffin JD (2002) Cloning and analysis of a mads-box gene expressed in Poa annua md-2 apices during floral meristem formation. Int. J. Plant Sci. 163: 43-50.
Liljegren SJ, Ditta GS, Eshed HY, Savidge B, Bowman JL, Yanofsky MF (2000) SHATTERPROOF MADS-box genes control seed dispersal in Arabidopsis. Nature 404: 766-770.
Mao L, Begum D, Chuang HW, Budiman MA, Szymkowiak EJ, Irish EE, Wing RA (2000) JOINTLESS is a MADS-box gene controlling tomato flower abscission zone development. Nature 406: 910-913.

Mariani C, Debeuckeleer M, Truettner J, Leemans J, Goldberg RB (1990) Induction of male-sterility in plants by a chimeric ribonuclease gene. Nature 347: 737-741.
Matzk F, Meister A, Schubert I (2000) An efficient screen for reproductive pathways using mature seeds of monocots and dicots. Plant J. 21: 97-108.
Mazzucato A, Barcaccia G, Pezzotti M, Falcinelli M (1995) Biochemical and molecular markers for investigating the mode of reproduction in the facultative apomict poa-pratensis L. Sex. Plant Reprod. 8: 133-138.
McWillam JR (1980) The development and significance of seed retention in grasses. In: Seed Production, Hebbelthwaite PD (Ed), pp 51-60. Butterworths, London.
Michaels SD, Amasino RM (1999) FLOWERING LOCUS C encodes a novel MADS domain protein that acts as a repressor of flowering. Plant Cell 11: 949-956.
Nakagawa H, Ebina M, Takahara M, Tsuruta S (2003) Linkage analysis and cDNA analysis toward the isolation of an apospory gene in guineagrass. In: Abstracts of Molecular Breeding of Forage and Turf, Third International Symposium, p.44. May 18-22, Dallas, Texas and Ardmore, Oklahoma, USA.
Nakagawa M, Shimamoto K, Kyozuka J (2002) Over-expression of RCN1 and RCN2, rice TERMINAL FLOWER 1/CENTRORADIALIS homologues, confers delay of phase transition and altered panicle morphology in rice. Plant J. 29: 743-750.
Ormrod JC, Bernier G (1990) Cell-cycle patterns in the shoot apex of *Lolium-temulentum* l cv ceres during the transition to flowering following a single long day. J. Exp. Bot. 41: 211-216
Ortiz JPA, Pessino SC, Leblanc O, Hayward MD, Quarin CL (1997) Genetic fingerprinting for determining the mode of reproduction in Paspalum notatum, a subtropical apomictic forage grass. Theor. Appl. Genet. 95: 850-856.
Ozias-Akins P, Goel S, Akiyama Y, Gualtieri G, Conner JA, Morishige D, Mullet JE, Hanna WW (2003) Characterization of the genomic region associated with the transmission of apomixis in *Pennisetum* and *Cenchrus.* In: Abstracts of Plant & Animal Genome XI Conference, W12. January 11-15, San Diego, CA, USA.
Ozias-Akins P, Roche D, Hanna WW (1998) Tight clustering and hemizygosity of apomixis-linked molecular markers in Pennisetum squamulatum genetic control of apospory by a divergent locus that may have no allelic form in sexual genotypes. Proc. Natl Acad. Sci. USA 95: 5127-5132.
Padidam M, Gore M, Lu DL, Smirnova O (2003) Chemical-inducible, ecdysone receptor-based gene expression system for plants. Transgenic Res. 12: 101-109.
Pasumarty SV, Matsumura T Higuchi S and Yamada T (1993) Causes of low seed set in white clover (Trifolium repens L). Grass Forage Sci. 48: 79-83.
Pessino SC, Evans C, Ortiz JPA, Armstead I, Do Valle CB, Hayward MD (1998) A genetic map of the apospory-region in Brachiaria hybrids: identification of two markers closely associated with the trait. Hereditas 128: 153-158.
Pessino SC, Ortiz JPA, Hayward MD, Quarin CL (1999) The molecular genetics of gametophytic apomixis. Hereditas 130: 1-11.
Pessino SC, Ortiz JPA, Leblanc O, doValle CB, Evans C, Hayward MD (1997) Identification of a maize linkage group related to apomixis in Brachiaria. Theor. Appl. Genet. 94: 439-444.
Pnueli L, Carmel-Goren L, Hareven D, Gutfinger T, Alvarez J, Ganal M, Zamir D, Lifschitz E (1998) The SELF-PRUNING gene of tomato regulates vegetative to reproductive switching of sympodial meristems and is the ortholog of CEN and TFL1. Development 125: 1979-1989.
Pupilli F, Caceres ME, Quarin CL, Arcioni S (1997) Segregation analysis of RFLP markers reveals a tetrasomic inheritance in apomictic *Paspalum* simplex. Genome 40: 822-828.
Pupilli F, Labombarda P, Caceres ME, Quarin CL, Arcioni S (2001) The chromosome segment related to apomixis in Paspalum simplex is homoeologous to the telomeric region of the long arm of rice chromosome 12. Mol. Breed. 8: 53-61.

Ratcliffe OJ, Amaya I, Vincent CA, Rothstein S, Carpenter R, Coen ES, Bradley DJ (1998) A common mechanism controls the life cycle and architecture of plants. Development 125: 1609-1615.
Reynaerts A, Vandewiele H, Desutter G, Janssens J (1993) Engineered genes for fertility-control and their application in hybrid seed production. Scientia Horticulturae 55: 125-139
Roberts JA, Elliott KA, Gonzalez-Carranza ZH (2002) Abscission,dehiscence, and other cell separation processes. Ann. Rev. Plant Biol. 53: 131-158.
Roberts JA, Whitelaw CA, Gonzalez-Carranza ZH, McManus MT (2000) Cell separation processes in plants - Models, mechanisms and manipulation. Annals Bot. 86: 223-235.
Roche D, Cong PS, Chen ZB, Hanna WW, Gustine DL, Sherwood RT, Ozias-Akins P (1999) An apospory-specific genomic region is conserved between Buffelgrass (Cenchrus ciliaris L.) and Pennisetum squamulatum Fresen. Plant J. 19: 203-208.
Roche D, Conner JA, Budiman MA, Frisch D, Wing R, Hanna WW, Ozias-Akins P (2002) Construction of BAC libraries from two apomictic grasses to study the microcolinearity of their apospory-specific genomic regions. Theor. Appl. Genet. 104: 804-812.
Rosellini D, Lorenzetti F, Bingham ET (1998) Quantitative ovule sterility in *Medicago sativa*. Theor. Appl. Genet. 97: 1289-1295.
Rosellini D, Pezzotti M, Veronesi F (2001) Characterization of transgenic male sterility in alfalfa. Euphytica 118: 313-319.
Rosellini D, Ferranti F, Barone P, Veronesi F (2003) Expression of female sterility in alfalfa (*Medicago sativa* L.). Sexual Plant Reprod. 15: 271-279.
Sargent JA, Osborne DJ, Dunford SM (1984) Celle separation and ist hormonal control in the *Gramineae*. J. Exp. Bot. 35:1663-1667.
Savidan Y (2000) Apomixis, the way of cloning seeds. Biofutur 2000: 38-43.
Simpson GG, Dean C (2002) Flowering - Arabidopsis, the rosetta stone of flowering time? Science 296: 285-289.
Simpson GG, Gendall AR, Dean C (1999) When to switch to flowering. Ann. Rev. Cell Dev. Biol. 15: 519-550.
Spangennberg G, Kalla R, Lidgett A, Sawbridge T, Ong EK, John U (2001) Transgenesis and genomics in molecular breeding of forage plants. Proc. XIX International Grassland Congress, 11-21 February, Sao Pedro, Sao Paulo, Brasil.
Spillane C, MacDougall C, Stock C, Kohler C, Vielle-Calzada JP, Nunes SM, Grossniklaus U, Goodrich J (2000) Interaction of the Arabidopsis Polycomb group proteins FIE and MEA mediates their common phenotypes. Curr. Biol. 10: 1535-1538.
Srivastava V, Ow DW (2001) Single-copy primary transformants of maize obtained through the co-introduction of a recombinase-expressing construct. Plant Mol. Biol. 46: 561-566.
Sung ZR, Chen LJ, Moon YH, Lertpiriyapong K (2003) Mechanisms of floral repression in Arabidopsis. Curr. Opin. Plant Biol. 6: 29-35.
Tavoletti S, Bingham ET, Yandell BS, Veronesi F, Osborn TC (1996) Half tetrad analysis in alfalfa using multiple restriction fragment length polymorphism markers. Proc. Natl Acad. Sci. USA 93: 10918-10922.
Tavoletti S, Pesaresi P, Barcaccia G, Albertini E, Veronesi F (2000) Mapping the jp (jumbo pollen) gene and QTLs involved in multinucleate microspore formation in diploid alfalfa. Theor. Appl. Genet. 101: 372-378.
Taylor JE, Whitelaw CA (2001) Signals in abscission. New Phytologist 151: 323-339.
Thorogood D, Kaiser WJ, Jones JG, Armstead I (2002) Self-incompatibility in ryegrass 12. Genotyping and mapping the S and Z loci of Lolium perenne L. Heredity 88: 385-390.
Vielle-Calzada JPV, Crane CF, Stelly DM (1996) Botany - Apomixis: The asexual revolution. Science 274: 1322-1323.
Vielle-Calzada JP, Nuccio ML, Budiman MA, Thomas TL, Burson BL, HUSSEY MA, Wing RA (1996) Comparative gene expression in sexual and apomictic ovaries of Pennisetum ciliare (L) Link. Plant Mol. Biol. 32: 1085-1092.
Vyshniakova MA (1991) Callose as an indicator of sterile ovules. Phytomorphology 41: 245-252.

Wang ZY, Hopkins A, Mian R (2001) Forage and turf grass biotechnology. Crit. Rev. Plant Sci. 20: 573-619.
White JGH (1990) Herbage seed production. In: Pastures, Their Ecology and Management, Langer RHM (ed.). Oxford University Press, Oxford.
Williams ME (1993) In: Serida WF (Ed) Abstr. 35th Annu. Maize Genet. p 46. Conf. University of North Dakota.
Zuo JR, Chua NH (2000) Chemical-inducible systems for regulated expression of plant genes. Curr. Opin. Biotechnol. 11: 146-151.

Identification and Mapping of Self-incompatibility Related Genes in *Lolium perenne*

Inge Van Daele[1,2], Erik Van Bockstaele[1,2] and Isabel Roldán-Ruiz[1]
[1]Department of Plant Genetics and Breeding, Agricultural Research Centre-Gent. Caritasstraat 21, B9090 Melle. [2]Department of Plant Production, University of Gent. Coupure Links 653, B9000 Gent.(Email: i.roldan-ruiz@clo.fgov.be)

Key words: EST, mapping, self-incompatibility, ryegrass, cDNA-AFLP

Abstract:

Self-incompatibility (SI) is a genetically controlled mechanism to prevent inbreeding in flowering plants. Ryegrasses (*Lolium* spp.) display a gametophytic self-incompatibility response controlled by the two multi-allelic loci S and Z. Our objective is to identify genes involved in the self-incompatibility response in perennial ryegrass, using the cDNA-AFLP technique. The cDNA-AFLP fragments displaying differential gene expression in non-pollinated, cross-pollinated and self-pollinated pistils will be mapped in a population, which is characterised for the self-incompatibility genotypes.

A. Hopkins, Z.Y. Wang, R. Mian, M. Sledge and R.E. Barker (eds.), Molecular Breeding of Forage and Turf, 127-131.

1. INTRODUCTION

Ryegrasses (*Lolium* spp.), like all members of the *Poaceae* studied to date, display gametophytic SI controlled by two multi-allelic loci (S and Z). Self-fertilisation is prevented when both, the S and Z alleles, present in the pollen are matched in the style. The SI system in ryegrass is not complete and some seed can be produced after self-fertilisation in absence of compatible pollen. Although *S* and *Z* determine the recognition process between pollen and stigma, it is assumed that, similarly to what has been reported for other plant families, many other genetic factors are involved in the cascade of events that is triggered by an incompatible reaction in the pistil.

To date only one gene putatively involved in the SI-mechanism of the *Poaceae* has been isolated in *Phalaris coerulescens* (Li et al. 1994). This gene, coding for a protein with thioredoxin activity, was initially supposed to represent the *S*-gene of the *Poaceae* (*S*-thioredoxin, Li et al. 1994). However, further research demonstrated that this thioredoxin gene is not the *S*-gene, but is located 2 cM apart from the S-gene in *P. coerulescens*. According to these authors the thioredoxin gene could be involved in the SI response in the *Poaceae*, even if it does not represent the *S*-gene (Langridge et al. 1999). The SI response has also been studied in *Secale* cereale. Wehling et al. (1994) amplified a fragment of about 280 bp in rye with primers derived from a conserved region of the *Brassica SLG-13* gene. Polymorphisms in this fragment were correlated with the SI-genotypes. So they mentioned that protein phosphorylation and Ca^{2+}, as constituents of a signal transduction, are involved in the SI response in rye. More recently, Thorogood et al. (2002) identified the map positions of *S* and *Z* in *L. perenne*, using the ILGI reference population. *S* and *Z* mapped to chromosome 1 and chromosome 2 respectively (Thorogood et al. 2002). Although knowledge about the map-position of S and Z can help to develop strategies for the isolation of these genes, nothing is still known about the putative function or genomic organisation of other factors involved in the SI-response in the *Poaceae*.

It is beyond any doubt that a better understanding of the recognition and rejection processes between stigma and pollen would be of relevance for the development of improved breeding strategies for this commercially important crop. Therefore, the main objective of this work is to study the molecular basis of the SI mechanism in ryegrass. Genes which are differentially expressed in non-pollinated, cross-pollinated and self-pollinated pistils, are identified by using the cDNA-AFLP technique (Bachem et al. 1996). In addition, those cDNA fragments displaying interesting differential expression patterns are transformed into polymorphic

DNA-markers (ESTs) which are mapped in the ILGI reference population and the DvP-CLO reference population (Hilde et al. 2002).

2. MATERIALS AND METHODS

Two different *L. perenne* genotypes (T_3 and T_6) were used in the study. RNA was extracted from the following materials:

- T_6 self-pollinated pistils
- T_6 non-pollinated pistils
- T_6 pistils cross-pollinated with T_3 pollen
- T_3 pollen grains; T_6 pollen grains
- Leaf material.

The self- and cross-pollinated pistils were harvested at four different times after pollination (1, 4, 8 and 24 hours). This resulted in a total of 12 samples to compare. All pollinations were carried out in the greenhouse, under controlled conditions. For cross-pollination, emasculated flowers were used. mRNA was extracted using the 'guandininium thiocynate method' described by Chomczynski and Sacchi (1987). Double stranded cDNA was synthesised from the mRNA using a biotinalyted oligo$(dT)_{25}$ primer. AFLP reactions were performed with *Bsty*I/*Mse*I primer couples (and using a total of four selective bases). Fragment separation was carried out in sequencing gels and radioactive labelling and detection methods were used. The cDNA fragments which showed a differential expression pattern were excised from the gel, reamplified and sequenced. The sequences were then entered in the database of plants ('National Centre for Biotechnology Information, NCBI) and the program 'Fasta' was used to search for homologies with known genes and to assign putative functions to these fragments.

Fragment sequences were also used to develop specific PCR primer pairs. The presence of polymorphisms in the PCR-amplified fragments was checked in two populations (the ILGI reference population and the DvP-CLO reference *L. perenne* population). First, PCR-fragment length polymorphisms were searched for. For those PCR primer pairs unable to identify length polymorphisms in the mapping populations studied, restriction-site polymorphisms were searched for.

3. RESULTS AND DISCUSSION

To date all 256 cDNA-AFLP primer combinations that are possible with a set of two restriction enzymes when 2+2 selective nucleotides are used have been tested. A total of 155 cDNA-AFLP fragments displaying differential expression patterns in self-pollinated, cross-pollinated and non-pollinated pistils have been excised from the gel. The size of the fragments selected fluctuates between 100 and 700 base pairs. Some of these fragments display homology with members of different gene families (hydrolases, glucosidases, RNA-polymerases, SRK kinases). But the majority (70%) display homology only with genes of unknown function.

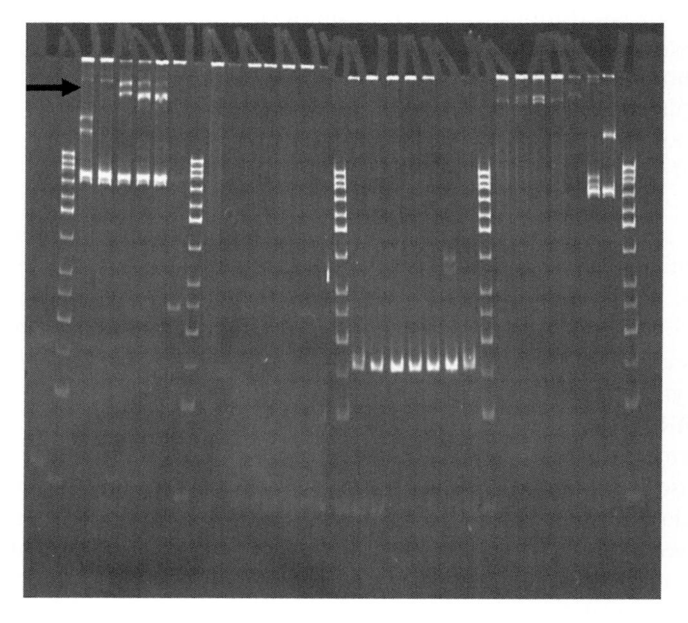

Figure 1. Development of polymorphic markers from cDNA-AFLP fragments in the ILGI population. Lane 1: 50 bp ladder, Lane 2: parent plant, Lane 2-5: progeny plants, lane 6: cDNA as a positive control. Lane 2-7, cdb 26-440, Lane 9-15: cdb 27-300, Lane 17-23: cdb 49-280, Lane 25-31 thioredoxin gene.

For 41 out of the 155 cDNA-AFLP fragments selected, primer pairs have been designed and tested. Twenty-eight primer pairs were able to amplify single PCR-products. Five of them revealed length polymorphisms in at least one of the mapping populations used (lanes 2-6 in Figure 1). These markers are currently being mapped. For those primer pairs which amplified fragments of similar molecular weights in all the individuals of the mapping populations studied (lanes 17-23 in Figure 1), restriction site polymorphisms are being tested (lanes 15-27 in Figure 2).

This kind of information will help us to analyse the genomic organisation of the factors involved in SI-response in ryegrasses.

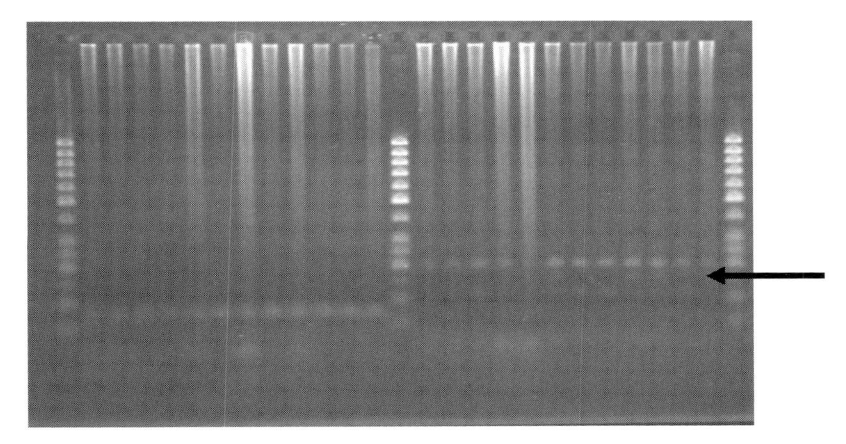

Figure 2. Search for restriction site polymorphisms in cDNA-AFLP fragments in the ILGI population (lanes 2-6 and 15-19) and the DvP-CLO reference population (lanes 7-13 and 20-26). Lane 15: parent of the ILGI population, Lane 16-19: progeny of the ILGI population, Lane 20 and 21: parents of the DvP-CLO reference population, Lane 22-26: progeny of the DvP-CLO reference population. Lane 15-26: cdb 30-400 restricted with RsaI.

ACKNOWLEDGEMENTS

This work is supported by a grant of the Flemish Institute for the promotion of the scientific and technologic research in the industry (I.W.T.) and by the European project GRASP (QLRT-2001-00862). We thank Ariane Staelens and Carina Pardon for excellent technical assistance.

REFERENCES

Bachem CWB., van der Hoeven RS, de Bruijn SM, Vreugdenhil D, Zabeau M and Visser RGF (1996) Visualization of differential expression using a novel method of RNA fingerprinting based on AFLP: Analysis of gene expression during potato tuber development. Plant J. 9: 745-753.

Chomczynski P and Sacchi N (1987) Single-step method of RNA isolation by acid guanidinium thiocyanate-phenol-chloroform extraction. Anal. Biochem. 162: 156-159.

Langridge P, Baumann U, Juttner J (1999) Revisiting and revising the self-incompatibilty genetics of *Phalaris coerulescens*. Plant Cell 11: 1826.

Li X, Nield J, Hayman D, Langridge P (1994) Cloning of a putative self-incompatibility gene from the pollen of the grass *Phalaris coerulescens*. Plant Cell 6: 1923-1932

Muylle H, Van Bockstaele E, Roldán-Ruiz I (2002). Genetic dissaction of crown rust resistance in a *Lolium perenne* mapping population. Proceed. Eucarpia meeting 2002 (in press).

Thorogood D, Kaiser WJ, Jones JG, Armstead I (2002) Self-incompatibility in ryegrass 12. Genotyping and mapping the S and Z loci of *Lolium perenne* L. Heredity 88: 385-390.

Wehling P, Hackauf B, Wricke G (1994) Identification of S-locus linked PCR fragments in rye (*Secale cereale* L) by denaturing gradient gel electrophoresis. Plant J. 5: 891-893.

Functional Analysis of the Perennial Ryegrass – *Epichloë* Endophyte Interaction

Barry Scott

Centre for Functional Genomics, Institute of Molecular BioSciences, Massey University, Private Bag 11 222, Palmerston North, New Zealand. (Email: d.b.scott@massey.ac.nz).

Key words: *Epichloë/Neotyphodium* endophytes, alkaloids, lolitrems, primary metabolites, secondary metabolites

Abstract: *Epichloë/Neotyphodium* endophytes are a group of clavicipitaceous fungi that form symbiotic associations (symbiota) with temperate grasses such as perennial ryegrass and tall fescue. The fungus derives nutrients from the host and a means of dissemination through the seed. The plant acquires increased tolerance to biotic and abiotic stresses such as herbivory and water insufficiency. Fungal synthesis of anti-insect and anti-mammalian secondary metabolites appears to be the mechanism for protection of the symbiotum from herbivory. Genes for the biosynthesis of lolines, ergot alkaloids and indole-diterpenes have recently been cloned from *Epichloë/Neotyphodium* endophytes and shown to be organised in gene clusters. In most cases these genes are not expressed in axenic culture but are highly up-regulated in the grass host. A molecular analysis of the regulation of these genes will provide fundamental insights into the molecular interaction between grass host and symbiont. Interaction between host and symbiont also involves exchange of primary metabolites. A gene for vitamin B1 biosynthesis has been deleted in *E. typhina* to examine the effect of this change on fungal compatibility with the host. Vitamin B1 auxotrophs were as effective as wild-type in both host colonisation and stromata formation.

A. Hopkins, Z.Y. Wang, R. Mian, M. Sledge and R.E. Barker (eds.), Molecular Breeding of Forage and Turf, 133-144.

1. INTRODUCTION

Epichloë/Neotyphodium endophytes are a group of clavicipitaceous fungi (Clavicipitaceae, Ascomycota) that form symbiotic associations (symbiota) with temperate grasses of the sub-family Pooideae. At least nine different sexual species are recognised including *Epichloë typhina*, a broad host range species (Schardl and Wilkinson 2000), and *E. festucae*, a natural symbiont of *Festuca spp.* (Leuchtmann et al. 1994) that is also capable of forming compatible associations with perennial ryegrass, *Lolium perenne* (Christensen et al. 1997). The asexual *Neotyphodium spp.* are predominantly inter-specific hybrids that form mutualistic associations with their host (Schardl et al. 1994; Tsai et al. 1994; Moon et al. 2000). *N. lolii*, an asexual symbiont of perennial ryegrass, is one of the few species that naturally appears to be haploid (Christensen et al. 1993; Schardl et al. 1994).

The *Epichloë/Neotyphodium* endophytes are biotrophic fungi that systemically colonise the intercellular spaces of leaf primordia, leaf sheaths, and leaf blades of vegetative tillers and the inflorescence tissues of reproductive tillers. The major benefits to the fungal symbiont are access to nutrients and a means of dissemination through the seed. Benefits to the host include increased tolerance to both biotic (e.g. insect and mammalian herbivory) and abiotic stresses (e.g. drought) (Scott 2001a).

2. METABOLIC INTERACTIONS: FUNGAL SYNTHESIS OF BIOPROTECTIVE SECONDARY METABOLITES

The ability of *Epichloë/Neotyphodium* endophytes to synthesize a range of secondary metabolites *in planta* constitutes a major ecological benefit for the symbiotum (Schardl 1996). Metabolites identified to date include both anti-insect (peramine and lolines) and anti-mammalian (ergot alkaloids and indole-diterpenes) compounds (Bush et al. 1997).

Peramine, classified as a pyrrolopyrazine alkaloid, appears to be unique to the *Epichloë/Neotyphodium* genera. Peramine production has been reported in cultures of *N. lolii* (Rowan 1993) and *E. typhina* (Schardl et al. 1999). There is no experimental information on the biosynthesis of this compound but analysis of its structure (Figure 1) suggests it is the product of a two module non-ribosomal peptide synthase (Schardl et al. 1999).

Loline alkaloids are saturated 1-aminopyrrolizidine alkaloids (Figure 1), most likely derived from aspartate and ornithine (Bush et al. 1993; Schardl et al. 1999). Lolines can accumulate in grass tissues to levels as high as 2% of the plant's dry mass (Craven et al. 2001), far in excess of the fungal biomass and the levels of other alkaloids reported (Spiering et al. 2002). A recent study reported loline (loline, *N*-formylloline and *N*-acetylloline) production in axenic cultures of *N. uncinatum* growing on minimal media (Blankenship et al. 2001), demonstrating for the first time that these compounds are fungal products. However, lolines were not detected in cultures of three other endophytes isolated from symbiota known to produce lolines, including *E. festucae* from *Lolium giganteum*, *N. coenophialum* from *Lolium arundinaceum* and *N. siegelii* from *Lolium pratense* (Blankenship et al. 2001).

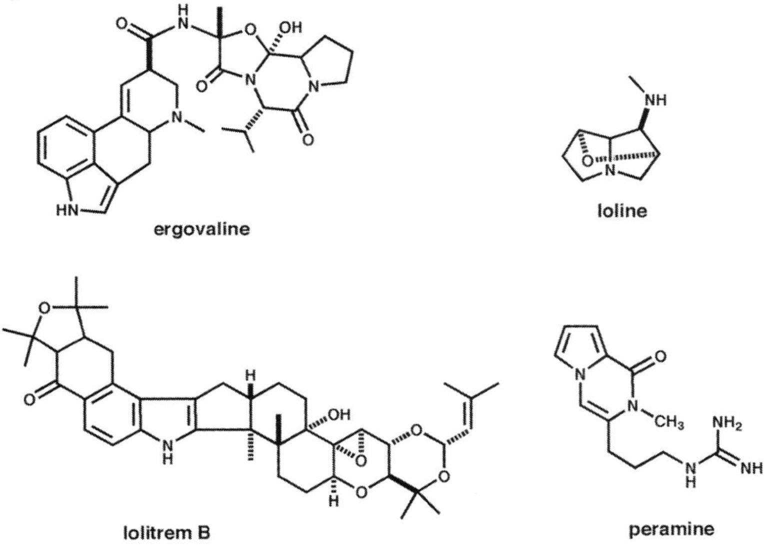

Figure 1. Chemical structures of loline, peramine, lolitrem B and ergovaline.

The biosynthesis of ergot alkaloids is the best understood of the endophyte alkaloid biosynthetic pathways, principally because of research carried out with *Claviceps purpurea*, the fungus responsible for ergotism from contaminated rye (Socic and Gaberc-Porekar 1992). The first committed step in this pathway is synthesis of dimethylallyl tryptophan (DMAT), a reaction catalysed by DMAT synthase (Gebler and Poulter 1992). DMAT is subsequently converted via several intermediates to the clavine alkaloids, such as lysergic acid, which are characterised by the presence of an ergolene ring system. The ergolene acids and alcohols are further transformed into complex ergopeptine derivatives such as ergotamine by non-ribosomal peptide synthetases. Ergotamine synthesis is catalysed by two non-ribosomal peptide synthetases, LPS2 (lysergyl peptide synthetase 2) and LPS1, a peptide synthetase that sequentially adds the amino acids alanine, phenylalanine and proline to the activated lysergic acid (Riederer et al. 1996; Walzel et al. 1997). Ergovaline is the most abundant ergopeptine detected in *Epichloë/Neotyphodium*-grass associations and is similar in structure to ergotamine except that it contains a valine instead of a phenylalanine in the second position of the tripeptide. Under natural conditions, ergot alkaloid synthesis by *C. purpurea* is restricted to sclerotial tissue formed following advanced colonisation of rye ovarian tissue (Tudzynski et al. 1999). However, several deregulated mutant strains have been isolated that produce significant amounts of ergot alkaloids in axenic culture (Tudzynski et al. 1999). Similarly, ergot alkaloids are abundant in *Epichloë/Neotyphodium*-grass symbiota but are difficult to detect in axenic culture although the synthesis of ergot alkaloids in *N. coenophialum* has been reported (Bacon and Robbins 1979; Bacon 1988).

Indole-diterpenes are a substantial and structurally diverse group of fungal metabolites, many of which are potent tremorgenic mammalian mycotoxins (Steyn and Vleggaar 1985). Characteristic of this group of compounds is the presence of a cyclic diterpene skeleton derived from four isoprene units and an indole moiety derived from tryptophan or a tryptophan precursor, such as anthranilic acid (Byrne et al. 2002). While biosynthetic schemes have been proposed on the basis of radiolabelling studies and the identification of likely intermediates from related filamentous fungi (Mantle and Weedon 1994; Munday-Finch et al. 1996), very little is known about the nature of the intermediates or the enzymology of their biosynthesis. The best known group of these fungal tremorgens are the lolitrems produced by *Epichloë* and *Neotyphodium spp.* in association with tall fescue (*Festuca arundinacea*) and perennial ryegrass (*Lolium perenne*). There have been two reports of indole-diterpene biosynthesis in axenic cultures of *Neotyphodium spp.* grown on agar plates and in submerged culture (Penn et al. 1993; Reinholz and Paul 2001). Penn et al. (1993) detected paxilline, lolitrem B and lolitriol in cultures of *N. lolii*, *N. coenophialum* (FaTG-1), *Neotyphodium sp.* (FaTG-2), *Neotyphodium sp.* (FaTG-3), *E. festucae* and *N. uncinatum*. Reinholz and Paul (2001) detected Lolitrem B in agar plate cultures of *N. lolii* and the non-endophyte fungus, *Penicillium paxilli*.

3. GENETICS AND MOLECULAR CLONING OF A LOLINE BIOSYNTHETIC GENE CLUSTER

Genetic crosses between Lol^{+} and Lol^{-} parents of *E. festucae* showed that the ability to synthesize lolines (Lol^{+}) segregated as a single genetic locus (Wilkinson et al. 2000). Two AFLP polymorphic markers were identified that co-segregated with the *lol* locus (Wilkinson et al. 2000), providing useful markers for isolation of the genes by map-based cloning (Kutil et al. 2003). Using the technique of suppression subtractive hybridisation PCR

(Diatchenko et al. 1996), two transcripts, *lolA* (encoding an aspartate kinase) and *lolC* (homocysteine synthase), were shown to be up-regulated in loline-producing cultures of *N. uncinatum* (Spiering et al. 2002). Long-range PCR established that these genes were linked and part of a cluster of nine putative genes for loline biosynthesis spanning about 25 kb of the genome (Spiering et al. 2003). *N. uncinatum* has two allelic clusters for loline biosynthesis (Spiering et al. 2003), a result consistent with the proposed inter-specific origin (*E. typhina* x *E. bromicola*) for this species (Craven et al. 2001). *E. festucae* has a single cluster for loline biosynthesis with a high degree of synteny (both order and orientation) with the *N. uncinatum* clusters (Kutil et al. 2003). Confirmation that these genes are indeed responsible for loline biosynthesis requires a gene-disruption/silencing experiment.

4. MOLECULAR CLONING OF ENDOPHYTE ERGOT ALKALOID BIOSYNTHETIC GENES

A gene encoding DMAT synthase (*dmaW*), the first committed step in ergot alkaloid biosynthesis, was first cloned from *Claviceps fusiformis* using reverse genetics (Tsai et al. 1995) and differential cDNA (Arntz and Tudzynski 1997) approaches. Subsequently, a *dmaW* homologue was cloned from *C. purpurea* and shown to be closely linked to the gene encoding the peptide synthetase (LPS1) required for the biosynthesis of ergotamine (Tudzynski et al. 1999; Panaccione et al. 2001). Using a degenerate PCR approach Wang and Schardl (2001) successfully amplified *dmaW*-like products from *N. coenophialum*, *Neotyphodium sp.* Lp1, *N. inebrians*, and *Balansia obtecta*. Southern blot analysis of genomic digests from various *Epichloë* and *Neotyphodium* species showed *dmaW* homologues were present in all ergot alkaloid producing strains but were also found in *E. glyceriae* and *E. clarkii*, species not known to synthesize ergopeptines. However, it is possible that these species, like *C. fusiformis* (Tudzynski et al. 1999), produce only clavines.

Using low stringency hybridisation a homologue of the *C. purpurea* peptide synthetase gene (*lps1*/*lpsA*) was cloned from *N. lolii* (Panaccione et al. 2001). This gene, designated *lpsA*, for lysergyl peptide synthetase, is present as a single copy in the perennial ryegrass endophyte strain Lp1. Targeted replacement of *lpsA* generated transformants of Lp1 that were unable to synthesize ergovaline in perennial ryegrass (Panaccione et al. 2001). Deletion of this gene had no effect on the ability of the mutant to colonise perennial ryegrass when compared with the wild-type strain. This result indicates that the synthesis of ergovaline has no role in the molecular interaction between the symbiont and the grass host (Panaccione et al. 2001).

5. MOLECULAR CLONING AND GENETIC ANALYSIS OF A GENE CLUSTER FOR PAXILLINE BIOSYNTHESIS

The recent cloning from *Penicillium paxilli* of a cluster of genes for paxilline biosynthesis (Figure 2) provides an insight into the enzymes required for indole-diterpene biosynthesis (Young et al. 2001). A combination of gene deletion and chemical complementation studies has confirmed that at least five genes are required for paxilline biosynthesis (Young et al. 2001; McMillan et al. 2003).

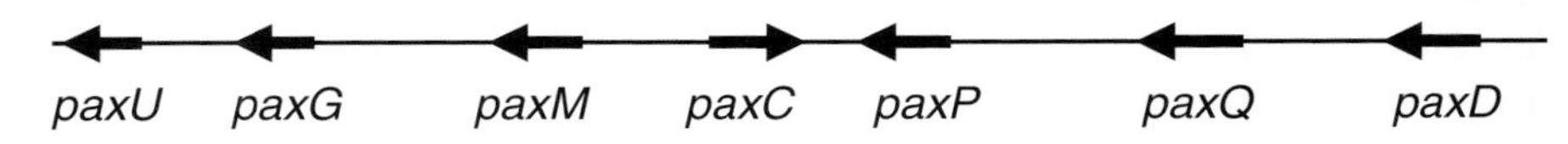

Figure 2. Gene cluster for paxilline biosynthesis in *Pencillium paxilli*.

The first committed step in paxilline biosynthesis is proposed to be catalysed by a geranylgeranyl diphosphate (GGPP) synthase, encoded by *paxG* (Fig. 3). Deletion of this gene gives rise to a paxilline negative phenotype (Young et al. 2001). Candidate genes for early steps in the pathway are *paxM,* a FAD-dependent monooxygenase, and *paxC*, a prenyl transferase. Deletion of either *paxM* or *paxC* results in mutants that lack the ability to synthesize any identifiable indole-diterpene intermediate (Scott et al. 2003, unpublished results). On the basis of these results, we have proposed that the first stable indole-diterpene product formed is paspaline (Figure 3) (Scott et al. 2003).

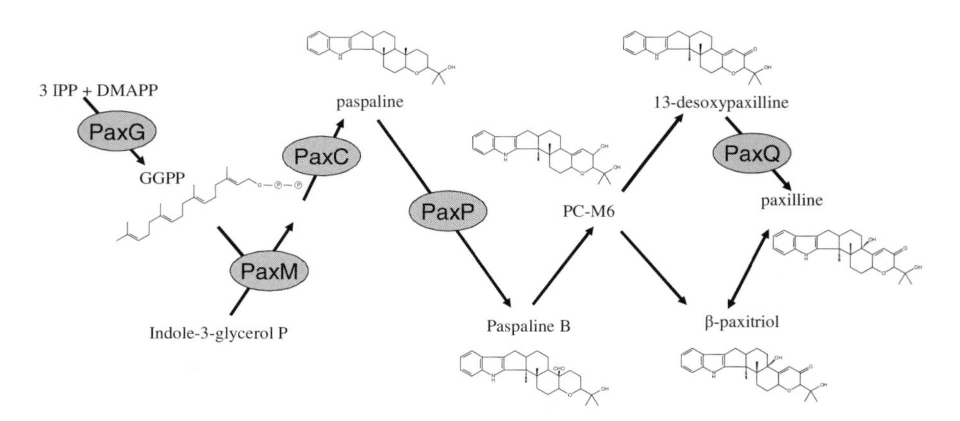

Figure 3. Proposed pathway for the biosynthesis of paxilline in *Penicillium paxilli*.

At least three other genes are required for paxilline biosynthesis (Figure 3). Deletion of *paxP* and *paxQ*, genes that encode cytochrome P450 monooxygenase enzymes, gave rise to mutants that accumulate paspaline and 13-desoxypaxilline respectively (McMillan et al. 2003). These compounds are proposed intermediates in a metabolic grid for the biosynthesis of paxilline and other indole-diterpenes (Munday-Finch et al. 1996). This metabolic grid suggests that the penultimate substrate for paxilline biosynthesis could be either 13-desoxypaxilline or β-paxitriol. If this is correct, then PC-M6 may also be a substrate for PaxQ (Figure 3). Similarly, the conversion of β-paxitriol to paxilline and PC-M6 to 13-desoxypaxilline both involve oxygenation at C-10, suggesting a single dehydrogenase enzyme may catalyse both reactions. The gene encoding this enzyme has yet to be identified.

The cloning and characterisation of this gene cluster makes it possible to clone orthologous gene clusters from other filamentous fungi, including *Epichloë/Neotyphodium* endophytes.

6. MOLECULAR CLONING AND GENETIC ANALYSIS OF A GENE CLUSTER FOR LOLITREM BIOSYNTHESIS

Using degenerate primers designed to conserved regions of fungal GGPP synthases, including both the primary and secondary genes from *P. paxilli* (Young et al. 2001), the *N. lolii* orthologue (*ltmG*) of *paxG* has been isolated (Young et al. 2003 unpublished results). Sequence analysis of a 25-kb genomic region around *ltmG* identified orthologues of *paxM* (*ltmM*) and *paxP* (*ltmP*) (Figure 4). RT-PCR analysis showed the *ltm* genes are weakly expressed in culture, including conditions where cultures are de-repressed for carbon or nitrogen metabolism, but highly expressed *in planta*. Sequences adjacent to the *ltm* genes have strong similarities to remnants of retro-elements. It is possible these retro-element platforms separate *ltmG*, *ltmM* and *ltmP* from the remaining genes required for lolitrem biosynthesis. These retroelements are very abundant and highly dispersed within the genomes of both *N. lolii* and *E. festucae* (Scott and Young 2003). A targeted deletion of *ltmM* has been constructed in *Epichloë festucae* and artificial associations of wild-type and mutant strains established with perennial ryegrass. Analysis of the lolitrem phenotype of these associations confirmed this gene is necessary for lolitrem biosynthesis.

The cloning of indole-diterpene genes now allows us to screen field isolates for the presence, distribution and expression of these genes (Scott 2001b). Molecular screening will provide the certainty required for prediction of the toxin phenotype of naturally occurring endophytes, such as those that lack the ability to synthesise the mammalian toxins, lolitrem B and ergovaline. Such isolates are desirable for use in pastoral agricultural systems to overcome animal toxicoses problems associated with endophyte synthesis of mammalian toxins, yet retain other bioprotective features such as the ability to resist insect herbivory (Fletcher 1999; Popay et al. 1999).

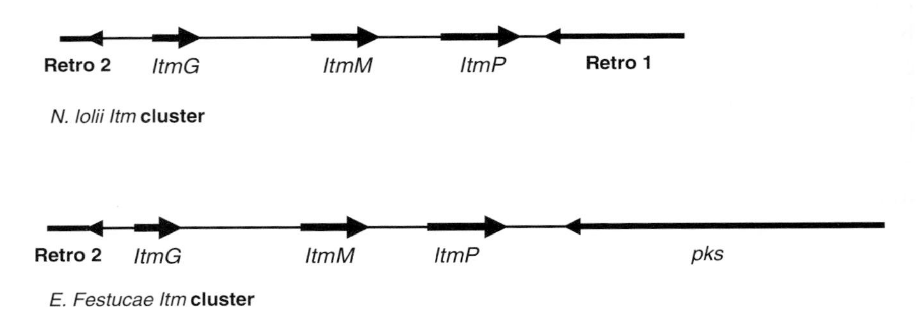

Figure 4. Gene cluster for lolitrem biosynthesis in *Neotyphodium lolii* and *Epichloë festucae*.

The availability of cloned toxin biosynthetic genes will also allow us to test how the expression of these genes in the symbiosis is regulated in response to both biotic and abiotic stress. Significant differences in the levels of both lolitrem B and ergovaline were detected between water sufficient and water insufficient perennial ryegrass symbiota grown under controlled environmental conditions (Hahn et al. 2003, unpublished results).

7. METABOLIC INTERACTIONS: FUNGAL REQUIREMENTS FOR PRIMARY METABOLITES

While microscopy studies have confirmed a very close association between endophytes and their plant hosts in the apoplastic space (Christensen et al. 2002), very little is known about metabolic exchanges that occur between each partner. To gain insight into the metabolic interface between host and symbiont and to test whether there is mutual dependency for metabolites, as has been shown recently for the *Rhizobium*-legume symbiosis (Lodwig et al. 2003), the cloning and analysis of a selection of genes involved in endophyte primary metabolism has been undertaken.

As a first step toward understanding the role of primary metabolism in the interaction between endophytes and their grass hosts we have chosen to examine the role of endophyte thiamine (vitamin B1) biosynthesis in this interaction. We have targeted this pathway because thiamine is a coenzyme for many steps in primary cellular metabolism. The orthologue (*thi1*) of *Saccharomyces cerevisiae THI4* (Praekelt et al. 1994) was isolated from *N.*

lolii and *E. typhina* by PCR using degenerate primers designed to conserved regions of known thiazole biosynthetic genes (Zhang et al. unpublished results). This gene is strongly expressed in culture and *in planta* and shows alternative splicing, with distinct patterns of the isoforms expressed under different nutritional conditions. A 'knockout' of the *E. typhina thi1* gene was generated and shown to have reduced hyphal density and branching compared to the wild-type on defined media lacking thiamine. Both thiamine and thiazole complemented this defect. No differences in infectivity were observed between wild-type and mutant strains in their ability to establish stable artificial associations with perennial ryegrass. However, some differences were observed in host colonisation, with the mutant strain behaving more like the asexual *N. lolii*. However, both the wild-type and mutant strains formed stromata on reproductive tillers.

8. CONCLUSIONS

The cloning of key secondary metabolite genes from endophytes is a major advance in our understanding of the interaction between endophytes and their grass hosts as these represent key genes associated with mutualism. We are now in a position to determine the spatial and temporal expression of these genes in the symbiosis, determine whether their expression changes in response to biotic and/or abiotic stresses and begin to identify the molecular basis of signalling pathways that occur between each partner. The availability of deletion mutants will also allow direct confirmation of the biological effects of these metabolites and lead to a better understanding of plant bioprotection mechanisms.

ACKNOWLEDGEMENTS

The unpublished research from my laboratory described in this paper was carried out by Lisa McMillan, Jonathan Astin, Rohan Lowe, Carolyn Young, Andrea Bryant, Heike Hahn, Michelle Bryant and Xiuwen Zhang in collaboration with Emily Parker and Geoffrey Jameson from the Institute of Fundamental Sciences, Massey University and Michael Christensen, Gregory Bryan, Bryan Tapper, Sarah Finch and Christopher Miles at AgResearch, New Zealand. I am also grateful to Michelle Bryant and Brendon Monahan for proof reading this manuscript. This research was supported by grants from the Foundation for Research, Science and Technology (MAUX0217) and the Royal Society of New Zealand (Marsden MAU010) to Barry Scott. Heike Hahn is supported by a fellowship from the German Academy of Natural Sciences Leopoldina.

REFERENCES

Arntz C, Tudzynski P (1997) Identification of genes induced in alkaloid producing cultures of *Claviceps sp*. Curr. Genet. 31: 357-360.

Bacon CW (1988) Procedure for isolating the endophyte from tall fescue and screening the isolates for ergot alkaloids. Appl. Env. Microbiol. 54: 2615-2618.

Bacon CW, Robbins JD (1979) Ergosine, ergosinine and chanaclavine I from *Epichloë typhina*. J. Agric. Food Chem. 27: 595-598.

Blankenship JD, Spiering MJ, Wilkinson HH, Fannin FF, Bush LP, Schardl CL (2001) Production of loline alkaloids by the grass endophyte, *Neotyphodium uncinatum*, in defined media. Phytochemistry 58: 395-401.

Bush LP, Fannin FF, Siegel MR, Dahlman DL, Burton HR (1993) Chemistry, occurrence and biological effects of saturated pyrrolizidine alkaloids associated with endophyte-grass interactions. Agric. Ecosyst. Environ. 44: 81-102.

Bush LP, Wilkinson HH, Schardl CL (1997) Bioprotective alkaloids of grass-fungal endophyte symbioses. Plant Physiol. 114: 1-7.

Byrne KM, Smith SK, Ondeyka JG (2002) Biosynthesis of nodulisporic acid A: precursor studies. J. Chem. Soc. Chem. Commun. 124: 7055-7060.

Christensen MJ, Ball OJ-P, Bennett RJ, Schardl CL (1997) Fungal and host genotype effects on compatibility and vascular colonization by *Epichloë festucae*. Mycolog. Res. 101: 493-501.

Christensen MJ, Bennett RJ, Schmid J (2002) Growth of *Epichloë/Neotyphodium* and p-endophytes in leaves of *Lolium* and *Festuca* grasses. Mycolog. Res. 106: 93-106.

Christensen MJ, Leuchtmann A, Rowan DD, Tapper BA (1993) Taxonomy of *Acremonium* endophytes of tall fescue (*Festuca arundinacea*), meadow fescue (*F. pratensis*) and perennial rye-grass (*Lolium perenne*). Mycolog. Res. 97: 1083-1092.

Craven KD, Blankenship JD, Leuchtmann A, Hignight K, Schardl CL (2001) Hybrid fungal endophytes symbiotic with the grass *Lolium pratense*. Sydowia 53: 44-73.

Diatchenko L, Lau Y-FC, Campbell AP, Chenchik A, Moqadam F, Huang B, Lukyanov S, Lukyanov K, Gurskaya N, Sverdlov ED, Siebert PD (1996) Suppression subtractive hybridization: a method for generating differentially regulated or tissue-specific cDNA probes and libraries. Proc. Natl. Acad. Sci. USA 93: 6025-6030.

Fletcher LR (1999) "Non-toxic" endophytes in ryegrass and their effect on livestock health and production. In: Ryegrass endophyte: an essential New Zealand symbiosis, Woodfield DR, Matthew C (eds.), pp. 133-139. Napier, New Zealand.

Gebler JC, Poulter CD (1992) Purification and characterisation of dimethylallyl tryptophan synthase from *Claviceps purpurea*. Arch. Biochem. Biophys. 296: 308-313.

Kutil BL, Spiering MJ, Schardl CL, Wilkinson HH (2003) Evolution of a secondary metabolite gene cluster implicated in loline alkaloid biosynthesis of grass-endophytes (*Epichloë* and *Neotyphodium spp*.). Fungal Genetics Newsletter (suppl.) 50:41

Leuchtmann A, Schardl CL, Siegel MR (1994) Sexual compatibility and taxonomy of a new species of *Epichloë* symbiotic with fine fescue grasses. Mycologia 86: 802-812.

Lodwig EM, Hosie AHF, Bourdes A, Findlay K, Allaway D, Karunakaran R, Downie JA, Poole PS (2003) Amino-acid cycling drives nitrogen fixation in the legume-*Rhizobium* symbiosis. Nature 422: 722-726.

Mantle PG, Weedon CM (1994) Biosynthesis and transformation of tremorgenic indole-diterpenoids by *Penicillium paxilli* and *Acremonium lolii*. Phytochemistry 36: 1209-1217.

McMillan LK, Carr RL, Young CA, Astin JW, Lowe RGT, Parker EJ, Jameson GB, Finch SC, Miles CO, McManus OB, Schmalhofer WA, Garcia ML, Kaczorowski GJ, Goetz MA, Tkacz JS, Scott B (2003) Molecular analysis of two cytochrome P450 monooxygenase genes required for paxilline biosynthesis in *Penicillium paxilli* and effects of paxilline intermediates on mammalian maxi-K ion channels. Mol Gen Genom (in press).

Moon CD, Scott B, Schardl CL, Christensen MJ (2000) The evolutionary origins of *Epichloë* endophytes from annual ryegrasses. Mycologia 92: 1103-1118.

Munday-Finch SC, Wilkins AL, Miles CO (1996) Isolation of paspaline B, an indole-diterpenoid from *Penicillium paxilli*. Phytochemisty 41: 327-332.

Panaccione DG, Johnson, RD, Wang J, Young CA, Damrongkool P, Scott B, Schardl CL (2001) Elimination of ergovaline from a grass-*Neotyphodium* endophyte symbiosis by genetic modification of the endophyte. Proc. Nat. Acad. Sci. USA 98: 12820-12825.

Penn J, Garthwaite I, Christensen MJ, Johnson CM, Towers NR (1993) The importance of paxilline in screening for potentially tremorgenic *Acremonium* isolates. In: Proceedings of the Second International Symposium on *Acremonium*/Grass Interactions, Hulme DE, Latch GCM, Easton HS (eds.), pp. 88-92. AgResearch, Grasslands Research Centre, Palmerston North, New Zealand.

Popay AJ, Hume DE, Baltus JG, Latch GCM, Tapper BA, Lyons TB, Cooper BM, Pennell CG, Eerens JPJ, Marshall SL (1999) Field performance of perennial ryegrass (*Lolium perenne*) infected with toxin-free fungal endophytes (*Neotyphodium spp.*). In: Ryegrass endophyte: an essential New Zealand symbiosis, Woodfield DR, Matthew C (eds.), pp. 113-122. Napier, New Zealand.

Praekelt UM, Byrne KL, Meacock PA (1994) Regulation of *THI4* (*MOL1*), a thiamine-biosynthetic gene of *Saccharomyces cerevisiae*. Yeast 10: 481-490.

Reinholz J, Paul VH (2001) Toxin-free *Neotyphodium*-isolates achieved without genetic engineering - a possible strategy to avoid "ryegrass staggers". In: 4th International Neotyphodium/grass interactions Symposium, Paul VH, Dapprich PD (eds.), pp. 261-271. Soest, Germany,

Riederer B, Han M, Keller U (1996) D-lysergyl peptide synthetase from the ergot fungus *Claviceps purpurea*. J. Biol. Chem. 271: 27524-27530.

Rowan DD (1993) Lolitrems, peramine and paxilline: mycotoxins of the ryegrass/endophyte interaction, Agriculture, Ecosystems & Environment, pp. 103-122. Elsevier Science Publishers B.V., Amsterdam.

Schardl CL (1996) *Epichloë* species: fungal symbionts of grasses. Ann. Rev. Phytopathol. 34: 109-130.

Schardl CL, Leuchtmann A, Tsai H-F, Collett MA, Watt DM, Scott DB (1994) Origin of a fungal symbiont of perennial ryegrass by interspecific hybridization of a mutualist with the ryegrass choke pathogen, *Epichloë typhina*. Genetics 136: 1307-1317.

Schardl CL, Wang J, Wilkinson HH, Chung K-R (1999) Genetic analysis of biosynthesis and roles of anti-herbivore alkaloids produced by grass endophytes. In: Proceeding of International Symposium of Mycotoxicology '99. Mycotoxin contamination: health risk and prevention project, Kumagai S GT, Kawai K, Takahashi H, Yabe K, Yoshizawa T, Koga H, Kamimura H, Akao M (eds.), pp. 118-125. Chiba, Japan.

Schardl CL, Wilkinson HH (2000) Hybridization and cospeciation hypotheses for the evolution of grass endophytes. In: Microbial Endophytes, Bacon CW, White JF, Jr (eds), pp. 63-83, Marcel Dekker, Inc., New York.

Scott B (2001a) *Epichloë* endophytes: symbionts of grasses. Curr. Opin. Microbiol. 4: 393-398.

Scott B (2001b) Molecular interactions between *Lolium* grasses and their fungal symbionts. In: Molecular Breeding of Forage Crops, Spangenberg G (ed.), pp. 261-274. Kluwer Academic Publishers, Dordrecht.

Scott B, Young C (2003) Genetic manipulation of clavicipitalean endophytes. In: Clavicipitalean fungi: evolutionary biology, chemistry, biocontrol and cultural impacts, White JF, Jnr., Bacon CW, Hywel-Jones NL, Spatafora JW (eds.), pp425-443. Marcel Dekker Inc., New York (in press).

Scott DB, Jameson GB, Parker EJ (2003) Isoprenoids: gene clusters and biosynthetic puzzles. In: Advances in fungal biotechnology for industry, agriculture and medicine, Tkacz JS, Lange L (eds.). Kluwer Academic/Plenum Publishers, New York (in press).

Socic H, Gaberc-Porekar V (1992) Biosynthesis and physiology of ergot alkaloids. In: Fungal biotechnology, Arora DK, Elander RP, Mukerji KG (eds.), pp. 123-155. Marcel Dekker, New York.

Spiering MJ, Moon CD, Schardl CL (2003) Gene clusters associated with production of 1-aminopyrrolizidine (loline) alkaloids in the grass endophytes *Neotyphodium uncinatum*. Fungal Genetics Newsletter (suppl.) 50: 41.

Spiering MJ, Wilkinson HH, Blankenship JD, Schardl CL (2002) Expressed sequence tags and genes associated with loline alkaloid expression by the fungal endophyte *Neotyphodium uncinatum*. Fungal Genet. Biol. 36: 242-254.

Steyn PS, Vleggaar R (1985) Tremorgenic mycotoxins. Prog Chem Organic Natural Products 48: 1-80.

Tsai H-F, Liu J-S, Staben C, Christensen MJ, Latch GCM, Siegel MR, Schardl CL (1994) Evolutionary diversification of fungal endophytes of tall fescue grass by hybridization with *Epichloë* species. Proc. Natl. Acad. Sci. USA 91: 2542-2546.

Tsai H-F, Wang H, Gebler JC, Poulter CD, Schardl CL (1995) The *Claviceps purpurea* gene encoding dimethylallyltryptophan synthase, the committed step for ergot alkaloid biosynthesis. Biochem. Biophys. Res. Commun. 216: 119-125.

Tudzynski P, Hölter K, Correia T, Arntz C, Grammel N, Keller U (1999) Evidence for an ergot alkaloid gene cluster in *Claviceps purpurea*. Mol. Gen. Genet. 261: 133-141.

Walzel B, Riederer B, Keller U (1997) Mechanism of alkaloid cyclopeptide formation in the ergot fungus *Claviceps purpurea*. Chem. Biol. 4: 223-230.

Wang J, Schardl CL (2001) Association of *Neotyphodium coenophialum* dimethylallyltryptophan synthase gene with ergot alkaloid biosynthesis. In: 4th International *Neotyphodium*/grass interactions Symposium, Paul VH, Dapprich PD (eds.) pp. 301-305. Soest, Germany.

Wilkinson HH, Siegel MR, Blankenship JD, Mallory AC, Bush LP, Schardl CL (2000) Contribution of fungal loline alkaloids to protection from aphids in a grass-endophyte mutualism. Mol. Plant-Microbe Int. 13: 1027-1033.

Young CA, McMillan L, Telfer E, Scott B (2001) Molecular cloning and genetic analysis of an indole-diterpene gene cluster from *Penicillium paxilli*. Mol. Microbiol. 39: 754-764.

Gene Discovery and Microarray-Based Transcriptome Analysis in Grass Endophytes

S. Felitti, K. Shields, M. Ramsperger, P. Tian, T. Webster[1], B. Ong[1], T. Sawbridge[1,2] and G. Spangenberg[1,2]
Plant Biotechnology Centre, Department of Primary Industries, La Trobe University, Bundoora, Victoria 3086, Australia and CRC for Molecular Plant Breeding, [1]Victorian Microarray Technology Consortium, [2]Victorian Bioinformatics Consortium, Australia. (Email: German.Spangenberg@dpi.vic.gov.au).

Key words: *Neotyphodium* and *Epichloë* endophytes, grass-endophyte symbiosis, gene discovery, expressed sequence tags, microarrays, transcriptome analysis

Abstract:

A genomic resource of 9411 expressed sequence tags (ESTs) representing 5038 genes has been established for *Neotyphodium* and *Epichloë* endophytes. A cDNA spotted microarray interrogating 4195 *Neotyphodium* and 920 *Epichloë* genes (EndoChip) has been generated. It provides a tool for high-throughput transcriptional profiling of endophytes, genome-specific gene expression analysis and expression profiling of novel endophyte genes, as well as transcriptome analysis in the grass-endophyte association. Microarray-based gene expression analysis using RNA from *in vitro* grown *N. coenophialum* and *N. lolii* endophytes demonstrated its utility in comparative transcriptome analysis of both endophyte species. Additional proof of concept was obtained in using the EndoChip to analyse genome wide-gene expression in endophytes grown *in vitro* and *in planta* and for gene expression profiling of novel genes based on coordinated expression with genes of known cellular functions. Northern hybridisation analysis allowed for validation of microarray-based gene expression data. In combination with established unigene microarrays for the host grasses it provides a tool for the dissection of the grass-endophyte association at the transcriptome level.

A. Hopkins, Z.Y. Wang, R. Mian, M. Sledge and R.E. Barker (eds.), Molecular Breeding of Forage and Turf, 145-153.

1. INTRODUCTION

Neotyphodium lolii, *N. coenophialum* and *Epichloë festucae* are frequent endophytic symbiotic fungi of the temperate pasture grasses, perennial ryegrass (*Lolium perenne* L.), tall fescue (*Festuca arundinacea* Schreb.) and red fescue (*Festuca rubra* L.), respectively.

The presence of the endophyte has been shown to improve seedling vigour, persistence and drought tolerance in marginal environments as well as provide protection against some insect pests. However, endophyte-infected grasses may be toxic to livestock because the fungus produces a wide range of chemicals, many of which have a high degree of biological activity against mammalian systems. The most thoroughly studied compounds are alkaloids, including ergopeptine alkaloids, indole-isoprenoid lolitrems, pyrrolizidine alkaloids, and pyrrolopyrazine alkaloids.

Endophyte-infected grasses are more aggressive, can acclimate to severe conditions more quickly, recover from stressful situations more rapidly, and resist insects and fungal pathogens more effectively than non-infected grasses. Endophytes differ, however, in their ability to confer herbivore resistance and in their insect deterrent properties. Furthermore, endophyte isolates differ in their alkaloid toxin profiles. This ability may be controlled by the host-endophyte interaction and thus may be subjected to selection, as has been demonstrated for the production of ergopeptine alkaloids in tall fescue. Endophyte-infected grasses may exhibit enhanced drought tolerance, presumably through alterations in carbohydrate partitioning, osmotic adjustment and enhanced root growth.

In contrast to the information on alkaloids and animal toxicosis, the beneficial physiological aspects of the endophyte/grass interactions have not been well characterised in any system. Very little is known regarding the factors important in host colonization or nutrient exchange between plant and fungus. The physiological mechanisms which lead to increased plant vigour and enhanced tolerance to abiotic stresses unrelated to the reduction in pest damage to endophyte-infected grasses are unknown. Similarly, genes involved in the endophyte/grass host interaction have not yet been isolated. Their isolation and characterisation would be critical for an effective manipulation of grass/endophyte interactions.

Until recently, no comprehensive genomic resource for *Neotyphodium* and *Epichloë* species has been generated (Spangenberg et al. 2000), with only 700 gene sequences including three *N. uncinatum* cDNAs encoding candidate genes putatively involved in loline biosynthesis, three *N.*

coenophialum genes (two peptide synthase genes and one *in planta*-expressed gene), highly conserved gene sequences such as actin, β tubulin, elongation factor-1 and ribosomal RNA genes as well as several cDNA and genomic clones from *Neotyphodium* and *Epichloë* deposited in GeneBank.

2. ENDOPHYTE GENE DISCOVERY

Expressed sequence tags (ESTs) provide the starting point for elucidating the function of thousands of endophyte genes. A world's first large-scale integrated grass endophyte functional genomics program (Symbio-Genomics) based on EST discovery has been established for *Neotyphodium* and *Epichloë* species (Spangenberg et al. 2000). This symbio-genomics program targets the discovery of genes involved in host colonization, nutrient exchange between plant and fungus, biosynthesis of active secondary metabolites involved in the grass-endophyte interaction and physiological mechanisms leading to increased plant vigour and enhanced abiotic stress tolerance.

Ten cDNA libraries have been constructed from *N. coenophialum*, *N. lolii* and *E. festucae* endophytes grown *in vitro* and *in planta*. The *in planta* enriched endophyte cDNA libraries were constructed from infected plant material at three developmental stages: imbibed seeds, vegetative stage and reproductive phase. A total of 9411 ESTs were generated representing 7.74 Mb of *Neotyphodium* and *Epichloë* gene sequences.

2.1 Functional Categorisation of *Neotyphodium* Genes

Clustering of ESTs from both *Neotyphodium* species produced 3806 clusters, which represent the minimum number of genes present in the EST collection. Functional categories were assigned to EST clusters based on significant matches following BLASTn and BLASTx analyses.

3. TRANSCRIPTOME ANALYSIS USING cDNA MICROARRAYS

With the advent of high-throughput DNA sequencing, one method of choice for expression studies is the analysis of multiple genes in parallel by hybridisation of indirectly labelled mRNA to single gene probes on a solid support. This allows the steady state mRNA levels of multiple probes (called targets) to be assessed in a single experiment. In order to achieve high efficiency, targets are printed as DNA spots on glass slides using high precision robotic systems. This allows thousands of spots to be placed on one slide. The mRNA sample to be analysed is indirectly labelled with a fluor and the fluorescence from the hybridised sample is measured using high precision slide scanner. Microarray-based reverse northern analysis requires the production of cDNA microarrays, and suitable labelling and hybridisation conditions to give sufficient, reproducible, linearly responsive fluorescent signal for reliable analysis.

3.1 Development of a Grass Endophyte Unigene Microarray

A fungal endophyte unigene microarray (EndoChip) for transcriptome analysis has been generated. This cDNA spotted microarray contains 4195 *in vitro* grown *Neotyphodium* cDNAs (1618 from *N. lolii* and 2477 from *N. coenophialum*), 920 *E. festucae* cDNAs, 210 *in planta* grown *Neotyphodium* cDNAs and 403 host grass cDNAs (19 from *L. perenne* and 384 from *F. arundinacea*). It also contains a Lucidea control set (Amersham Biosciences). All cDNAs were spotted in 3 replicates per slide.

3.2 Microarray-based Transcriptome Analysis in *N. lolii*

Microarray-based transcriptome analysis with total RNA from *N. lolii* grown *in vitro* was used to study endophyte gene expression under different culture conditions. Hierarchical cluster analysis (Figure 1) and time series plots (Figure 2) showed differential gene expression patterns for *N. lolii* grown in liquid and solid culture.

3.3 Microarray-based Novel Gene Expression Profiling in *Neotyphodium*

The application of the EndoChip in the expression profiling of novel *Neotyphodium* genes is shown (Figure 3). It allowed the identification of novel genes coordinately expressed with genes of known cellular function. The endophyte unigene microarray also allowed to study transgene expression in genetically modified endophytes grown *in vitro*. RNA from *Neotyphodium* endophytes transformed with chimeric *hph* and *gusA* genes was used in microarray analysis. Transgene expression in transgenic endophyte was demonstrated (Figure 4).

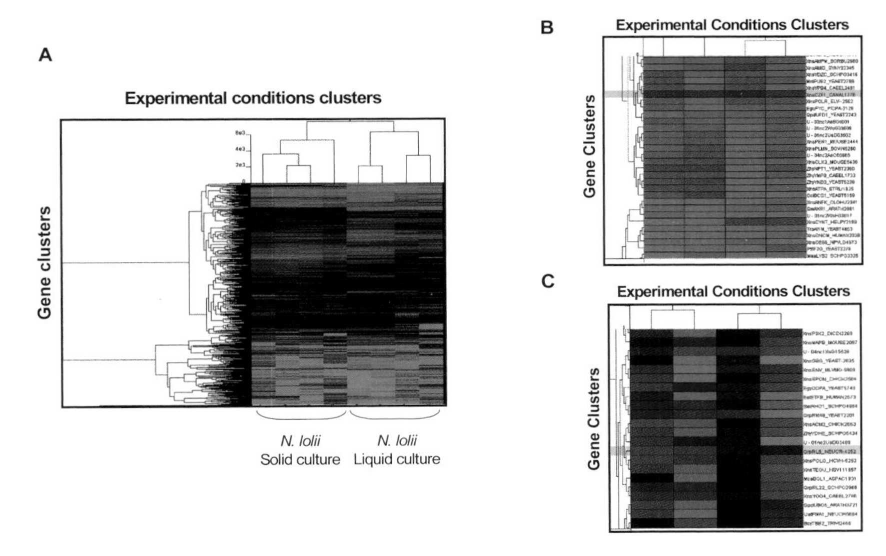

Figure 1. Hierarchical clustering of different signal values (Euclidean distance) from *N. lolii* grown in liquid culture vs. *N. lolii* grown in solid culture using GeneSight™ software. A) Hierarchical clustering of mean signal values. B) Hierarchical clustering of ratio of median signal values showing *N. lolii* genes up-regulated in liquid culture. C) Hierarchical clustering of ratio of median signal values showing *N. lolii* genes with similar expression levels in both culture conditions.

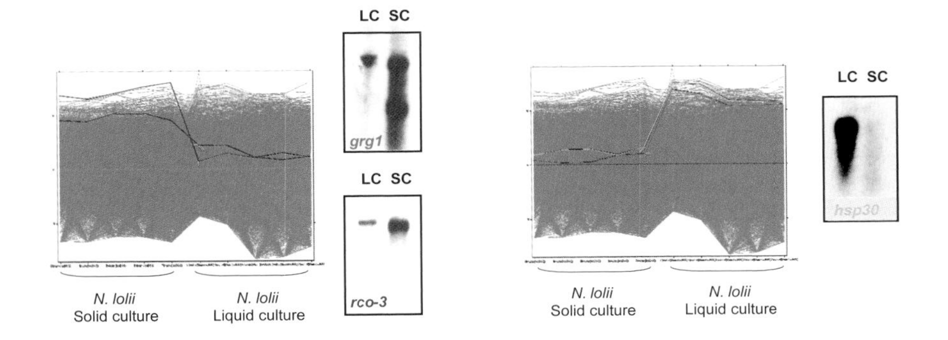

Figure 2. N. lolii transcriptome analysis. Time series plots of mean signal values. *grg1*: homologue to a glucose-repressible gene, *rco-3*: homologue to a glucose transporter gene, and hsp30: homologue to a 30 kDa heat shock protein from *Neurospora crassa. N. lolii* cDNAs were used as probes in northern hybridisation analysis. This analysis allowed for validation of microarray-based gene expression data.

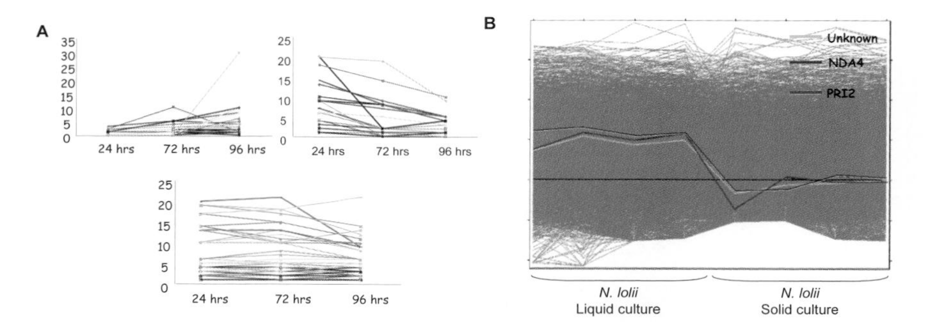

Figure 3. A) Time series plots of median signal values showing numerous *N. lolii* gene classes expression levels. B) Time series plot of median signal values showing a gene of unknown function coordinately expressed with cell cycle genes.

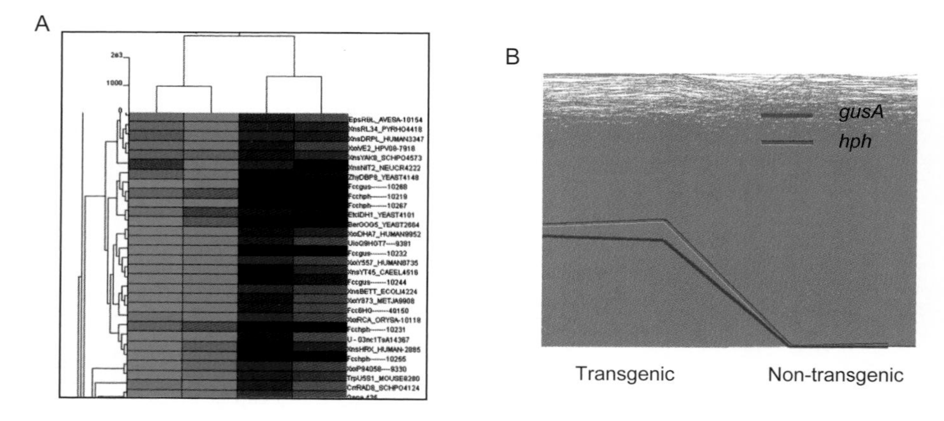

Figure 4. Neotyphodium in vitro gene expression on the EndoChip containing *gus*A and *hph* genes. A) Hierarchical clustering of mean signal values showing differential gene expression in different endophyte genotypes at target loci. B) Time series plot of mean signal values showing the expression levels of both transgenes in the transgenic and wild-type strains.

3.4 Microarray-based Transcriptome Analysis in *Neotyphodium* and *Epichloë* Species

Microarray-based gene expression analysis with RNA from *in vitro* grown *N. coenophialum* and *N. lolii* demonstrated its utility in comparative transcriptome analysis of both endophyte species (Figure 5).

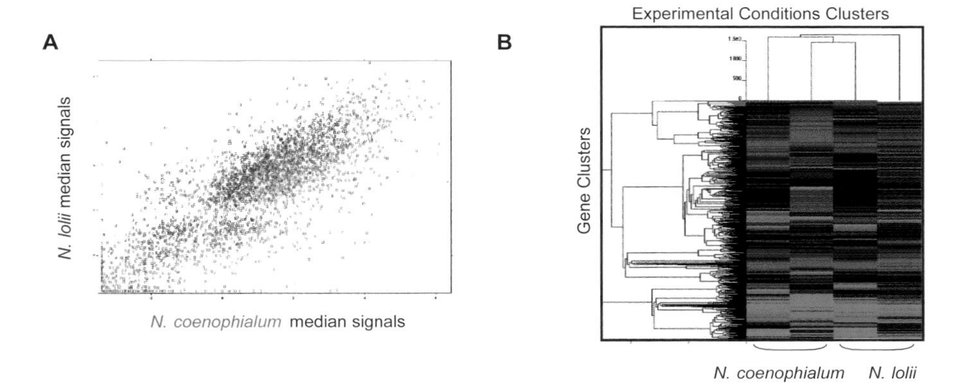

Figure 5. Comparative transcriptome analysis of *N. coenophialum* and *N. lolii*. A) Scatter plot of median signal values showing *N. coenophialum* and *N. lolii* prevalent expression. B) Hierarchical clustering of ratio median signal values showing species-prevalent gene expression.

Microarray-based transcriptome analysis with total RNA from *N. lolii* and *E. festucae* grown *in vitro* allowed to study genome-specific gene expression in asexual and sexual forms of grass endophytes (Figure 6).

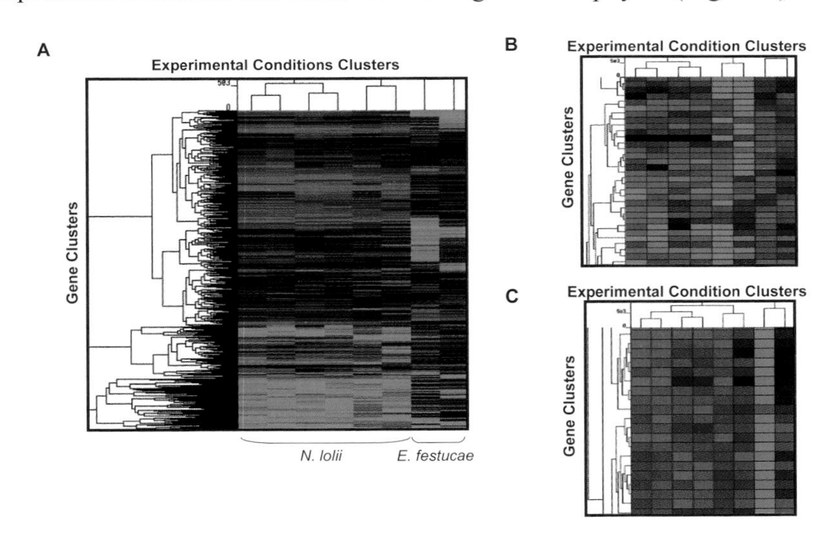

Figure 6. Hierarchical clustering of mean signal values (Euclidean distance) from *N. lolii* and *E. festucae* grown in liquid culture using GeneSight™ software. The genes that are shown in red are highly expressed, while the ones that are shown in green have low expression levels. A) *N. lolii* and *E. festucae*-prevalent gene expression. B) *E. festucae*-prevalent gene expression showing predominantly amino acid metabolism genes. C) *N. lolii*-prevalent gene expression showing numerous gene classes.

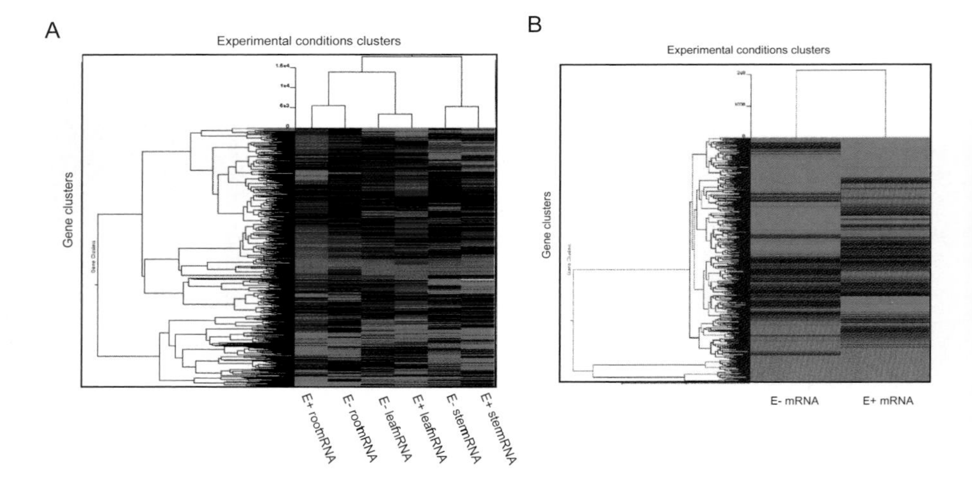

Figure 7. A) Hierarchical clustering of mean signal values (Euclidean distance) from infected and uninfected ryegrass organs on the Endophyte unigene microarray. B) Endophyte-infected stem prevalent gene expression.

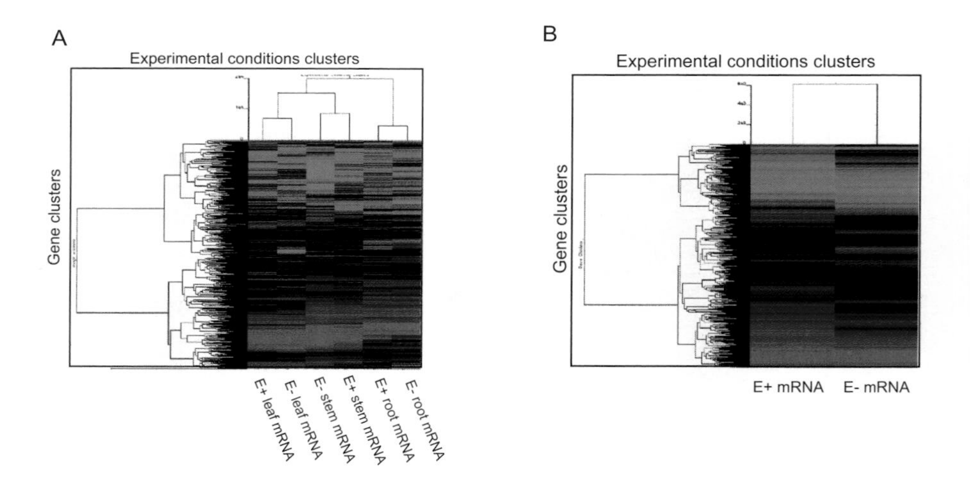

Figure 8. A) Hierarchical clustering of mean signal values (Euclidean distance) from infected and uninfected ryegrass organs on the 15K Ryegrass unigene microarray. B) Endophyte-infected stem prevalent gene expression.

2. ANALYSIS OF GENETIC DIVERSITY IN FUNGAL ENDOPHYTES

2.1 Background

Initial analysis of isolates from reference collections (van Zijll de Jong et al. 2003a,b) detected low levels of genetic variation within both *N. lolii* and *N. coenophialum*. *Neotyphodium* endophytes are asexual in reproductive mode, and so the limited genetic variation detected within *N. lolii* and *N. coenophialum* is not unexpected. Higher levels of genetic variation were detected within *E. festucae* which has the ability to undergo sexual reproduction. However, many of these reference isolates were obtained from the same grass variety or grass varieties with putative common ancestry, hence the analysis of additional isolates is required.

2.2 Detection of Endophytes *in planta*

EST-SSR markers have been developed for the efficient *in planta* detection of endophyte genetic diversity in large grass populations. Previous studies have demonstrated the use of a small number of genomic DNA-derived SSR markers for *in planta* detection (Groppe and Boller 1997; Moon et al. 1999). SSR markers with different sized products were labelled with different fluorochromes (FAM, HEX and NED) and assembled in duplex or triplex combinations for fluorescent-based detection on the MegaBACE 1000 capillary sequencer system (Amersham Biosciences, Little Chalfont, U.K.). Consistent detection of endophytes *in planta* was achieved with 20 of the 22 SSR markers tested. The SSR markers were able to detect endophytes in a number of grass species including *N. coenophialum* in tall fescue, *N. uncinatum* in meadow fescue and *N. lolii* in perennial ryegrass (Figure 1A). The sensitivity of SSR markers to detect endophytes *in planta* was tested with *in vitro*-composed admixtures of genomic DNA isolated from a cultured endophyte and an endophyte-free grass genotype. Some differences were observed in the sensitivity of different SSR markers, with selected markers able to detect endophyte at endophyte:grass DNA mass ratios as low as 1:5000. The SSR markers were able to detect endophytes in the upper leaf blade of the grass, but the signals were generally weak. Stronger endophyte signals were detected in the leaf sheaths of grasses, in which endophytes are known to be most abundant (Keogh et al. 1996).

Whole genome amplification methods such as the bacteriophage ø129-mediated multiple displacement amplification (MDA) technique (Dean et al. 2002), which is commercialised as the GenomiPhi DNA Amplification system (Amersham Biosciences), may be used to increase the quantity of

DNA derived from endophyte-infected grasses without adversely affecting the detection of endophytes *in planta*. Amplification of template DNA from isolated endophyte and from both endophyte-infected and uninfected grass yielded microgram quantities of high molecular weight products from nanogram inputs. Products of the expected molecular size were obtained with MDA from isolated endophyte and endophyte-infected grass with the relevant SSR markers (Figure 1B). These products remained absent in the uninfected grass. The threshold of detection of endophytes *in planta* with the SSR markers was not altered for *in vitro* assembled admixtures of endophyte and endophyte-free grass DNA that had undergone MDA.

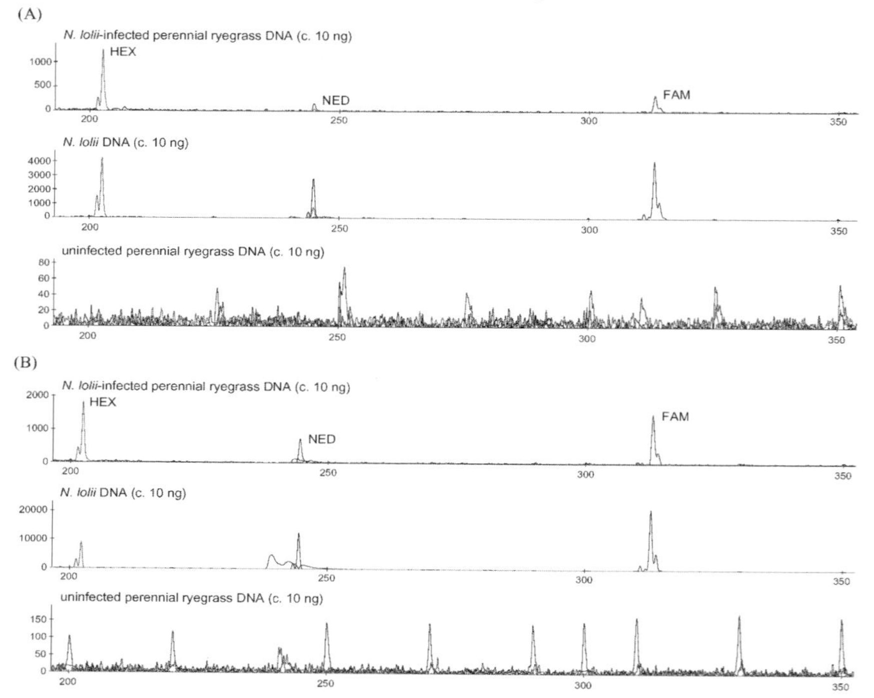

Figure 1. Detection of endophyte from purified template DNA (A) and template DNA obtained with MDA (B) using three EST-SSR markers (NLESTA1TA10-FAM, NCESTA1CC10-HEX, NCESTA1FH03-NED). In each case, c. 10 ng of template DNA has been derived from *N. lolii*-infected perennial ryegrass, cultured *N. lolii* and uninfected perennial ryegrass. The peaks visualised using the uninfected grass samples are derived from the ROX-labelled in-capillary size standards.

The possibility of cross-amplification from other fungi present in grasses with EST-derived SSR markers, due to the conserved nature of coding sequences, made it important to test marker specificity. There was no cross-amplification of products from genomic DNA of the grass powdery mildew pathogen *Erysiphe graminis*, but three EST-SSR markers showed weak cross-amplification from genomic DNA of the fungal saprophytes *Aspergillus nidulans* (2/3), *Penicillium marneffei* (2/3) and the grass crown rust pathogen *Puccinia coronata* f.sp. *lolii* (1/3). Two of the SSR loci showed sequence identity with known protein-coding genes of *A. nidulans* (a carboxypeptidase and a nuclear movement protein), although only primer pairs designed to the former produced amplification products from this species. The third SSR locus showed sequence identity with an ADP-ribosylation factor binding protein present in yeast, mouse and humans. The signals detected in *A. nidulans*, *P. marneffei* and *P. coronata* are probably too weak to be detected *in planta*.

2.3 Genetic Diversity in the Perennial Ryegrass Endophyte *Neotyphodium lolii*

Genetic diversity in *N. lolii* was investigated with the 20 SSR markers that consistently detected endophytes *in planta*. The plant material that was used was sub-sampled from a globally-distributed perennial ryegrass germplasm collection of 477 wild and cultivated accessions, as well as from Australian farms with variable incidence of endophyte-related livestock toxicosis and a variety (Grasslands Samson) containing an endophyte (AR1) with reduced toxicity effects. Endophyte viability in the germplasm collection may have been affected by sub-optimal seed storage conditions (Welty et al. 1987). A total of 132 accessions were examined cytologically for the presence of endophyte, with only 21 identified as containing at least one endophyte-positive plant.

The genetic variation detected among perennial ryegrass endophytes was low (van Zijll de Jong et al. 2003c). Most of the diversity detected was within the range of diversity previously detected in reference *N. lolii* isolates (Figure 2). The more diverse endophytes in genotypes from Morocco and Tunisia may belong to different endophyte taxa. In a larger phenogram developed to include data from other reference *Neotyphodium* and *Epichloë* isolates, the Moroccan endophyte was found to group closely with an unidentified *Neotyphodium* isolate and *E. festucae* isolates. The Tunisian endophyte grouped most closely with *N. lolii* isolates. This endophyte may belong to *Lolium perenne* taxonomic group 2 (LpTG-2). Endophytes in this group are believed to be interspecific hybrids of *N. lolii* and *E. typhina* (Schardl et al. 1994). Multiple amplification products were often detected in the Tunisian endophyte, suggestive of a hybrid origin. These results need to be confirmed with additional SSR markers.

The AR1 endophyte, which is unable to produce the animal toxins lolitrem B and ergovaline, was found to be genetically divergent from most other endophytes in perennial ryegrass. However, endophytes from farms with a major incidence of livestock toxicosis were not found to be genetically distinct from endophytes from farms with no incidence or a minor incidence of livestock toxicosis.

(A)

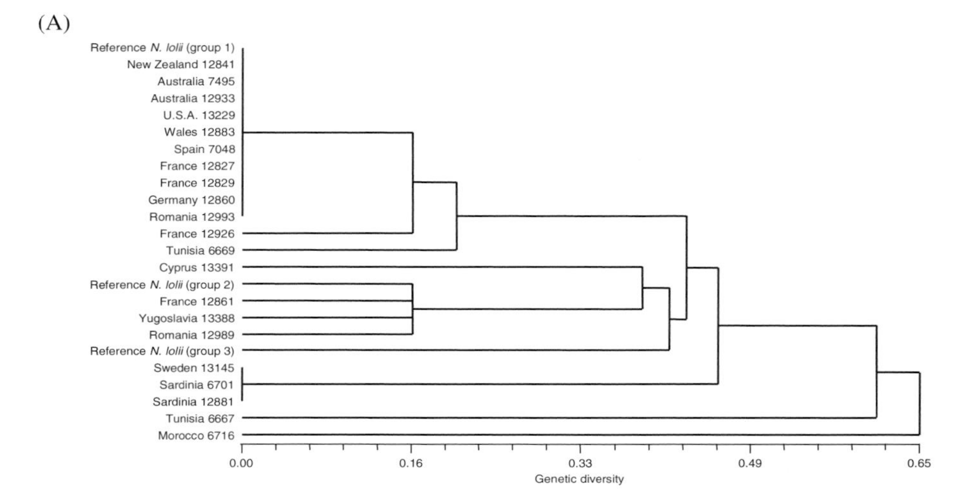

(B)

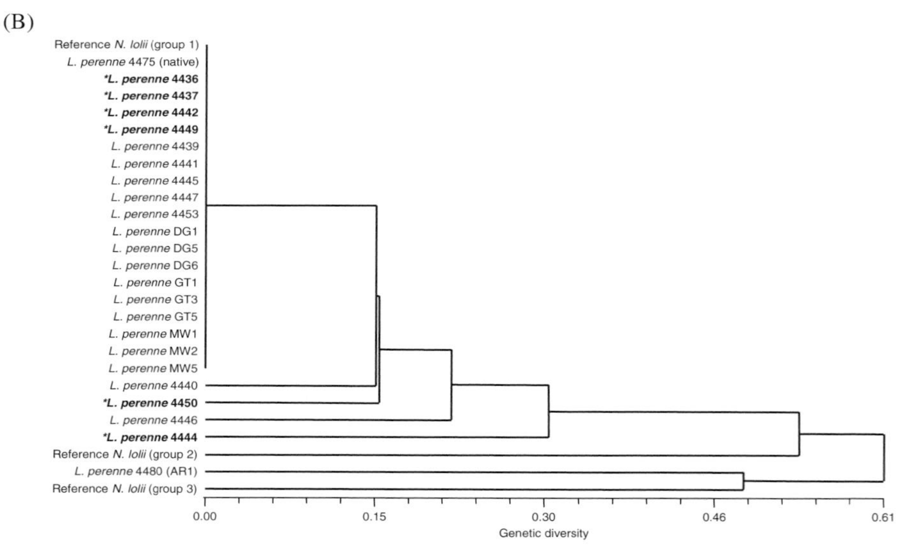

Figure 2. Genetic diversity of endophytes in (A) a globally-distributed perennial ryegrass germplasm collection and (B) in perennial ryegrass plants collected from Australian farms with variable incidence of endophyte-related livestock toxicosis. Plants from farms with an incidence of livestock toxicosis are in bold and labelled with an asterisk. Plants of the variety (Grasslands Samson) with the AR1 endophyte or the native endophyte in this variety are labelled as (AR1) or (native).

Genetic variation among host grasses was higher than among endophytes, suggesting that quantitative variability in endophyte-related effects may largely result from variation in the host genotype. A recent study has found evidence of host genetic control of endophyte toxin levels (Easton et al. 2002). By contrast, the qualitative variation associated with AR1 may be attributable to the endophyte genotype.

The geographic pattern of diversity in *N. lolii* shows some similarities with the proposed dispersion of perennial ryegrass from its putative near-Eastern centre of origin. A study of the variation in *Lolium* species of chloroplast DNA, which like the endophyte is maternally inherited, identified three clusters of populations which were not geographically randomly distributed (Balfourier et al. 2000). The three clusters appeared to have originated in the Middle East and to have spread following the migration of the first human farmers along three routes: a north-eastern one (Danubian movement), a south-western one, including the British Isles (Mediterranean movement) and a North African continental route. Endophytes in accessions along these three putative migration routes were, in broad terms, found to group together on the phenogram. Although additional genotypes from each accession need to be analysed to confirm these results, this data supports the model of a monophyletic origin for *N. lolii* (van Zijll de Jong et al. 2003c).

2.3 Genetic Diversity in the Meadow Fescue Endophyte *Neotyphodium uncinatum*

The EST-SSR markers developed for *in planta* genetic diversity analysis in *N. lolii* are also applicable to other endophyte taxa. The genetic diversity of *N. uncinatum* in nine meadow fescue genotypes (that displayed differences in cold tolerance and resistance to snow mould) was assessed. Low levels of genetic variation were detected among *N. uncinatum* endophytes. Only two of the 20 SSR markers detected variation, which separated the endophytes into two groups (Figure 3). Isozyme studies also detected limited polymorphism within *N. uncinatum* and the separation of isolates into two isozyme phenotypes (Christensen et al. 1993; Leuchtmann 1994). The diversity among *N. uncinatum* endophytes reflected in part the genetic relationships, based on varietal origin, of the host grasses. Analysis with additional SSR markers may lead to the detection of further polymorphism between *N. uncinatum* endophytes. In previous studies (van Zijll de Jong et al. 2003a), some of the polymorphism detected between reference *N. coenophialum* isolates was not accounted for with the 20 SSR markers used here for *in planta* analysis.

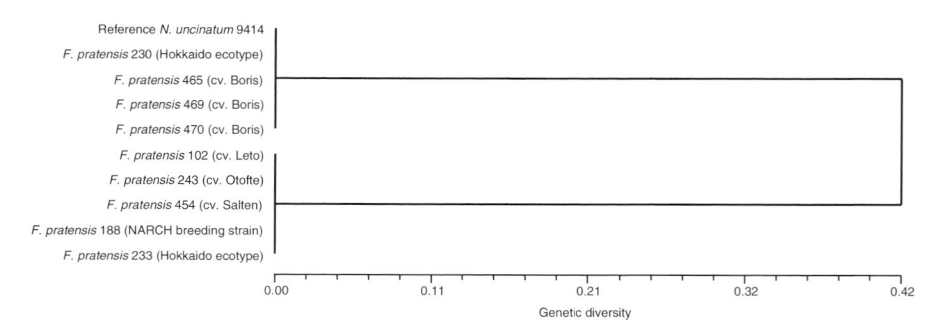

Figure 3. Genetic diversity of *N. uncinatum* genotypes detected *in planta* in meadow fescue genotypes displaying differences in cold tolerance and resistance to snow mould.

3. CONCLUSIONS

The development of SSR markers to detect endophytes *in planta* permitted the genetic diversity of endophytes in larger grass populations to be assessed. Low levels of genetic variation were detected in endophytes of perennial ryegrass and meadow fescue. These results suggest that variation for expression of endophyte characters may be more associated with host grass genotypic differences. The quantitative trait locus (QTL) analysis of endophyte-specific phenotypic effects in full-sib genetic mapping families containing a single resident endophyte genotype provides the means for genetic dissection of such effects. Families with this structure (Guthridge et al. 2003) are the subject of intensive study in our current work. Extensive perennial ryegrass germplasm collections also provide the resource for analysis of simultaneous genetic variation of both host and endophyte. Although the specific EST-SSR markers described here and genomic-DNA derived SSR markers for perennial ryegrass (Jones et al. 2001) provide ideal tools for such analysis, sequence-independent co-genotyping methods such as diversity array technology (DArT: Jaccoud et al. 2001; van Zijll de Jong et al. 2003d) may also be of value.

ACKNOWLEDGEMENTS

This work was supported by the Victorian Department of Primary Industries and the Cooperative Research Centre for Molecular Plant Breeding. Endophyte EST sequences were generated by Karen Fulgueras, Jutta Nagel, Marc Ramsperger and Kate Shields of the Plant Biotechnology Centre. Reference endophyte isolates were kindly provided by Dr. Adrian Leuchtmann (ETH, Zürich, Switzerland) and Nola McFarlane (DPI-PVI,

Hamilton, Victoria, Australia). Dr. Kevin Reed (DPI-PVI, Hamilton, Victoria, Australia) and Dr. Toshihiko Yamada (NARCH, Hitsujigaoka, Sapporo, Japan) kindly provided some of the plant material used in this study.

REFERENCES

Balfourier F, Imbert C, Charmet G (2000) Evidence for phylogeographic structure in *Lolium* species related to the spread of agriculture in Europe. A cpDNA study. Theor. Appl. Genet. 101: 131-138.

Breen JP (1993) Enhanced resistance to three species of aphids (*Homoptera, Aphididae*) in *Acremonium* endophyte-infected turfgrasses. J. Econ. Entomol. 86: 1279-1286.

Christensen MJ, Easton HS, Simpson WR, Tapper BA (1998) Occurrence of the fungal endophyte *Neotyphodium coenophialum* in leaf blades of tall fescue and implications for stock health. N.Z. J. Agr. Res. 41: 595-602.

Christensen MJ, Leuchtmann A, Rowan DD, Tapper BA (1993) Taxonomy of *Acremonium* endophytes of tall fescue (*Festuca arundinacea*), meadow fescue (*F. pratensis*) and perennial ryegrass (*Lolium perenne*). Mycol. Res. 97: 1083-1092.

Dean FB, Hosono S, Fang L, wu X, Fawad Faruqi A, Bray-Ward P, Sun Z, Zong Q, Du Y, Du J, Driscoll M, Song W, Kingsmore SF, Egholm M, Lasken RS (2002) Comprehensive human genome amplification using multiple displacement amplification. Proc. Natl. Acad. Sci USA 99: 5261-5266.

Easton HS, Latch GCM, Tapper BA, Ball OJP (2002) Ryegrass host genetic control of concentrations of endophyte-derived alkaloids. Crop Sci. 42: 51-57.

Elbersen HW, West CP (1996) Growth and water relations of field-grown tall fescue as influenced by drought and endophyte. Grass Forage Sci. 51: 333-342.

Gallagher RT, Hawkes AD, Steyn PS, Vleggaar R (1984) Tremorgenic neurotoxins from perennial ryegrass causing ryegrass staggers disorder of livestock: structure elucidation of lolitrem B. J. Chem. Soc. Chem. Comm. 9: 614-616.

Groppe K, Boller T (1997) PCR assay based on a microsatellite-containing locus for detection and quantification of *Epichloë* endophytes in grass tissue. Appl. Env. Microbiol. 63: 1543-1550.

Guthridge KM, McFarlane NM, Ciavarella TA, Batley J, Jones ES, Smith KF, Forster JW (2003) Molecular marker-based analysis of morphological traits in perennial ryegrass (*Lolium perenne* L.). In: Abstracts of Molecular Breeding of Forage and Turf, Third International Symposium, p.58. May 18-22, Dallas, Texas and Ardmore, Oklahoma, USA.

Jaccoud D, Peng K, Feinstein D, Kilian A (2001) Diversity arrays: a solid state technology for sequence information-independent genotyping. Nucl. Acids. Res. 29(4): e25.

Jones ES, Dupal MP, Kölliker R, Drayton MC, Forster JW (2001) Development and characterisation of simple sequence repeat (SSR) markers for perennial ryegrass (*Lolium perenne* L.). Theor. Appl. Genet. 102: 405-415.

Keogh RG, Tapper BA, Fletcher RH (1996) Distributions of the fungal endophyte *Acremonium lolii*, and of the alkaloids lolitrem B and peramine, within perennial ryegrass. N.Z. J. Agr. Res. 39: 121-127.

Leuchtmann A (1994) Isozyme relationships of *Acremonium* endophytes from 12 *Festuca* species. Mycol. Res. 98: 25-33.

Malinowski DP, Belesky DP (1999) *Neotyphodium coenophialum*-endophyte infection affects the ability of tall fescue to use sparingly available phosphorus. J. Plant. Nutr. 22: 835-853.

Moon CD, Scott B, Schardl CL, Christensen MJ (2000) The evolutionary origins of *Epichloë* endophytes from annual ryegrasses. Mycologia 92: 1103-1118.

Moon CD, Tapper BA, Scott B (1999) Identification of *Epichloë* endophytes *in planta* by a microsatellite-based PCR fingerprinting assay with automated analysis. Appl. Env. Microbiol. 65: 1268-1279.

Prestidge RA, Gallagher RT (1988) Endophyte fungus confers resistance to ryegrass: Argentine stem weevil larval studies. Ecol. Entomol. 13: 429-435.

Rowan DD, Gaynor DL (1986) Isolation of feeding deterrents against Argentine stem weevil from ryegrass infected with the endophyte *Acremonium loliae*. J. Chem. Ecol. 12: 647-658.

Schardl CL, Leuchtmann A, Tsai HF, Collett MA, Watt DM, Scott DB (1994) Origin of a fungal symbiont of perennial ryegrass by interspecific hybridization of a mutualist with the ryegrass choke pathogen, *Epichloë typhina*. Genetics 136: 1307-1317.

Siegel MR, Latch GCM, Bush LP, Fannin FF, Rowan DD, Tapper BA, Bacon CW, Johnson MC (1990) Fungal endophyte infected grass: alkaloid accummulation and aphid response. J. Chem. Ecol. 16: 3301-3316.

Tsai HF, Liu JS, Staben C, Christensen MJ, Latch GCM, Siegel MR, Schardl CL (1994) Evolutionary diversification of fungal endophytes of tall fescue grass by hybridization with *Epichloë* species. Proc. Natl. Acad. Sci. U.S.A. 91: 2542-2546.

van Zijll de Jong E, Guthridge KM, Spangenberg GC, Forster JW (2003a) Development and characterization of EST-derived simple sequence repeat (SSR) markers for pasture grass endophytes. Genome 46: 277-290.

van Zijll de Jong E, Guthridge KM, Batley J, Edwards D, Spangenberg GC, Forster JW (2003b) EST-SSR sequence variation in *Neotyphodium* grass endophytes. Genomics of plant-symbiont relations. P5. MBFT2003, Dallas, Texas

van Zijll de Jong E, Bannan NR, Batley J, Spangenberg GC, Smith KF, Forster JW (2003c) Genetic diversity in the perennial ryegrass fungal endophyte *Neotyphodium lolii*. In: Abstracts of Molecular Breeding of Forage and Turf, Third International Symposium, p.7. May 18-22, Dallas, Texas and Ardmore, Oklahoma, USA.

van Zijll de Jong E, Kilian A, Batley J, Spangenberg GC, Forster JW (2003d) Diversity array technology for co-genotyping of the perennial ryegrass-*Neotyphodium lolii* interaction. Genomics of plant-symbiont relations. In: Abstracts of Molecular Breeding of Forage and Turf, Third International Symposium, p.3. May 18-22, Dallas, Texas and Ardmore, Oklahoma, USA.

Welty RE, Azevedo MD, Cooper TM (1987) Influence of moisture content, temperature and length of storage on seed germination and survival of endophytic fungi in seeds of tall fescue and perennial ryegrass. Phytopathology 77: 893-900.

White JF, Morgan-Jones G, Morrow AC (1993) Taxonomy, life cycle, reproduction and detection of *Acremonium* endophytes. Agr. Ecosyst. Environ. 44: 13-37.

Wilkinson HH, Siegel MR, Blankenship JD, Mallory AC, Bush LP, Schardl CL (2000) Contribution of fungal loline alkaloids to protection from aphids in a grass-endophyte mutualism. Mol. Plant. Microbe. In. 13: 1027-1033.

Yates SG, Plattner RD, Garner GB (1985) Detection of ergopeptine alkaloids in endophyte infected, toxic Ky-31 tall fescue by mass spectrometry/mass spectrometry. J. Agr. Food Chem. 33: 719.

Molecular Breeding for Animal, Human and Environmental Welfare

M. O. Humphreys and M. T. Abberton
Institute of Grassland and Environmental Research, Plas Gogerddan, Aberystwyth, SY23 3EB, UK. (Email: mervyn.humphreys@bbsrc.ac.uk).

Key words: ryegrass, clover, resource-use efficiency, fatty acids, carbohydrates, protein utilisation, QTL, marker –assisted selection, transformation

Abstract:

Since farmers started to cultivate land and deliberately sow seed, selection has led to the development of domesticated crops suitable for a wide range of agro-ecological and socio-cultural conditions. Based on increasing understanding of genetics and genes, new prospects for molecular breeding have emerged. Both forward and reverse genetic approaches and techniques such as marker-assisted selection and introgression, mutation breeding and transformation are now available. Increased precision in genetic manipulation promotes the application of more targeted breeding strategies to a wider range of traits serving a broader range of objectives. These include enhanced prospects for multifunctional land-use with improvements in the efficient use of inputs such as nitrogen, phosphate and water; in soil quality and protection against erosion; in pest and disease resistance; in enhanced animal nutrition and improved livestock product quality; in reduced pollution risks and health problems; and in potential for alternative use of grassland such as for amenity purposes, industrial processing and pharmaceuticals.

A. Hopkins, Z.Y. Wang, R. Mian, M. Sledge and R.E. Barker (eds.), Molecular Breeding of Forage and Turf, 165-180.

1. INTRODUCTION

There is increasing emphasis on the sustainability of farming systems with on-farm produced feed assuming greater importance. Forage legumes are an essential component of these systems because they fix nitrogen, have a high nutritive value and can improve soil quality. As well as supporting agricultural production grassland has value for recreational and landscape use and helps to protect land from soil erosion and pollution. However intensively stocked grasslands may lose substantial amounts of nitrate to drainage water, particularly after cultivation and emissions of ammonia and nitrous oxide to the atmosphere from grassland agriculture can also be significant. This reflects the low efficiency with which grazing animals utilise forage protein. If not required to support livestock farming, forage crops can be refined to produce high value products and energy. Commercial refining of forage grasses is at present limited to small-scale production of protein, fibre and biogas from ryegrass and clover in Switzerland and Germany (www.2bio.ch). However, the industrial use of crops is expanding and diversifying. A range of feedstock chemicals for industry may be produced by grass fermentation (Lin et al. 1985). Lactic acid is one product that is now being used to produce the biodegradable plastic polylactic acid (Vink et al. 2003). Alcohols derived from the fermentation of grass carbohydrate could even be a major source of hydrogen for use as a clean vehicle fuel (Cortright et al. 2002). Many temperate forage grasses are unsuitable for direct combustion because of a high water content (Wilkins 1997). A possible exception is reed canary grass (*Phalaris arundinacea* L.) if field dried in freezing conditions (Andersen and Lindvall 1997). Greater understanding of ecological and social interactions combined with an appreciation of economic benefits are needed to determine how best to maintain profitable land-use while protecting the landscape and maintaining desirable levels of biodiversity. Decisions have to be made on what species are most fit for purpose; on the extent to which desirable levels of genetic improvement can be achieved; the most appropriate breeding strategies to use; and whether commercial seed production is viable.

2. IMPROVEMENT TARGETS

2.1 Reduced Pollution

The potential to reduce environmental pollution in agricultural and 'amenity' grasslands requires knowledge of genotype interactions with a range of environmental factors. This includes nitrogen (in all its reactive forms), phosphorus, sulphur, persistent organic pollutants, heavy metals and particulate pollutants including transmission to waters and exchange with the atmosphere. It is also relevant to consider carbon, both from the perspective of sequestration and cycling as well as its impact on 'driving' other nutrient transformations and utilisation. Legumes have a unique role through the fixation and transfer of atmospheric nitrogen. Leaching of nitrogen under defoliated clover swards can be considerable because of major inputs of nitrogen into the soil from the senescence and decomposition of roots and nodules (Dubach and Russelle 1994). Inefficient use of high protein forage in the rumen or silo can increase land and water pollution, acidification and greenhouse gas emissions.

2.2 Water Quantity and Quality

Climate change scenarios indicate that a key element will be the use and availability of water. Grasslands in the Europe currently use 33% of all available water and it is predicted that areas of the UK will suffer summer droughts and increasing temperatures with climate change. It is estimated that a 50mm water deficit can reduce dry-matter yield of ryegrass by 1 tonne. Water deficit in SE England is, already, frequently four times this level. Other factors affecting water quality include increased winter rainfall causing flooding with greater run-off and industrial/agricultural pollution. Prospects for breeding improved drought and flood tolerance are good in grasses and forage legumes. Plant interactions with agricultural pollutants (e.g. nutrients, micro-organics and pathogens) may also provide breeding targets.

2.3 Improved Resource Use Efficiency by Forages and Ruminants

Improved nitrogen-use efficiency (NUE) increases the amount of dry matter produced per unit of available N. A 10% higher NUE could reduce nitrogen surpluses by 35% without loss of production (van Loo and Vellinga 1994). Carbon resource allocation also has a role in several aspects of sustainable land use including sward persistency, interactions with soil microflora, plant protein utilisation by ruminants and the production of 'greenhouse' gasses. The efficient utilisation of plant protein by ruminants requires readily available energy and/or reduced rates of protein degradation. High sugars (WSC) in grasses can dramatically improve the efficiency of protein use by animals so that more protein is converted into milk and meat constituents and less nitrogen is excreted in the urine (Miller et al. 2001). It also improves silage quality with better protein preservation (Davies et al. 2002). Starch, as a fermentable carbohydrate, may also provide a significant target in forage improvement. Starch accounts for less than 10-15% of the reserves in vegetative tissue of grasses even when total carbon reserves are high. It is unclear why starch does not accumulate to higher concentrations in leaves of the fructan grasses. The regulation of starch metabolism in grass leaves is distinct from the regulatory mechanisms governing fructan synthesis and is markedly different from that in plants such as clover, where starch is the dominant reserve. Studies of starch metabolism in the fructan grasses is providing fundamental information about the regulation of assimilate partitioning and help to identify assimilate-responsive genes and promoters associated with carbohydrate metabolism. The availability of structural carbohydrates in cell walls is constrained by lignin deposition and provides a range of targets with regard to improving digestibility (Chen et al. 2002).

Improved forage protein utilisation with reduced nitrogen wastage and reduced methane production may also be achieved by means of plant tannins that reduce rates of protein degradation. Tannins are polymerised units of flavan-3-ols arising from a poorly defined branch at the lower end of the flavonoid pathway and have anti-oxidant and protein-precipitating properties that can improve forage quality, human health and plant pest resistance (Robbins et al. 2002). There is increasing information on the molecular mechanisms that control the biosynthetic pathway to condensed tannins (proanthocyanidins) (Robbins et al. 1999). Work is in progress to determine whether the pathway can operate in the foliar tissues of white clover and lucerne and produce bloat-safe field crops with enhanced nitrogen efficiency when grazed by ruminant livestock (Aerts et al. 1999). *Lotus* species naturally contain high levels of tannins in vegetative tissues and there is potential to improve the capacity of *Lotus* to establish in swards and compete with other legumes and grasses. White and red clover contain tannins in the flowers although there is little in vegetative tissues. However this could be quite important with regard to the use of prolific flowering *T. nigrescens* hybrids in swards (Marshall et al. 1998).

There is also an increasing need to consider plant and animal utilisation of phosphate (a major agricultural pollutant) throughout production cycles. Although phytate is not a significant part of forage leaf tissues and its non-availability is more relevant to monogastrics, it is an important issue in relation to higher input diets with seed based concentrates for dairy cows. Increasing P availability in herbage varieties may be a way to reduce phytate:P ratio in diets. Italian ryegrass is better than maize and several other species in extracting P from heavily manured soils. Both P content and dry matter yield are key factors (Pederson et al. 2002, Thelier-Huche et al. 1994).

2.4 Deeper Rooting Forages for Improved Mineral Content and Animal Welfare

Altering the content and ratios of minerals in herbage to prevent metabolic disorders provide possible breeding objectives. Grass tetany is caused by low levels of magnesium in the blood of cattle or sheep. Varieties of Italian ryegrass and tall fescue with markedly higher levels of magnesium have proved to be very effective in maintaining levels of blood magnesium in grazing sheep (Moseley and Baker 1991) and cattle (Crawford et al. 1998) respectively. Milk fever, caused by low blood calcium, produces animal welfare and production problems that could be addressed by reducing potassium content of forage without reducing calcium and magnesium concentrations (Sanchez et al. 1994). Relevant work is being carried out on the relationship between carbohydrate resource allocation and root growth traits. Initial work looking at a crude measure of root amount suggests that there are several regions of the genome with important influence on root traits. Further work is required on traits such as root amount, root profile, root initials and rooting capacity and root penetrability through impedance layers.

2.5 Plant Toxins and Animal Welfare

The accumulation of toxins in grasses can drastically reduce the productivity of animals and adversely affect their welfare. A major source of alkaloid toxins is *Neotyphodium* endophytic fungi that occur mainly in the leaf sheaths of ryegrasses and fescues but also grow up the flowering stems to infest the seed. Loss of muscular co-ordination by sheep grazing ryegrass in New Zealand (ryegrass staggers) and toxicosis in cattle grazing tall fescue in the USA were found to be linked to infection by endophytic fungi (Fletcher and Harvey 1981; Hoveland et al. 1983). Eczema in lambs grazing meadow fescue in the Netherlands also was linked to endophyte infestation (Huizing et al. 1991). Endophytes can be removed easily by treatment of plants with systemic fungicide before seed production, although this is often at the cost of persistency because of reduced resistance to insects. There has been considerable progress in developing strains of endophyte that deter insect attack but do not produce the alkaloids that are the main causes of toxicity to grazing animals. Endogenous production of alkaloids by grasses is also common. Leaves of endophyte-free meadow fescue and ryegrass/fescue hybrids can contain high levels of perloline, which is known to inhibit ruminal digestion (Bush 2001). Nine different alkaloids are found in endophyte-free reed canary grass and there is considerable genetic variation in their concentrations (Marum et al. 1979). There have been no reports of significant concentrations of alkaloids in milk and meat of animals grazing endophyte-infected or intrinsically high alkaloid grasses. Alkaloids are the prime suspects if animals perform very differently when grazing grass varieties with similar DMD and fibre content, especially if this is related to signs of poor health.

2.6 Improving the Functional Feed Value of Forage for Healthier Outputs

Possibilities exist to improve meat and milk quality and hence human health through forage breeding. Plant lipids contain a high proportion of polyunsaturated fatty acids, associated with the thylakoid membranes of chloroplasts, and are the primary source of beneficial fatty acids in the food chain. Better utilisation of forage in ruminant diets combined with breeding to increase the delivery of fatty acids into ruminant products is an important long-term objective. It appears that some genes associated with plant lipid metabolism are sugar-inducible and initial work has identified genetic variation in a number of leaf fatty acids. There is variation among ryegrass species and varieties for total fatty acid content and the proportion of linoleic acids (Dewhurst et al. 2001). Increasing the voluntary intake of grasses through improved DMD or reduced fibre content, thereby increasing the proportion of grass in the diet, would also increase the intake of linoleic acids indirectly.

3. MOLECULAR BREEDING STRATEGIES FOR GENETIC IMPROVEMENT

All the targets described previously are amenable to improvement through molecular breeding procedures. Conventional breeding approaches using DNA markers can be used to improve trait components or a range of traits simultaneously through marker-assisted introgression and selection (MAS). Traditionally, breeders have relied on visible phenotypic traits to select improved varieties. This can be very costly particularly if several years are required to measure the trait (e.g. persistency) or if animals are involved (e.g. intake). MAS relies on identifying DNA markers associated with a desired trait and selecting on the basis of the marker (genotype) rather than the trait (phenotype).

3.1 Use of DNA Markers in Forage Improvement

Genome mapping using various DNA markers is well-advanced in annual grasses such as barley, wheat, rice and maize. Microsatellite markers are potentially the most useful for marker-assisted selection, being highly repeatable, co-dominant and amenable to automation (Gupta and Varshney 2000). Single nucleotide polymorphisms (SNPs) are also being used as more sequence information becomes available (Henry 2001). Markers that are closely linked to desirable alleles associated with Quantitative Trait Loci (QTL) can be identified and used to aid introgression and selection programmes. Traditionally, major gene and polygenic variation has been analysed in different ways, but QTL analysis now allows a more integrated approach in dissecting complex traits and assessing gene effects.

Genome mapping of forage grasses has lagged behind those of the major cereal crops, but considerable progress has been made in mapping the ryegrass genome using RFLP, AFLP and microsatellite markers (Jones et al. 2001; Armstead et al. 2002; Forage Grass Database FoggDB; www.ukcrop.net/perl/ace/search/FoggDB). Mapping in other grasses is less well developed but work is progressing in tall fescue (Xu et al. 1995) and meadow fescue (Rognli pers.comm.). QTL associated with water-soluble carbohydrate (WSC) accumulation in the leaves and leaf sheaths of ryegrass have been identified (Humphreys and Turner 2001; Turner et al. 2001). QTL have also been identified for other traits including regrowth (reserve-driven growth), constitutive leaf extension rate (photosynthate-driven growth), leaf chlorophyll content, plant size (vigour) and heading date. Some QTL overlie or fall close to the location of genes with known function. In ryegrass a major tiller base sucrose QTL overlies the location of an alkaline invertase gene (Turner et al. 2001) and other candidate genes associated with carbon metabolism (Lidgett et al. 2002) and digestibility (McInnes et al. 2002) have been isolated. A major heading date QTL located in perennial ryegrass lies close to a gene highly homologous to the *Arabidopsis* flowering time gene *CONSTANS* (*CO*) (Donnison et al. 2002). Allelic variation in this gene correlates with differences in heading date and may account for up to 25% of the total variation. QTL for crown rust resistance have also been identified in perennial and Italian ryegrasses (Roderick et al. 2003; Dumsday et al. 2003; Muylle et al. 2003; Fujimori et al. 2003). Exploiting genetic variation in *stay-green* characteristics of plants is a major turfgrass breeding objective and there is considerable quantitative variation within perennial ryegrass. Analysis of leaf pigment loss in detached leaves of perennial ryegrass has revealed 6 large QTL (Thorogood and Laroche 2001). Introgression from *F. pratensis* into *L. perenne* with the aid of DNA markers and genomic *in-situ* hybridisation (GISH) also produced an improvement of high temperature rust resistance (Roderick et al. 2000).

Marker-assisted selection can be used to increase gains from recurrent combined phenotypic and family selection within elite breeding populations.

A possible approach is to identify markers in mother plants that are associated with improved plot performance and select progeny plants with the most valuable markers as parents for the next generation of family selection. Initial results look promising (Wilkins et al., 2003). Of 25 RFLP markers known to map to individual loci in ryegrass, 7 were significantly associated with various plot traits including yield at the main silage cut, yield during vegetative growth, herbage DMD and WSC, and ground cover.

In forage legumes the majority of work on QTL analysis and MAS has been carried out in alfalfa (*Medicago sativa*). On a world basis alfalfa is the most important temperate forage legume although tetrasomic inheritance causes difficulties for breeders and diploid varieties have been produced via haploids. Using molecular markers such as RAPDS and RFLPS, linkage maps have been produced at the diploid level (Brummer et al., 1993; Echt et al., 1993; Kiss et al., 1993). More recently, SSR (microsatellite) markers have been used and trait mapping carried out on tetraploid alfalfa (Diwan et al., 2000; Skinner et al., 2000). Linkage mapping and other genetic studies in alfalfa studies have been hampered by inbreeding depression and markers have been used to investigate heterosis (Osborn et al. 1997).

White clover (*Trifolium repens*) is an important perennial forage legume in Northern Europe, New Zealand and parts of Australia. A white clover map based on AFLPs and microsatellites has been developed using self-compatible inbred lines (Abberton et al. 2000). RAPDS have been used to assess the extent of genetic diversity between inbred lines and associated heterosis in crosses between them (Joyce et al. 1999). Targets for MAS in white clover include a range of agronomic, physiological and quality traits (Webb et al. 2003).

So far there has been little use of molecular markers in red clover breeding programmes (Taylor and Quesenberry 1996) and a map has not been published. Isozyme and RAPDs have been used for genetic diversity studies (Kongkiatnam et al. 1995). RAPD markers have been developed for disease resistance (Page et al. 1997) and isozyme heterozygosity linked to morphological characteristics (Xie and Mosjidis 2001). There have also been studies of gene expression (Nelke et al. 1999).

Introgression in forage legumes has concentrated on alfalfa and white clover and both molecular and cytological markers are beginning to aid this work (Abberton and Marshall 2002). Interspecific hybrids have been produced between red clover and some of its relatives but with little impact. Introgression between species of *Medicago* is limited by polyploidy, inbreeding depression and the lack of an easily characterised karyotype although a range of hybrids have been generated (Osborn et al. 1997; Arcioni et al. 1997). The most successful introgression has been with *M. dzhawakhetica* (2n =4x=32) and genome-specific markers have been developed using RAPDs and RFLPs (Osborn et al. 1997). In white clover work has focused on introgressions using *T. nigrescens* and *T. ambiguum*. *T. nigrescens* Viv. is an annual diploid species that is sexually compatible with white clover and is believed to be one of its ancestral genomes. Introgression has been used to increase seed production (Marshall et al. 1998) and to improve resistance to clover cyst nematode (Hussain et al. 1997). Introgressions from *T. ambiguum* Bieb has produced drought resistant material with both stolons and rhizomes (Abberton et al. 2000). BSA-AFLP has been used to identify markers for the rhizomatous habit in plants of backcross families (Abberton and Marshall 2002). In New Zealand virus resistance has also been transferred from *T. ambiguum* (Woodfield and Brummer 2001).

Despite the promising potential of MAS, poor precision in locating QTL, bias of individual gene effects and risks of identifying 'false-positive' associations have generated disappointing responses (Kearsey and Farquhar 1998). However, based on QTL information for tiller base polymeric fructan content, marker selections were made on Linkage Groups (LG) 1 and 2 in perennial ryegrass. Mean differences in %WSC between alternative homozygous genotypes on LG1 and LG2 were 6.1 and 6.9 for % tiller base fructan and 4.7 and 8.1 for total % WSC. Heading dates of selections were similar. Progeny from the contrasting selections on both Linkage Groups demonstrated significant differences in %WSC in the direction expected from the selection imposed. Similar success with marker selection has been obtained by combined selection on several NUE QTL (Dolstra et al. 2003).

cDNA derived macro and micro-arrays can also aid precision and success in MAS. They are increasingly available from a range of species (particularly rice and barley for *Lolium* and *Medicago* for *Trifolium*) and can be used to identify genes and determine their expression. A good example is provided by recent investigations into the genomics of the phenylpropanoid pathway in *Medicago truncatula* that is important in plant defence mechanisms (Dixon et al. 2002). Regulatory regions of the genome affecting the expression of many genes may be highly polymorphic. cDNAs from genes associated with carbohydrate metabolism, lignin/cell wall biosynthesis and stress tolerance are providing new information on gene expression patterns. Although work has concentrated mainly on improving forage quality characteristics, traits are being mapped that confer enhanced soil physico-chemical quality and microbial diversity, including soil porosity and aeration, functional microbial activity (mineralisation potential) and root exudation of

C and N and tolerance to, or hyper-accumulation of, heavy metals (Zn, Pb, Cd). Transcriptome analysis of populations adapted to contrasting environments such as high salt concentrations in the soil, flooding or growth in soils polluted with heavy metals will provide valuable information on candidate genes and their expression relevant to sustainable land-use.

In forage legumes transcriptome changes are being investigated that trigger nodule and root senescence in response to defoliation and hence determine genotypic variation for leaching propensity. Functional genomics approaches can also be applied to plant responses involved in rhizobial and arbuscular mycorrhizal (AM) symbiotic associations. Optimal functioning of legume symbioses, such as rhizobium and mycorrhizal fungi (AMF), are central to the successful development of extensive agricultural systems. These symbiotic associations not only provide environmental and nutritional benefits by permitting reduced inputs of N and P (Harrison et al. 2002), but successful plant-AMF symbioses also offer non-nutritional benefits by improving disease resistance, stress tolerance, soil structure, by aiding phytoremediation and by supporting both plant and soil bio-diversity. Preliminary studies indicate that white clover cultivars vary in their response to colonization by single isolates of AMF. Recent advances in genomics of AM symbiosis have targeted AMF colonization of roots. However, simple colonization of the plant does not ensure optimal functioning of the plant-AMF interaction. Functional phenotypes of near-isogenic lines (NILs) of white clover, which vary in AMF colonisation and effectivity (Eason et al. 2001) provide a major resource in identifying genes expressed during symbiotic functioning. Genes expressed during symbioses can be used to identify genotypes with fully functional and effective symbioses. The functioning of selected, cloned promoters and genes will be elucidated using a combination of molecular and physiological approaches.

3.2 Genetic Engineering Approaches in Molecular Breeding

Transgenic crops are now well established outside Europe, occupying some 53 M ha (James 2001). So far biolistics has been the most successful means of transforming grasses. The soluble carbohydrate composition of Italian ryegrass has been altered by transformation with the *Bacillus subtilis* sacB gene (Ye et al. 2001), the resistance of perennial ryegrass to ryegrass mosaic virus (RMV) has been increased by transformation with an RMV coat protein gene (Xu et al. 2001), and a sulphur-rich albumin gene from sunflower has been expressed in leaves of transgenic tall fescue (Wang et al. 2001). Maize genes that modulate related anthocyanin and condensed tannin pathways have been introduced into *Lotus* and alfalfa using transgene technology (Robbins et al. 2003 Gruber et al. 2001). Cattle digested transgenic alfalfa 12-15% more slowly in the first six hours after ingestion, which could help to prevent bloat although protein digestion was normal by 24 hours.

Most perennial forages are outbreeders which makes the practical exploitation of transgenes difficult in countries where related wild species are prevalent. Without strict control of pollination during seed multiplication transgenes may spread and testing seed lots of non-transgenic varieties for the presence of the transgene will increase seed production costs. Control of flowering can lead to improvements in leafiness (and hence quality) but is also highly desirable to prevent transgene spread. This also requires the development of gene-switch technology to allow normal seed production when required. Apomixis combined with male sterility could also be used to enable seed production of transgenic grasses without gene flow to other varieties (Hayward 2001). However such a system would have to be absolutely reliable (van Dijk and van Damme 2000) to avoid changes in the genetic architecture of natural populations.

The pace of future advances will depend to some extent on further developments in transformation technology. *Agrobacterium* transformation of grasses would provide efficient single copy gene transfer with segregation of unlinked T-DNA sequences but this requires development of new vectors and higher efficiency. Development of transformation systems that will handle agronomic traits requires the delivery and integration of large DNA fragments (BAC (bacterial artificial chromosome) clones fine mapped to agronomic traits that would be functionally equivalent to mini-introgression segments). This would be further aided by the availability of heritable, self replicating plant artificial chromosomes based on self-replicating centromeric sequences.

Precise genetic manipulation requires accurate targeting of transgene expression either to different cell types or to specific intercellular compartments (endoplasmic reticulum/vacuole /apoplast/golgi). Some progress has been made using cell specific promoters in grasses and legumes (but many more cell specific promoters are needed) and for compartment targeting using 5' and 3' signal sequences in grasses associated with specific wall degrading enzymes (Morris et al. 2001). Plastid transformation gives higher expression of prokaryotic genes and may facilitate "containment" of transgenes within the transformed plants, but there are marked species dependent differences in the level of paternal inheritance of plastid DNA.

Forage crops may also be transformed with a view to Bio-Farming and they could compete with more "traditional" industrial crops such as maize, wheat or potatoes. The fact that forage is both a direct and major component of the ruminant diet means that forages also offer the potential for direct delivery of therapeutic proteins (antibodies, vaccines, viral inhibitors, etc) to improve animal health. The incorporation of such agents into forages via genetic manipulation is theoretically possible, scientifically attractive but currently untested. There is considerable generic interest in the expression of pharmacologically active compounds in crops for human consumption (pharming) but there has been opposition on both ethical and safety grounds. However, improvement of animal feeds by the incorporation of bioactive molecules may offer opportunities to test the overall validity of the approach whilst avoiding direct human ingestion.

REFERENCES

Abberton MT, Marshall AH, Michaelson-Yeates TPT, Williams TA, Thornley W, Prewer W, White C, Rhodes I (2000) An integrated approach to introgression breeding for stress tolerance in white clover. In: Crop Development for the Cool and Wet Regions of Europe: Achievements and Future Prospects; Proceedings of the Final COST Action 814 Conference, Parente G, Frame J (eds.), pp.311-314. Office for Official Publications of the European Communities.

Abberton MT, Michaelson-Yeates TPT, Marshall AH (2000) Molecular marker analysis in white clover. In: New Approaches and Techniques in Breeding Sustainable Fodder Crops and Amenity Grasses; Proceedings of the 22nd Eucarpia Fodder Crops and Amenity Grasses Section Meeting, Provorov NA, Tichonovich IA, Veronesi F (eds.), pp.192-195. All-Russia Institute for Agricultural Microbiology, St Petersburg.

Abberton MT, Marshall AH (2002) Marker assisted introgression in white clover. In: 1st International Conference on Legume Genomics and Genetics: Translation to Crop Improvement, Minneapolis-St Paul, MN, June 2-6 2002 pp. 109.

Aerts RJ, Barry TN, McNabb WC (1999) Polyphenols and agriculture: beneficial effects of proanthocyanidins in forages. Agr. Ecosyst. Environ. 75: 1-2.

Andersen B, Lindvall E (1997) Use of biomass from reed canarygrass (*Phalaris arundinacea* L.) as raw material for production of paper and pulp and fuel. In: Proceedings of the XVIII International Grassland Congress, Canada. 1. pp. 3. Calgary: B. R. Christie.

Arcioni S, Damiani F, Mariani A, Pupilli F (1997) Somatic hybridisation and embryo rescue for the introduction of wild germplasm. In: Biotechnology and the Improvement of Forage Legumes. Biotechnology in Agriculture Series, No. 17, McKersie BD, Brown DCW (eds.), pp. 61-89. CAB International.

Armstead IP, Turner LB, King IP, Cairns AJ, Humphreys MO (2002) Comparison and integration of genetic maps generated from F2 and BC1-type mapping populations in perennial ryegrass (*Lolium perenne* L.). Plant Breed. 121: 501-507

Brummer EC, Bouton JH, Kochert G (1993) Development of an RFLP map in diploid alfalfa. Theor. Appl. Genet. 83: 329-332.

Bush LP (2001) Perloline, the forgotten plant alkaloid. In: Proceedings of the XIX International Grassland Congress, Gomide JA, Mattos WRS, da Silva SC (Eds.), pp. 461-462. São Paulo, Brazil.

Cortright RD, Davda RR, Dumesic JA (2002) Hydrogen from catalytic reforming of biomass-derived hydrocarbons in liquid water. Nature 418: 964-967.

Chen L Auh C, Chen F, Cheng X, Aljoe H, Dixon R, Wang Z (2002) Lignin deposition and associated changes in anatomy, enzyme activity, gene expression and ruminal degradability in stems of tall fescue at different developmental stages. J. Agri. Food Chem. 50: 5558-5565.

Crawford RJ, Massie MD, Sleper DA, Mayland HF (1998) Use of an experimental high-magnesium tall fescue to reduce grass tetany in cattle. J. Production Agri. 11: 491-496.

Davies DR, Leemans DK, Merry RJ (2002) Ensiling either high or low sugar containing perennial ryegrasses with or without red. In: Multi-Function Grasslands: Quality Forages, Animal Products and Landscapes; Proceedings 19th General Meeting of the European Grassland Federation (EGF), Durand JL, Emile JC, Huyghe C, Lemaire G, (eds.), pp.194-195. La Rochelle, France.

Dewhurst RJ, Scollan ND, Youell SJ, Tweed JKS, Humphreys MO (2001) Influence of species, cutting date and cutting interval on the fatty acid composition of grasses. Grass Forage Sci. 56: 68-74.

Diwan N, Bouton JH, Kochert G, Cregan PB (2000) Mapping of simple sequence repeat (SSR) DNA markers in diploid and tetraploid alfalfa. Theor. Appl. Genet. 101: 165-172.

Dixon R, Achnine L, Kota P, Liu C, Reddy M and Wang L (2002) The phenylpropanoid pathway and plant defence - a genomics perspective. Mol. Plant Pathol. 3: 371-390.

Dolstra O, Boucoiran CFS, Dees D, Denneboom C, Groot PJ, De Vos ALF, Van Loo EN (2003) Genome-wide search of genes relevant to nitrogen-use-efficiency in perennial ryegrass. In: Abstracts of Plant & Animal Genome XI Conference, pp.47. January 11-15, San Diego, CA, USA.

Donnison IS, Cisneros P, Montoya T, Armstead IP, Thomas B, Thomas AM, Jones RN, Morris P (2002) The floral transition in model and forage grasses. Flowering Newsletter 33: 42-48.

Dubach M, Russelle MP (1994) Forage legume roots and nodules and their role in nitrogen transfer. Agron. J. 86: 259-266.

Dumsday JL, Smith KF, Forster JW, Jones ES (2003) SSR-based genetic linkage analysis of resistance to crown rust (*Puccinia coronata* Corda f. sp. *lolii*) in perennial ryegrass (*Lolium perenne* L.). Plant Pathol. (in press).

Eason WR, Webb KJ, Michaelson-Yeates TPT, Abberton MT, Griffith GW, Culshaw CM, Hooker JE, Dhanoa MS (2001) Effect of genotype of *Trifolium repens* on mycorrhizal symbiosis with *Glomus mosseae*. J. Agri. Sci. 137: 27-36.

Echt CC, Kidwell KK, Knapp SJ, Osborn TC, McCoy TJ (1993) Linkage mapping in diploid alfalfa (*Medicago sativa*). Genome 37: 61-71.

Fletcher LR, Harvey IC (1981) An association of a *Lolium* endophyte with ryegrass staggers. New Zealand Vet. J. 29: 185-186.

Fujimori M, Hayashi K, Hirata M, Mizuno K, Fujiwara T, Akiyama F, Mano Y, Komatsu T, Takamizo T (2003) Linkage analysis of crown rust resistance in Italian ryegrass (*Lolium multiflorum* Lam.). In: Abstracts of Plant & Animal Genome XI Conference, pp.46. January 11-15, San Diego, CA, USA.

Gruber MY, Ray H, Blahut-Beatty (2001) Genetic Manipulation of Condensed Tannin Synthesis in Forage Crops. In: Molecular Breeding of Forage Crops; Proceedings 2nd International Symposium, Spangenberg G (ed.), pp.189-201. Kluwer Academic Publishers, Dordrecht.

Gupta PK, Varshney RK (2000) The development and use of microsatellite markers for genetic analysis and plant breeding with emphasis on bread wheat. Euphytica 113: 163-185.

Harrison M, Dewbre G, Liu J (2002) A phosphate transporter from *Medicago truncatula* involved in the acquisition of phosphate released by arbuscular mycorrhizal fungi. Plant Cell 14: 2413–2429.

Hayward MD (2001). The future of molecular breeding of forage crops. In: Molecular Breeding of Forage Crops; Proceedings 2nd International Symposium, Spangenberg G (ed.), pp.325-337. Kluwer Academic Publishers, Dordrecht.

Henry RJ (2001) Plant Genotyping - The DNA Fingerprinting of Plants, pp.325. CABI Publishing, Wallingford.

Hoveland, CS, Schmidt, SP, King, CC, Odom, JW, Clark, EM, Mcguire, JA, Smith, LA, Grimes, HW, Holliman, JL (1983). Steer performance and association of *Acremonium coenophialum* fungal endophyte on tall fescue pasture. Agron. J. 75: 821-824.

Huizing HJ, Van Der Molen W, Kloek W, Den Nijs APM (1991). Detection of lolines in endophyte-containing meadow fescue in the Netherlands and the effect of elevated temperature on induction of lolines in endophyte-infected perennial ryegrass. Grass Forage Sci. 46: 441-445.

Humphreys MO, Turner LB (2001) Molecular markers for improving nutritive value in perennial ryegrass. In: Abstracts 16th Eucarpia Congress, Plant Breeding: Sustaining the Future, P1.17. September 10-14, 2001, Edinburgh.

Hussain SW, Williams WM, Mercer CF, White DWR (1997) Transfer of clover cyst nematode resistance from *Trifolium nigrescens* Viv. to *T. repens* L .by interspecific hybridisation . Theor. Appl. Genet. 95: 1274-1281.

James C (2001) Global review of commercialized transgenic crops 2001. International Service for the Acquisition of Agribiotech Applications in Crop Biotechnology. Brief No. 24.

Joyce TA , Abberton MT, Michaelson-Yeates TPT, Forster WJ (1999) Relationships between genetic distance measured by RAPD-PCR and heterosis in inbred lines of white clover (*Trifolium repens*). Euphytica 107: 159-165.

Jones ES, Mahoney NL, Hayward MD, Armstead IP, Jones JG, Humphreys MO, King IP, Kishida T, Yamada T, Balfourier F, Charmet G, Forster JW (2002) An enhanced molecular marker based genetic map of perennial ryegrass (*Lolium perenne*) reveals comparative relationships with other *Poaceae* genomes. Genome 45: 282-295.

Kearsey MJ, Farquhar AGL (1998) QTL analysis in plants; where are we now? Heredity 80: 137-142.

Kiss GB, Csanadi G, Kalman K, Kalo P, Okresz, L (1993) Construction of a basic genetic map for alfalfa using RFLP, RAPD and isozyme and morphological markers. Mol. Gen. Genet. 238: 129-137.

Kongkiatnam P, Waterway M, Fortin MG, Coulman BE (1995) Genetic variation within and between two cultivars of red clover (*Trifolium pratense* L.) Comparison of morphological, isozyme and RAPD markers. Euphytica 84: 237-246.

Lidgett A, Jennings K, Johnson X, Guthridge K, Jones E, Spangenberg G (2002) Isolation and characterisation of a fructosyltransferase gene from perennial ryegrass (*Lolium perenne*). J. Plant Physiol. 159: 1037-1043.

Lin KW, Patterson JA, Ladisch MR (1985) Anaerobic fermentation: microbes from ruminants. Enzyme Microbial Technol. 7: 98-107.

Marshall AH, Holdbrook-Smith K, Michaelson-Yeates TPT, Abberton MT, Rhodes I (1998) Growth and reproductive characteristics in backcross hybrids derived from *Trifolium repens* L. X *T. nigrescens* Viv. interspecific crosses. Euphytica 104: 61-66.

Marum P, Hovin AW, Marten GC (1979) Inheritance of three groups of indole alkaloids in reed canary grass. Crop Sci. 19: 539-544.

McInnes R, Lidgett A, Lynch D, Huxley H, Jones E, Mahoney N, Spangenberg G (2002) Isolation and characterisation of a cinnamoyl-CoA reductase gene from perennial ryegrass (*Lolium perenne*). J. Plant Physiol. 159: 415-422.

Miller LA, Moorby JM, Davies DR, Humphreys MO, Scollan ND, Macrae JC, Theodorou, MK (2001) Increased concentration of water-soluble carbohydrate in perennial ryegrass (*Lolium perenne* L.): milk production from late-lactation dairy cows. Grass Forage Sci. 56: 383-394.

Morris P, Bettany AJE, Dalton SJ, Buanafina MM de O, Robbins MP (2001) Strategies for manipulating the chemical composition of forage grasses to improve their nutritional value. Abstracts 9th International Cell Wall Meeting, 2-7 September 2001. Toulouse, France.

Moseley, G, Baker, DH (1991). The efficacy of a high magnesium grass cultivar in controlling hypomagnesaemia in grazing animals. Grass Forage Sci. 46: 375-380.

Muylle H, Van Bockstaele E, Roldán-Ruiz I (2003) Identification of four genomic regions involved in crown rust resistance in a *L. perenne* population. In: Abstracts of Molecular Breeding of Forage and Turf, Third International Symposium, p.13. May 18-22, Dallas, Texas and Ardmore, Oklahoma, USA.

Nelke M, Nowak J, Wright JM, McLean NL, Laberge S, Castonguay Y, Vezina LP (1999) Enhanced expression of a cold-induced gene coding for a glycine-rich protein in regenerative somaclonal variants of red clover (*Trifolium pratense* L.). Euphytica 105: 211-217.

Osborn TC, Brouwer D, McCoy TJ (1997) Molecular marker analysis in Alfalfa. In: Biotechnology and the Improvement of Forage Legumes. McKersie BD, Brown DCW (eds.), pp. 91-109. CAB International.

Page D, Delclos D, Aubert G, Bonavent JF, Mousset-Declas C (1997) Sclerotinia rot resistance in red clover: Identification of RAPD markers using bulked segregant analysis. Plant Breed. 116: 73-78.

Pederson GA, Brink GE, Fairbrother TE (2002) Nutrient uptake in plant parts of sixteen forages fertilised with poultry litter: nitrogen, phosphorus, potassium, copper and zinc. J. Agron. 94: 895-904.

Robbins MP, Bavage AD, Morris P (1999) Designer tannins: using genetic engineering to modify levels and structures of condensed tannins in *Lotus corniculatus*. In: Plant Polyphenols, Chemistry, Biology, Pharmacology, Ecology, Gross GG, Hemingway RW, Yoshida T, (eds.), pp. 301-314. Kluwer Academic / Plenum Publishers, New York.

Robbins MP, Allison GG, Bettany AJE, Dalton SJ, Davies TE, Hauck B, Hughes JW, Timms EJ, Morris P (2002) Biochemical and molecular basis of plant composition determining the degradability of forage for ruminant nutrition. In: Multifunctional Grassland: Quality Forages, Animal Products and Landscapes, Proceedings 19th General Meeting of the European Grassland Federation, Durand JL, Emile JC, Huyghe C, Lemaire G (eds.). pp. 37-43. La Rochelle, France.

Robbins MP, Paolocci F, Hughes J-W, Turchetti V, Allison G, Arcioni S, Morris P, Damiani F(2003) Sn, a maize bHLH gene, modulates anthocyanin and condensed tannin pathways in *Lotus corniculatus*. Online at http://jxb.oupjournals.org/cgi/reprint/54/381/239.pdf. J. Exper. Bot. 54 : 239-248.

Roderick HW, Thorogood D, Adomako B (2000) Temperature-dependent resistance to crown rust infection in perennial ryegrass, *Lolium perenne*. Plant Breed. 119: 93-95.

Roderick HW, Humphreys MO, Turner LB, Armstea IP, Thorogood D (2003) Isolate specific trait loci for resistance to crown rust in perennial ryegrass. In: Proceedings 24th EUCARPIA Fodder Crops and Amenity Grasses Section Meeting. September 22-26, 2002, Braunschweig, Germany (Posselt, U. K., Greef, J. M., eds.) Vortage fur Pflanzenzuchtung 59: 244-247

Sanchez WK, Beede DK, Cornell JA (1994) Interactions of sodium, potassium, and chloride on lactation, acid-base status, and mineral concentrations. J. Dairy Sci. 77: 1661-1675.

Skinner DZ, Loughin T, Obert DE (2000) Segregation and conditional probability association of molecular markers with traits in autotetraploid alfalfa. Mol. Breed. 6: 295-306.

Taylor NL, Quesenberry KH (1996) Current Plant Science and Biotechnology in Agriculture,Vol. 28, Red Clover Science. Kluwer Academic Publishers.

Thelier-Huche L, Simon JL, Corre L, le Salette J (1994) Valorisation, sur prarie et mais, de la fertilisation organique et minerale. Etude sur le long terme. Fourrages 138: 145-155.

Thorogood D, Laroche S (2001) QTL analysis of chlorophyll retention during leaf senescence in perennial ryegrass. In: Abstracts of the XVIth EUCARPIA Congress, Plant Breeding: Sustaining the Future, P1.27. September 10-14, 2001. Edinburgh, Scotland.

Turner LB, Armstead IP, Cairns AJ, Humphreys MO, Pollock CJ (2001) A physiological and biochemical analysis of the genetic basis of soluble carbohydrate storage in temperate grasses. J. Exper. Bot. (Supplement), Abstracts Society for Experimental Biology Annual Meeting, P7.19. April 2-6, 2001. Canterbury, Kent.

van Dijk P, van Damme J (2000) Apomixis technology and the paradox of sex. Trends Plant Sci. 5: 81-84.

van Loo, Vellinga (1994) Prospects of grass breeding for improved forage quality to improve dairy farm income and reduce mineral surpluses. Proceedings of the 15th General Meeting of the European Grassland Federation, Workshop Proceedings, p.124-127.

Vink ETH, Rábago KR, Glassner D, Gruber PR (2003) Applications of life cycle assessment to NatureworksTM polylactide (PLA) production. Polymer Degradation and Stability (in press)

Wang ZY, Ye XD, Nagel J, Potrykus I, Spangenberg G (2001) Expression of a sulphur-rich sunflower albumin gene in transgenic tall fescue (*Festuca arundinacea* Schreb.) plants. Plant Cell Rep. 20: 213-219.

Webb KJ, Abberton MT, Young SR (2003) Molecular genetics of white clover Focus on Biotechnology. In: Applied Genetics of Leguminosae Biotechnology, Jaiwal PK, Singh RP (eds.), pp. 263-279. Kluwer Academic Publishers, Dordrecht.

Wilkins PW (1997) Preliminary evaluation of some forage grasses as biomass crops for summer harvest in the UK. Aspects of Applied Biology, Biomass and Energy Crops, AAB Conference, pp.247-250. April 7-8, 1997. Silsoe College.

Wilkins PW, Armstead I, Deakin C, Dhanoa MS, Davies RW (2003) Marker-assisted selection for herbage yield and quality in an advanced breeding population of perennial ryegrass. In: Grass for Food & Grass for Leisure; Proceedings of the Eucarpia Fodder Crops and Amenity Grass Section Meeting for 2002. Braunschweig, Germany (Posselt, U. K., Greef, J. M., eds.) Vortage fur Pflanzenzuchtung 59: 74-79

Woodfield DR, Brummer EC (2001) Integrating molecular techniques to maximise the genetic potential of forage legumes. In: Molecular Breeding of Forage Crops, Spangenberg G (ed.), pp.51-65. Kluwer Academic Publishers, Dordrecht.

Xie C, Mosjidis JA (2001) Inheritance and linkage study of isozyme loci and morphological traits in red clover. Euphytica 119: 253-257.

Xu WW, Sleper DA, Chao S (1995) Genome mapping of polyploid tall fescue (*Festuca arundinacea* Schreb.) with RFLP markers. Theor. Appl. Genet. 91: 947-955.

Xu J, Schubert J, Altpeter F (2001) Dissection of RNA-mediated ryegrass mosaic virus resistance in fertile transgenic perennial ryegrass (*Lolium perenne* L.). Plant J. 26: 265-274.

Ye XD, Wu XL, Zhao H, Frehner M, Nösberger J, Potrykus I, Spangenberg G (2001) Altered fructan accumulation in transgenic Lolium multiflorum plants expressing a *Bacillus subtilis* sacB gene. Plant Cell Rep. 20: 205-212.

Improving Forage Quality of Tall Fescue (*Festuca arundinacea*) by Genetic Manipulation of Lignin Biosynthesis

Lei Chen, Chung-Kyoon Auh[1], Paul Dowling[2], Jeremey Bell and Zeng-Yu Wang

Forage Improvement Division, The Samuel Roberts Noble Foundation, 2510 Sam Noble Parkway, Ardmore, Oklahoma 73401, USA. [1]Current address: Department of Biological Sciences, SungKyunKwan University, Suwon 440-746, South Korea. [2]Current address: HiberGen Ltd., IDA Business Park, Bray, Co Wicklow, Ireland. (Email: zywang@noble.org).

Keywords: forage grass, digestibility, lignin biosynthesis, transgenic plant, tall fescue

Abstract:

In vitro dry matter digestibility is one of the most important characteristics of forages. Lignification of cell walls during plant development has been identified as the major factor limiting forage digestibility and concomitantly animal productivity. cDNA sequences encoding key lignin biosynthetic enzymes, cinnamyl alcohol dehydrogenase (CAD) and caffeic acid *O*-methyltransferase (COMT), were cloned from tall fescue (*Festuca arundinacea* Schreb.). Transgenic tall fescue plants carrying either sense or antisense CAD and COMT gene constructs were obtained by microprojectile bombardment of single genotype-derived embryogenic suspension cells. Severely reduced mRNA levels and significantly decreased enzymatic activities were found in four transgenic lines. These transgenic tall fescue plants had reduced lignin content, altered lignin composition and increased in vitro dry matter digestibility. No significant changes in cellulose, hemicellulose, neutral sugar composition, *p*-coumaric acid and ferulic acid levels were observed in the transgenic plants. This is the first time that a widely-grown forage grass species with modified lignin and improved digestibility was obtained by genetic engineering. Progresses in lignin modification of other plant species were also discussed.

A. Hopkins, Z.Y. Wang, R. Mian, M. Sledge and R.E. Barker (eds.), Molecular Breeding of Forage and Turf, 181-188.

1. INTRODUCTION

Lignin is an important chemical component of plant cell walls; however, it is essentially undigestible and negatively impacts livestock feed and paper pulp processing. The amount of digestible energy available to ruminant livestock from fermentation of cell wall polysaccharides is restricted by lignin, rendering much of the cellulose and hemicellulose inaccessible to rumen microorganisms. Lignification of cell walls during plant development has been identified as the major factor limiting digestibility of forage crops (Buxton and Russell 1988; Vogel and Jung 2001). Feeding and grazing studies have shown that small changes in forage digestibility can have a significant impact on animal performance (Casler and Vogel 1999).

Lignin in forage grasses comprises of guaiacyl (G) units derived from coniferyl alcohol, syringyl (S) units derived from sinapyl alcohol, and *p*-hydroxyphenyl (H) units derived from *p*-coumaryl alcohol. In addition to lignin content (or concentration), the composition of lignin is an important factor that influence cell wall degradability of forages (Buxton and Russell 1988; Vogel and Jung 2001). Natural or chemically induced low-lignin mutants have been isolated in maize, sorghum and pearl millet. Leaf and stem tissues of these brown-midrib mutants, which are characterized by decreased lignin contents and by structural modifications, are more digestible than those of the normal genotypes (Cherney et al. 1991).

Because of the large economic benefits that might be achieved, considerable research efforts have been made toward reducing lignin content (concentration) or modifying lignin composition by genetic engineering. However, reports on transgenic modification of lignin biosynthesis have so far been mainly on dicot species, such as tobacco, Arabidopsis, alfalfa and poplar (reviewed by Dixon et al. 2001). In monocot species, there have been only two reports on transgenic manipulation of lignin biosynthesis, one was on down-regulation of caffeic acid *O*-methyltransferase (COMT) in maize (Piquemal et al. 2002), and the other was on down-regulation of cinnamyl alcohol dehydrogenase (CAD) in tall fescue (Chen et al. 2003).

Tall fescue is a leafy, course-textured, vigorous perennial bunchgrass that reproduces vegetatively through tillering and by seed. Beside its excellent adaptation to diverse climate and edaphic conditions, its suitability and flexibility in beef production has contributed to its widespread use as a cool-season forage grass. Tall fescue forms the forage basis for beef cow-calf production in the east-central and southeast USA, supporting over 8.5 million beef cows and is used for sheep and horse production. Here we summarize our work on genetic manipulation of lignin biosynthesis in tall fescue.

2. LIGNIN DEPOSITION IN TALL FESCUE

We analyzed lignin deposition in stem tissues of tall fescue at three elongation (E1, E2, E3) and three reproductive (R1, R2, R3) stages (Chen et al. 2002). Anatomical analysis showed the deposition of G and S lignin during plant development, with S lignin increased with progressive maturity of stems from E1 to R3, while the relative content of G lignin decreased during the same period. One of the most obvious changes during stem lignification is the formation of a sclerenchyma ring. The ring become visible in the E2 stage and continued to increase in size and proportion of the cross-sectioned area through the later stages. Histochemical staining of different internodes at the R3 stage showed a decrease in sclerenchyma area and wall thickness from I1 to I4 internodes. Lignin content in stem tissues increased moderately during the elongation stage (E1 to E3), but the major increase in lignin content occurred when plants changed from elongation stage to reproductive stage. Lignin content at the reproductive stage was always much higher than that at the elongation stage, with 10 times more lignin deposited in the cell walls at the R3 stage than at the E1 stage. Cell wall digestibility of stems and internodes decreased with increasing maturity. S lignin content and S/G ratio were negatively correlated with digestibility. Lignin composition analyzed by gas chromatography/mass spectrometry revealed that S lignin content and S/G ratio increased with stem development, but contents of H and G lignins decreased during the same period. S lignin content and S/G ratio also increased from the younger upper internode down to the older basal internode of the stem, but G and H lignin decreased in parallel. Relative *O*-methyltransferase activities increased during stem development, and in parallel with the lignification process of stem. The pattern of enzyme activity during development varied with the choice of substrate, with highest activities seen when substrates were caffeoyl aldehyde and 5-OH ferulic acid, and lowest activities when caffeic acid and 5-OH coniferyl alcohol were used as substrates. The expression of caffeic acid *O*-methyltransferase and cinnamyl alcohol dehydrogenase genes increased during the stem elongation stage and remained at high levels during the reproductive stages. The changes at anatomical, metabolic and molecular levels during plant development were closely associated with lignification and digestibility (Chen et al. 2002).

3. GENETIC MANIPULATION OF CAD

Lignin biosynthesis comprised of a set of coordinated and regulated metabolic events, and many enzymes are involved in the pathway. Cinnamyl alcohol dehydrogenase (CAD) and caffeic acid *O*-methyltransferase (COMT) have been shown to play important roles in lignin biosynthesis (Halpin et al.

1998; Vailhe et al. 1998; Guo et al. 2001b). CAD catalyzes the last step in the biosynthesis of lignin precursors, which is the reduction of cinnamaldehydes to cinnamyl alcohols (Baucher et al. 1998). Transgenic manipulation of CAD by either antisense or sense strategy has been reported in a few dicot species. Reduction of CAD activity led to changes in S/G ratio or other phenolic compounds in tobacco (Baucher et al. 1996a; Bernard-Vailhe et al. 1998; Yahiaoui et al. 1998; Baucher et al. 1999), alfalfa (Baucher et al. 1999) and poplar (Baucher et al. 1996b; Lapierre et al. 1999; Pilate et al. 2002). Only a slight reduction in lignin content was observed in transgenic poplar (Lapierre et al. 1999; Pilate et al. 2002). Some of the changes resulted in increased lignin extractability or degradability (Baucher et al. 1996a; Bernard-Vailhe et al. 1996a; Yahiaoui et al. 1998; Baucher et al. 1999). Transgenic CAD plants with altered lignin showed normal development and architecture (Bernard-Vailhe et al. 1998; Baucher et al. 1999; Chabannes et al. 2001; Pilate et al. 2002).

We cloned four *CAD* cDNA sequences from tall fescue. These cDNA sequences were highly homologous with each other and showed very high similarity to that of other related monocot species such as perennial ryegrass. Recombinant tall fescue CAD expressed in *E. coli* exhibited highest V_{max}/K_m values when coniferaldehyde and sinapaldehyde were used as substrates. Transgenic tall fescue plants carrying either sense or antisense *CAD* gene constructs were obtained by microprojectile bombardment of single genotype-derived embryogenic suspension cells. Severely reduced levels of mRNA transcripts and significantly reduced CAD enzymatic activities were found in two transgenic plants carrying sense and antisense *CAD* transgenes, respectively. The transgenic lines showed significantly decreased lignin content and apparently altered lignin composition. No significant changes in cellulose, hemicellulose, neutral sugar composition, *p*-coumaric acid and ferulic acid levels were observed in the transgenic plants. Increases of *in vitro* dry matter digestibility of 7.2 to 9.5% were achieved in the CAD down-regulated lines (Chen et al. 2003). Consistent with some of the reports in tobacco and alfalfa (Halpin et al. 1994; Yahiaoui et al. 1998; Baucher et al. 1999), our study showed that the CAD down-regulated tall fescue plants had reduced S/G ratio. However, in contrast to the unchanged lignin quantity reported in some dicots species (Halpin et al. 1994; Higuchi et al. 1994; Hibino et al. 1995; Baucher et al. 1996a; Yahiaoui et al. 1998; Baucher et al. 1999), the transgenic tall fescue plants showed significantly reduced Klason lignin content as well as reduced amounts of G and S lignin monomers. This result is consistent with the enzyme activity data, in which the cloned tall fescue CAD was shown to be responsible for catalyzing the synthesis of both G and S lignin. Due to the lack of reports on transgenic manipulation of *CAD* in monocot species, the only comparable case we can find in a monocot species is the natural *bm1* mutants in maize. A decrease in CAD activity (by 60 – 70%), a reduction of Klason lignin content (by 20%), a reduction in the yield of G and S lignin monomers after thioacidolysis and an increase in digestibility were observed in maize *bm1* mutant plants (Halpin et al. 1998). Mapping studies strongly suggested that maize *bm1* directly affects expression of the *CAD* gene (Halpin et al. 1998). Our work on down-regulation of CAD in tall fescue (Chen et al. 2003) is the first report on genetic manipulation of CAD in monocot species.

4. GENETIC MANIPULATION OF COMT

COMT is a multi-specific enzyme that not only methylates caffeic acid to ferulic acid and 5-hydroxyferulic acid to sinapic acid, but also involved in the 3-*O*-methylation of monolignol precursors at the aldehyde or alcohol levels (Dixon et al. 2001; Humphreys and Chapple 2002). There have been several reports on the effects of down-regulation of COMT activity on lignin content and composition in transgenic dicot plants (Ni et al. 1994; Atanassova et al. 1995; Van Doorsselaere et al. 1995; Tsai et al. 1998; Jouanin et al. 2000; Zhong et al. 2000; Guo et al. 2001a; Guo et al. 2001b). Down-regulation of COMT in tobacco and poplar resulted in reduced lignin content (Ni et al. 1994; Jouanin et al. 2000) or altered lignin composition (Atanassova et al. 1995; Van Doorsselaere et al. 1995). Increased degradability of tobacco stems were observed in some further studies (Bernard-Vailhe et al. 1996b; Sewalt et al. 1997). Strong reduction of COMT in the tropical pasture legume *Stylosanthes humilis* resulted in no apparent reduction in lignin levels but in a significant reduction of S lignin (Rae et al. 2001). In vitro digestibility of stem material was increased by up to 10% in the transgenic plants. Similar experiments conducted in transgenic alfalfa led to decreases of as much as 30% in Klason lignin levels and 4% increase in dry matter digestibility (Guo et al. 2001a).

In monocot species, it has been shown that maize *bm3* mutant is severely deficient in OMT activity, with only 10% of the activity found in normal plants (Grand et al. 1985). By sequencing the COMT clones obtained from the *bm3-1 and bm3-2* maize, the *bm3-1* allele was found to arise from an insertional event producing a *COMT* mRNA altered in both size and amount, whereas the *bm3-2* was resulted from a deletion of part of the *COMT* gene (Vignols et al. 1995). Recently, transgenic maize plants were generated with a construct harboring a maize *COMT* cDNA in the antisense orientation. One transgenic line displayed a significant reduction in COMT activity (15%–30% residual activity) and barely detectable amounts of COMT protein. Biochemical analysis of this transgenic plant showed a strong decrease in Klason lignin content, a decrease in syringyl units, a lower *p*-coumaric acid content, and the occurrence of unusual 5-OH guaiacyl units. These results are reminiscent of some characteristics already observed for the maize *bm3* mutant (Piquemal et al. 2002).

We isolated four *COMT* cDNA clones from tall fescue by screening a cDNA library. To confirm that the cloned sequences encode corresponding enzymes and to investigate their kinetic properties and substrate specificity, the coding region of one of the *COMT* cDNAs (FaCOMT1b) was cloned into pET29a vector for expression in *E. coli*. The enzymatic properties of purified COMT recombinant protein were determined using different substrates. The preferred substrates for tall fescue recombinant COMT are 5-hydroxyferulic acid and caffeoyl aldehyde, consistent with enzymatic activity analysis using crude protein extracts from tall fescue stems. The coding region of FaCOMT1b was also used to construct sense and antisense chimeric transgenes, which were introduced back to tall fescue by biolistic transformation. After biochemical and molecular analyses, two co-suppressed sense transgenic tall fescue lines were identified as having down-regulated lignin biosynthesis. These transgenic plants showed substantially reduced levels of transcripts, significantly reduced enzymatic activities, significantly decreased lignin content, apparently altered lignin composition, and significantly increased dry matter digestibility. Similar to the maize *bm3* mutant, the transgenic tall fescue with down-regulated COMT also showed reduced S lignin level and reduced S/G ratio.

5. SUMMARY

Transgenic expression of lignin genes has provided a powerful tool to elucidate lignin biosynthetic pathways in dicot species. This is not yet the case in monocot species. The lack of reports on successful modification of lignin in forage grasses is mainly due to the difficulties in obtaining transgenics and identifying transgenic plants with changes in lignin. Our work on lignin modification in tall fescue is the first case that a widely-grown forage grass species with altered lignin and improved digestibility that might benefit livestock industry has been obtained by genetic engineering.

Genetic manipulation for increased in vitro dry matter digestibility of forage grasses can lead to rapid financial benefits to the agricultural community and society (Casler and Vogel 1999). Conventional breeding by phenotypic recurrent selection has resulted in the release of grass cultivars with improved dry matter digestibility. However, continued selection for increased dry matter digestibility might affect plant fitness. Genetic engineering allows us to target specific enzymes in the lignin biosynthetic pathway, thus offering an alternative and effective approach for improving forage digestibility of grasses.

REFERENCES

Atanassova R, Favet N, Martz F, Chabbert B, Tollier MT, Monties B, Fritig B, Legrand M (1995) Altered lignin composition in transgenic tobacco expressing *O*-methyltransferase sequences in sense and antisense orientation. Plant J. 8: 465-477.

Baucher M, Bernard Vailhe MA, Chabbert B, Besle JM, Opsomer C, Van Montagu M, Botterman J (1999) Down-regulation of cinnamyl alcohol dehydrogenase in transgenic alfalfa (*Medicago sativa* L.) and the effect on lignin composition and digestibility. Plant Mol. Biol. 39: 437-447.

Baucher M, Chabbert B, Pilate G, Tollier MT, Petit Conil M, Cornu D, Monties B, Inze D, Jouanin L, Boerjan W, Van Doorsselaere J, Van Montagu M (1996) Red xylem and higher lignin extractability by down-regulating a cinnamyl alcohol dehydrogenase in poplar. Plant Physiol. 112: 1479-1490.

Bernard-Vailhe MA, Cornu A, Robert D, Maillot MP, Besle JM (1996a) Cell wall degradability of transgenic tobacco stems in relation to their chemical extraction and lignin quality. J. Agric. Food Chem. 44: 1164-1169.

Bernard-Vailhe MA, Besle JM, Maillot MP, Cornu A, Halpin C, Knight M (1998) Effect of down-regulation of cinnamyl alcohol dehydrogenase on cell wall composition and on degradability of tobacco stems. J. Sci. Food Agric. 76: 505-514.

Bernard-Vailhe MA, Migne C, Cornu A, Maillot MP, Grenet E, Besle JM, Atanassova R, Martz F, Legrand M (1996b) Effect of modification of the *O*-methyltransferase activity on cell wall composition, ultrastructure and degradability of transgenic tobacco. J. Sci. Food Agric. 72: 385-391.

Buxton DR, Russell JR (1988) Lignin constituents and cell-wall digestibility of grass and legume stems. Crop Sci. 28: 553-558.

Casler MD, Vogel KP (1999) Accomplishments and impact from breeding for increased forage nutritional value. Crop Sci. 39: 12-20.

Chabannes M, Barakate A, Lapierre C, Marita JM, Ralph J, Pean M, Danoun S, Halpin C, Grima Pettenati J, Boudet AM (2001) Strong decrease in lignin content without significant alteration of plant development is induced by simultaneous down-regulation of CCR and CAD in tobacco plants. Plant J. 28: 257-270.

Chen L, Auh C, Chen F, Cheng XF, Aljoe H, Dixon RA, Wang ZY (2002) Lignin deposition and associated changes in anatomy, enzyme activity, gene expression and ruminal degradability in stems of tall fescue at different developmental stages. J. Agric. Food Chem. 50: 5558-5565.

Chen L, Auh C, Dowling P, Bell J, Chen F, Hopkins A, Dixon RA, Wang ZY (2003) Improved forage digestibility of tall fescue (*Festuca arundinacea*) by transgenic down-regulation of cinnamyl alcohol dehydrogenase. Plant Biotechnol. J. (in press).

Dixon RA, Chen F, Guo DJ, Parvathi K (2001) The biosynthesis of monolignols: a "metabolic grid" or independent pathways? Phytochemistry 57: 1069-1084.

Grand C, Parmentier P, Boudet A, Boudet AM (1985) Comparison of lignins and of enzymes involved in lignification in normal and brown midrib (*bm3*) mutant corn seedlings. Physiologie Vegetale 23: 905-911.

Guo DJ, Chen F, Wheeler J, Winder J, Selman S, Peterson M, Dixon RA (2001a) Improvement of in-rumen digestibility of alfalfa forage by genetic manipulation of lignin O-methyltransferases. Transgenic Res. 10: 457-464.

Guo DJ, Chen F, Inoue K, Blount JW, Dixon RA (2001b) Downregulation of caffeic acid *O*-methyltransferase and caffeoyl CoA *O*-methyltransferase in transgenic alfalfa: Impacts on lignin structure and implications for biosynthesis of G and S lignin. Plant Cell 13: 73-88.

Halpin C, Holt K, Chojecki J, Oliver D, Chabbert B, Monties B, Edwards K, Barakate A, Foxon GA (1998) Brown-midrib maize (*bm1*) - a mutation affecting the cinnamyl alcohol dehydrogenase gene. Plant J. 14: 545-553.

Halpin C, Knight ME, Foxon GA, Campbell MM, Boudet AM, Boon JJ, Chabbert B, Tollier MT, Schuch W (1994) Manipulation of lignin quality by downregulation of cinnamyl alcohol dehydrogenase. Plant J. 6: 339-350.

Hibino T, Takabe K, Kawazu T, Shibata D, Higuchi T (1995) Increase of cinnamaldehyde groups in lignin of transgenic tobacco plants carrying an antisense gene for cinnamyl alcohol dehydrogenase. Biosci. Biotechnol. Biochem. 59: 929-931.

Higuchi T, Ito T, Umezawa T, Hibino T, Shibata D (1994) Red-brown color of lignified tissues of transgenic plants with antisense CAD gene: wine-red lignin from coniferyl aldehyde. J. Biotechnol. 37: 151-158.

Jouanin L, Goujon T, deNadai V, Martin MT, Mila I, Vallet C, Pollet B, Chabbert B, PetitConil M, Lapierre C (2000) Lignification in transgenic poplars with extremely reduced caffeic acid *O*-methyltransferase activity. Plant Physiol. 123: 1363-1373.

Lapierre C, Pollet B, Petit Conil M, Toval G, Romero J, Pilate G, Leple JC, Boerjan W, Ferret V, Nadai Vd, Jouanin L, de Nadai V (1999) Structural alterations of lignins in transgenic poplars with depressed cinnamyl alcohol dehydrogenase or caffeic acid *O*-methyltransferase activity have an opposite impact on the efficiency of industrial kraft pulping. Plant Physiol. 119: 153-163.

Ni W, Paiva NL, Dixon RA (1994) Reduced lignin in transgenic plants containing an engineered caffeic acid O-methyltransferase antisense gene. Transgenic Res. 3: 120-126.

Pilate G, Guiney E, Holt K, PetitConil M, Lapierre C, Leple JC, Pollet B, Mila I, Webster EA, Marstorp HG, Hopkins DW, Jouanin L, Boerjan W, Schuch W, Cornu D, Halpin C (2002) Field and pulping performances of transgenic trees with altered lignification. Nat. Biotechnol. 20: 607-612.

Piquemal J, Chamayou S, Nadaud I, Beckert M, Barriere Y, Mila I, Lapierre C, Rigau J, Puigdomenech P, Jauneau A, Digonnet C, Boudet AM, Goffner D, Pichon M (2002) Down-regulation of caffeic acid *O*-methyltransf erase in maize revisited using a transgenic approach. Plant Physiol. 130: 1675-1685.

Rae AL, Manners JM, Jones RJ, McIntyre CL, Lu DY (2001) Antisense suppression of the lignin biosynthetic enzyme, caffeate O-methyltransferase, improves in vitro digestibility of the tropical pasture legume, Stylosanthes humilis. Aust. J. Plant Physiol. 28: 289-297.

Sewalt VJH, Ni WT, Jung HG, Dixon RA (1997) Lignin impact on fiber degradation: Increased enzymatic digestibility of genetically engineered tobacco (*Nicotiana tabacum*) stems reduced in lignin content. J. Agric. Food Chem. 45: 1977-1983.

Tsai CJ, Popko JL, Mielke MR, Hu WJ, Podila GK, Chiang VL (1998) Suppression of *O*-methyltransferase gene by homologous sense transgene in quaking aspen causes red-brown wood phenotypes. Plant Physiol. 117: 101-112.

Van Doorsselaere J, Baucher M, Chognot E, Chabbert B, Tollier MT, Petit Conil M, Leple JC, Pilate G, Cornu D, Monties B, Van Montagu M, Inze D, Boerjan W, Jouanin L (1995) A novel lignin in poplar trees with a reduced caffeic acid/5-hydroxyferulic acid *O*-methyltransferase activity. Plant J. 8:855-864.

Vignols F, Rigau J, Torres MA, Capellades M, Puigdomenech P (1995) The brown midrib3 (*bm3*) mutation in maize occurs in the gene encoding caffeic acid *O*-methyltransferase. Plant Cell 7: 407-416.

Vogel KP, Jung HJG (2001) Genetic modification of herbaceous plants for feed and fuel. Crit. Rev. Plant Sci. 20: 15-49.

Yahiaoui N, Marque C, Myton KE, Negrel J, Boudet AM (1998) Impact of different levels of cinnamyl alcohol dehydrogenase down-regulation on lignins of transgenic tobacco plants. Planta 204: 8-15.

Zhong R, Morrison WH, III, Himmelsbach DS, Poole FL, II, Ye Z, Zhong RQ, Ye ZH (2000) Essential role of caffeoyl coenzyme A *O*-methyltransferase in lignin biosynthesis in woody poplar plants. Plant Physiol. 124: 563-577.

Cloning of Red Clover and Alfalfa Polyphenol Oxidase Genes and Expression of Active Enzymes in Transgenic Alfalfa

Michael Sullivan[1], Sharon Thoma[2], Deborah Samac[3] and Ronald Hatfield[1]
[1]US Dairy Forage Research Center, USDA-ARS, 1925 Linden Drive West, Madison, WI, 53705, USA. [2]Dept. of Natural Science, Edgewood College, 1000 Edgewood College Dr., Madison, WI, 53711, USA. [3]USDA-ARS and Dept. of Plant Pathology, University of Minnesota, 1991 Upper Buford Circle, St. Paul, MN, 55108, USA. (Email: mlsulliv@facstaff.wisc.edu).

Key words: post-harvest proteolysis, ensiling, browning, forage legume

Abstract:

Red clover contains high levels of polyphenol oxidase (PPO) activity and *o*-diphenol substrates. This results in a characteristic post-harvest browning reaction associated with decreased protein degradation during ensiling. To define PPO's role in inhibition of post-harvest proteolysis, we are taking both biochemical and molecular approaches. We have cloned three unique PPO cDNAs from red clover leaves, RC PPO1-3. The cDNAs encode proteins that are predicted to be targeted to the chloroplast thylakoid lumen. RNA blotting and immunoblotting experiments indicate RC PPO1 is expressed predominantly in young leaf tissue, RC PPO2 is expressed most highly in flowers and petioles, and RC PPO3 is expressed in both leaves and flowers. We expressed the red clover cDNAs in alfalfa, which lacks both significant endogenous PPO activity and *o*-diphenol substrates in its leaves, to further characterize the individual proteins encoded by RC PPO1-3. The expressed proteins are active in alfalfa extracts and the individual enzymes show differences in substrate specificity, suggesting different functional roles. Additionally, we have cloned a PPO gene from alfalfa. Preliminary studies indicate alfalfa PPO expression is limited to flowers and seedpods. Expression of the cloned alfalfa PPO gene from a strong constitutive promoter in transgenic alfalfa results in low but measurable enzyme activity, suggesting a low specific activity compared to the red clover enzymes.

A. Hopkins, Z.Y. Wang, R. Mian, M. Sledge and R.E. Barker (eds.), Molecular Breeding of Forage and Turf, 189-195.

1. INTRODUCTION

Ensiling crops is a popular method of preserving forage for animal feed, particularly in the humid northern regions of the USA. Unfortunately, excessive proteolysis of ensiled forages can result in both economic losses to farmers (Rotz et al. 1993) and negative impacts on the environment since in the rumen, non-true protein nitrogen (ammonia, amino acids, and small peptides) is poorly utilized, most being excreted by the animal as urea.

Proteolytic losses in ensiled alfalfa are especially high, with degradation of 44-87% of the forage protein (Papadopoulos and McKersie 1983; Muck 1987). In contrast, red clover, a forage of protein content similar to alfalfa, has been found to have up to 90% less proteolysis than alfalfa during ensiling (Papadopoulos and McKersie 1983). This difference in the extent of post-harvest proteolysis is evident in extracts of clover and alfalfa leaves (Jones et al. 1995a,b,c). Red clover's lower extent of post-harvest proteolysis is not due to differences in its inherent proteolytic activity compared to alfalfa, but seems related to the presence of polyphenol oxidase (PPO) and *o*-diphenols based on several experimental observations (Jones et al. 1995a,b,c). Interestingly, alfalfa leaves and stems have little if any polyphenol oxidase activity or *o*-diphenol PPO substrates (unpublished data). Since PPO catalyzes the oxidation of *o*-diphenols to *o*-quinones (Figure 1), a possible mechanism whereby PPO acts to inhibit proteolysis could be via PPO-generated *o*-quinones binding directly to and inactivating endogenous proteases.

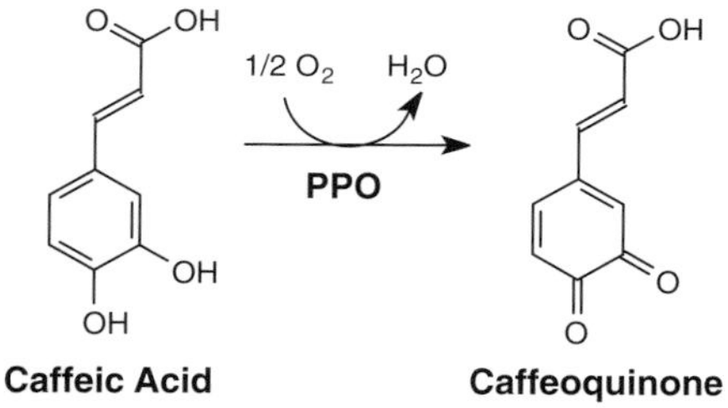

Figure 1. Polyphenol oxidase (PPO) catalyzed oxidation of an o-diphenol to an o-quinone (caffeic acid to caffeoquinone in this example).

In an effort to understand the role of PPO in inhibition of post-harvest proteolysis, we have been using both biochemical and molecular approaches. Here we report the isolation of three red clover PPO cDNAs, characterization of their expression pattern, and analysis of their enzymatic activities. Additionally, to understand the lack of PPO activity in alfalfa leaves and stems, we have isolated a genomic PPO clone from alfalfa and have begun to characterize its expression and encoded protein.

2. RESULTS

To clone red clover PPO genes, we took advantage of the conserved sequences of the copper binding motifs of several previously cloned plant PPOs (Figure 2) to design primers for reverse transcription and PCR of red clover leaf mRNA. Several PCR fragments were isolated, cloned, and sequenced. To obtain full-length clones, a 300 bp PCR fragment was used to screen a red clover leaf cDNA library. Three unique PPO cDNAs were identified and sequenced (Table 1).

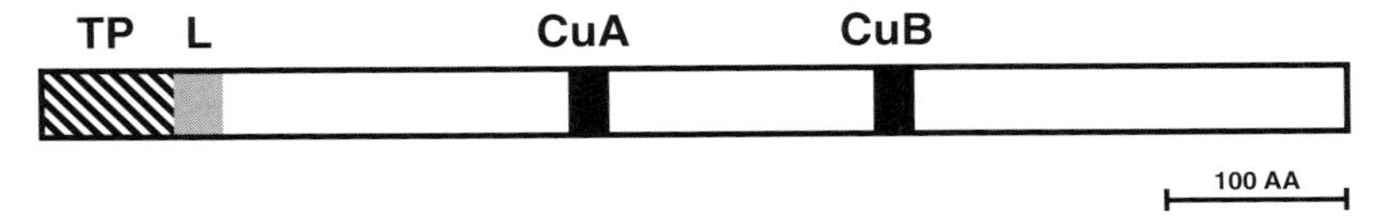

Figure 2. Schematic representation of PPO protein showing predicted transit peptide (TP), luminal targeting sequence (L), and conserved copper binding motifs (CuA and CuB).

Table 1. Characteristics of red clover and alfalfa PPO genes

Gene	Genbank Accession	Protein Length[a]	Expression Pattern
red clover PPO1	AY017302	605/511	leaf
red clover PPO2	AY017303	623/522	flower, petiole
red clover PPO3	AY017304	599/504	leaf, flower
alfalfa PPO	AY283062	607/512	flower, seed pod

[a]*Length in amino acids, predicted precursor/predicted mature.*

Sequence comparison among the three red clover PPO clones indicates they are 84-89% identical at the nucleotide level (76-80% identity at the amino acid level). A BLAST search of Genbank reveals sequence similarity to several previously cloned PPO genes. The red clover PPOs are most similar to that of the legume *Vicia faba* (Genbank accession Z11702, 60-62% amino acid identity). ChloroP and SignalP algorithms (www.cbs.dtu.dk/services) predict an N-terminal chloroplast transit peptide and thylakoid lumen targeting signal (Figure 2) (Peltier et al. 2000; Emanuelsson et al. 1999) which would result in processing to mature proteins of 504-522 amino acids in length. These predictions are consistent with the intracellular localization and processing sites of several other PPO enzymes (Steffens et al. 1994). In the case of RC PPO1, *E. coli*-expressed protein corresponding to the predicted mature form comigrates with a PPO protein present in red clover leaves, indicating that the predicted processing occurs (data not shown).

Expression of the various red clover PPO genes was examined by northern blotting and hybridization using RC PPO1-3 gene-specific probes (Figure 3 and data not shown). RC PPO1 is most highly expressed in unexpanded and young leaves, RC PPO2 is expressed most highly in flowers and petioles, and RC PPO 3 is expressed in both leaves and flowers. Although leaf expression for RC PPO2 is not apparent in the northern blot, recovery of the cDNA from a leaf library suggests it is expressed in leaves, albeit to a low level. Using antiserum raised against *E. coli*-expressed RC PPO1 on immunoblots of extracts of transgenic alfalfa expressing the individual RC PPO genes (described in more detail below), we found that the individual red clover PPOs could be distinguished by mobility on SDS-PAGE (Figure 3). Immunoblot analysis of red clover leaf and flower extracts show that RC PPO1 and 3 are most highly expressed in leaves whereas RC PPO2 is expressed most highly in flowers. These findings are consistent with the northern analysis.

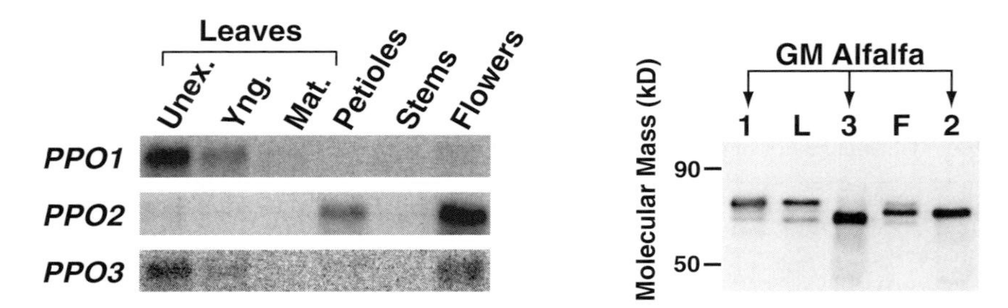

Figure 3. Red clover PPO genes show tissue-specific expression. (Left) Northern blots of RNA isolated from unexpanded (Unex.), young (Yng.) and mature (Mat.) leaves; petioles; stems; and flowers of red clover were hybridized with gene-specific probes as indicated. (Right) An Immunoblot of extracts of red clover leaves (L) and flowers (F) or leaves from transgenic (GM) alfalfa expressing RC PPO1-3 as indicated was probed with anti RC PPO1 antiserum.

Due to the low activity of RC PPO1 enzyme expressed in *E. coli* (data not shown), we decided to express the red clover PPO genes in alfalfa. Alfalfa is easily transformed and alfalfa leaves contain no detectible PPO activity or *o*-diphenol substrates. This lack of endogenous PPO and substrates would facilitate our analyses of the red clover gene products. The entire coding regions from RC PPO1-3 were inserted behind the cassava vein mosaic virus (CVMV) promoter in the pILTAB357 transformation vector (Verdaguer et al. 1996). The resulting constructs were transformed into alfalfa using *Agrobacterium*-mediated transformation (Austin et al. 1995). Several independent transformants were generated for each PPO gene, as well as control plants transformed with the pILTAB357 vector only. When the *o*-diphenol caffeic acid is added to extracts of PPO-expressing alfalfa leaves, a dramatic browning reaction is apparent within minutes, a hallmark of *o*-quinone production by PPO (Steffens et al. 1994). No such browning takes place in control alfalfa extracts, even after 48 hours. To quantify PPO activity we used a 5,5'-dithiobis-(2-nitrobenzoic acid) (DTNB) quinone trap assay (Esterbauer et al. 1977). For alfalfa expressing PPO1 and PPO3 and using a caffeic acid substrate, we found several independent alfalfa transformants with activities comparable to that seen in red clover leaves (i.e. 1-5 μmol/min/mg crude protein). One striking observation using caffeic acid as a substrate is that the activity of RC PPO2 (~0.1 μmol/min/mg) appears to be substantially lower than that of RC PPO1 and 3. For each of the red clover PPO genes, alfalfa plants showing the highest levels of PPO activity served as the enzyme source for additional analyses of enzyme activity. To determine if there are any differences in substrate utilization among the red clover PPOs, we measured PPO activity for several different *o*-diphenols in extracts of the transgenic alfalfa as well as in red clover leaf extracts (Figure 4).

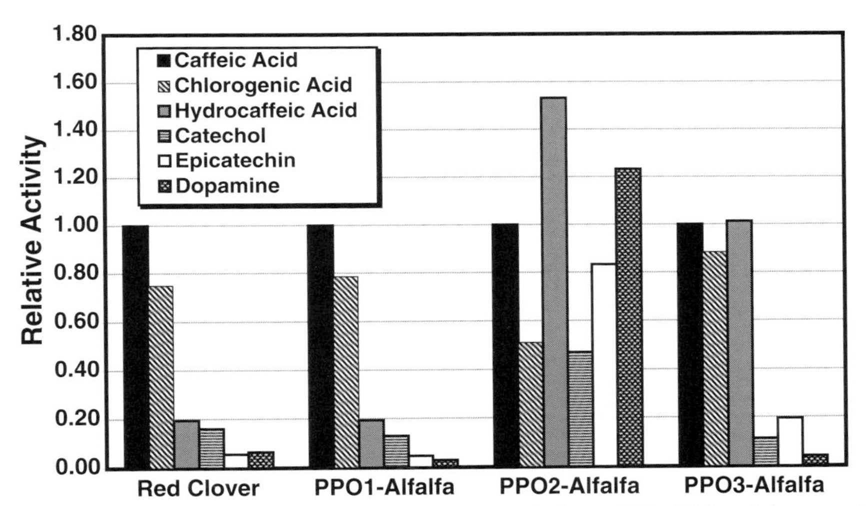

Figure 4. Differential utilization of o-diphenols among red clover PPOs. PPO activity towards various o-diphenol substrates was measured in extracts of red clover or alfalfa expressing individual RC PPO enzymes using a DTNB quinone trap assay (Esterbauer et al. 1977). To facilitate comparison among the different enzyme sources, activity is expressed relative to that for caffeic acid.

For RC PPO1 we found that the utilization of the *o*-diphenols tested is similar to that of red clover leaf extract, consistent with RC PPO1 being the major PPO of red clover leaves. RC PPO1-3 show substantial differences in their relative preferences for *o*-diphenol substrates. For example, both RC PPO2 and 3 utilize hydrocaffeic acid (3,4-dihydroxyhydrocinnamic acid) as well as or better than caffeic acid, while this is a mediocre substrate for RC PPO1. Also, RC PPO2 utilizes (-)-epicatechin and dopamine, relatively poor substrates for RC PPO1 and 3, as well as or better than caffeic acid. Additional studies of the red clover PPO enzymes are currently underway.

As noted above, we have detected little if any PPO activity in alfalfa leaves or stems. This lack of PPO activity could be due to lack of a functional gene or lack of PPO expression in the aerial tissues we have examined. Alternatively, PPO enzymes in alfalfa could have substrate specificities different from those of previously characterized PPOs and consequently not be detectible in standard activity assays. To distinguish among these possibilities, we cloned an alfalfa PPO gene by screening a genomic library (Gregerson et al. 1994) with a probe generated by PCR of alfalfa genomic DNA with PPO specific primers. We isolated and sequenced one clone containing an entire intronless alfalfa PPO gene along with 1200 bp of upstream sequence (Table 1). Of cloned PPO genes, alfalfa PPO is most similar to that of *Vicia Faba* (73% amino acid identity). The alfalfa enzyme shares only ~65% amino acid identity with the clover enzymes. Like most other plant PPOs, alfalfa PPO is predicted to be targeted to the chloroplast thylakoid lumen (Peltier et al. 2000; Emanuelsson et al. 1999; Steffens et al. 1994). Expression of alfalfa PPO was examined by northern blotting and hybridization under moderate stringency (to allow detection of transcripts from any related PPO genes). Of the tissues examined, PPO expression appears to be limited to flowers and seed pods (Figure 5).

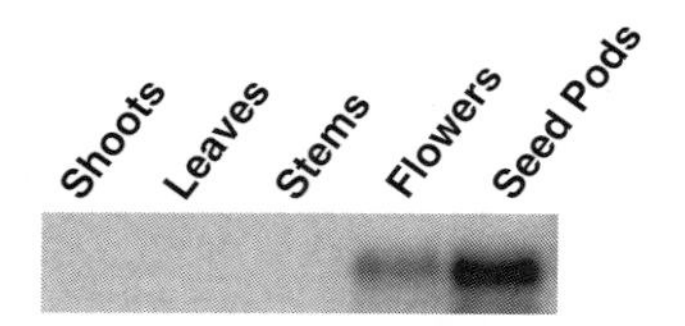

Figure 5. Northern blot analysis reveals alfalfa PPO expression is limited to flowers and seed pods.

We are currently examining the expression pattern of alfalfa PPO using a promoter::GUS fusion in transgenic alfalfa. Initial results indicate PPO is expressed in anthers, sepals, developing seeds and pods. To analyze the alfalfa PPO gene product, we are overexpressing the protein in transgenic alfalfa utilizing a CVMV promoter construct similar to those used for the red clover PPO genes. Extracts of the transgenic plants show significant browning when exogenous caffeic acid is supplied, although the browning reaction is slow compared to that of alfalfa expressing any of the red clover PPO genes (several hours compared to a few minutes). No browning is seen for control plants lacking the PPO transgene, even after several days. Since greater than twenty-five independent transformants have been examined, the slow browning reaction suggests that the specific activity of alfalfa PPO is low (at least for caffeic acid) compared to the red clover PPOs. We are currently carrying out more detailed studies of the enzymatic properties of alfalfa PPO.

ACKNOWLEDGEMENTS

We wish to thank Merici Evans and Mindy Dornbusch for excellent technical assistance and Dr. Jane Marita for comments on the manuscript.

REFERENCES

Austin S, Bingham ET, Matthews D, Shahan M, Will J, Burgess RR (1995) Production and field performance of transgenic alfalfa expressing alpha-amylase and manganese-dependent lignin peroxidase. Euphytica 85: 381-393.

Emanuelsson O, Nielsen H, von Heijne G (1999) ChloroP, a neural network-based method for predicting chloroplast transit peptides and their cleavage sites. Prot. Sci. 8: 978-984.

Esterbauer H, Schwarzl E, Hayn M (1977) A rapid assay for catechol oxidase and laccase using 2-nitro-5-thio benzoic acid. Anal. Biochem. 77: 486-494.

Gregerson RG, Miller SS, Petrowski M, Gantt JS, Vance CP (1994) Genomic structure, expression and evolution of the alfalfa aspartate aminotransferase genes. Plant Mol. Biol. 25: 387-399.

Jones BA, Hatfield RD, Muck RE (1995a) Characterization of proteolysis in alfalfa and red clover. Crop Sci. 35: 537-541.

Jones BA, Hatfield RD, Muck RE (1995b) Screening legumes forages for soluble phenols, polyphenol oxidase and extract browning. J. Sci. Food Agr. 67: 109-112.

Jones BA, Muck RE, Hatfield RD (1995c) Red clover extracts inhibit legume proteolysis. J. Sci. Food Agric. 67: 329-333.

Muck RE (1987) Dry matter level effects on alfalfa silage quality. I. Nitrogen transformation. Trans. ASAE 30: 7-14.

Papadopoulos YA, McKersie BD (1983) A comparison of protein degradation during wilting and ensiling of six forage species. Can. J. Plant Sci. 63: 903-912.

Peltier JB, Friso G, Kalume DE, Roepstorff P, Nilsson F, Adamska I, van Wijk KJ (2000) Proteomics of the chloroplast: Systematic identification and targeting analysis of lumenal and peripheral thylakoid proteins. Plant Cell 12: 319-341.

Rotz CA, Pitt RE, Muck RE, Allen MS, Buckmaster DR (1993) Direct-cut harvest and storage of alfalfa on the dairy farm. Trans. ASAE 36: 621-628.

Steffens JC, Harel E, Hunt MD (1994). Polyphenol oxidase. In: Genetic Engineering of Plant Secondary Metabolism, Ellis BE, Kuroki GW, Stafford HA (eds.), vol. 28, pp. 275-312. New York.

Verdaguer B, de Kochko A, Beachy RN, Fauquet C (1996) Isolation and expression in transgenic tobacco and rice plants, of the cassava vein mosaic virus (CVMV) promoter. Plant Mol. Biol. 31: 1129-1139.

Molecular Marker-Based Genetic Analysis of Pasture and Turf Grasses

John W. Forster[1,4], Elizabeth S. Jones[1,3,4], Jacqueline Batley[1,4] and Kevin F. Smith[2,4]

[1]Plant Biotechnology Centre, Primary Industries Research Victoria, Department of Primary Industries, La Trobe University, Bundoora, Victoria 3086, Australia. [2]Pastoral and Veterinary Institute, Primary Industries Research Victoria, Department of Primary Industries, Private Bag 105, Hamilton, Victoria 3300, Australia. [3]Present address: Crop Genetics Research and Development, Pioneer Hi-Bred International, 7300 NW 62nd Avenue, Johnston, Iowa 50131-1004, USA. [4]CRC for Molecular Plant Breeding, Australia. (Email: John.Forster@dpi.vic.gov.au).

Key words: ryegrass, fescue, molecular marker, genetic map, quantitative trait locus, candidate gene, single nucleotide polymorphism, germplasm collection

Abstract:

Molecular genetic marker systems have been developed for implementation in the breeding of a number of pasture and turf grass species. A reference genetic map for perennial ryegrass (*Lolium perenne* L.) has been generated through the activities of the International *Lolium* Genome Initiative (ILGI), containing conserved restriction fragment length polymorphism (RFLP) loci for comparative genetic analysis with other Poaceae species and simple sequence repeat (SSR) loci for map alignment across multiple pedigrees. Trait-specific genetic mapping in perennial ryegrass and related species is being performed in a number of ILGI-affiliated laboratories, allowing the detection of quantitative trait loci (QTLs) for a range of agronomic traits such as root and shoot morphology, floral development, herbage quality, nutrient assimilation, disease resistance and tolerance to abiotic stresses such as cold and drought. Specific novel strategies are required for effective molecular marker implementation in outbreeding grass species. Candidate gene-based markers provide the means for direct selection of superior alleles in parental genotypes of synthetic varieties and the effective screening and utilisation of large-scale germplasm resources. RFLP and SSR loci corresponding to functionally annotated expressed sequence tags (EST-RFLPs and EST-SSRs) from perennial ryegrass have been assigned to genetic map locations. Corresponding single nucleotide polymorphism markers (EST-SNPs) are under intensive development in our program, with the aim of providing highly informative systems capable of high-throughput automated analysis in breeding programs.

A. Hopkins, Z.Y. Wang, R. Mian, M. Sledge and R.E. Barker (eds.), Molecular Breeding of Forage and Turf, 197-238.

1. INTRODUCTION

1.1 Background

In the mid- to late-1990s, the status of molecular marker technology development for perennial ryegrass, the most important temperate pasture and turf grass species, was considerably less advanced than for many other major agricultural plant species. High-resolution genetic linkage maps, with a multiplicity of molecular marker types, had been constructed for major crop species such as tomato, soybean, maize, wheat, barley, rice, sorghum and others (e.g. Bhattramakki et al. 2000). Genetic maps had been utilised for QTL identification for multiple phenotypic characters associated with yield, disease resistance and abiotic stress tolerance (e.g. Jefferies et al. 1999). By contrast, only a rudimentary reference genetic map of perennial ryegrass based on a small number of RFLP and random amplification of polymorphic DNA (RAPD) markers was available (Hayward et al. 1994; Hayward et al. 1998). Apart from the interspecific pair-cross (p129) that was used for map construction, a small number of other pair-cross derived trait-specific mapping populations had been developed in the UK and Europe (Thomas et al. 1997). Limited QTL information was available for characters such as floral development traits and water-soluble carbohydrate content (Hayward et al. 1998; Humphreys et al. 1998). Genetic analysis using partial or genome-wide marker coverage had also been undertaken for a number of other forage grass species, with tall fescue (*Festuca arundinacea* Schreb.) providing the most advanced example (Xu et al. 1995). However, as for perennial ryegrass, the extent of map development and QTL analysis was limited, especially compared to other Poaceae species.

The key priorities for molecular marker research and development in outcrossing pasture grass species were:

- The development of efficient polymerase chain reaction (PCR)-based co-dominant (multiallelic) genetic marker systems in order to identify potential allelic complexity in pair-crosses, as well as for population diversity analysis. Simple sequence repeat (SSR) markers provided the best prospect of fulfilling this requirement.
- The development of reference genetic maps in order to coordinate readily transferred markers into framework sets for analysis of trait-specific mapping families.
- The development of trait-specific families based on pair-crosses between selected phenotypically divergent individuals.

- The development of accurate phenotypic assays in order to quantify the physiological and biochemical basis of breeders' characters.
- The mapping and tagging of genes/QTLs for key agronomic traits.
- The development of innovative strategies for the implementation of markers in the varietal improvement of outbreeding pasture grass species, given the existing paradigms for molecular marker-based breeding (such as recurrent back-crossing) had been implicitly or explicitly developed for inbreeding species.

A large coordinated research program was obviously necessary in order to address these priorities. It was clear that international collaboration would provide a highly efficient and desirable means to achieve rapid progress, especially as a number of different groups were interested in the development of a ryegrass genetic map and had identified similar priorities. The International *Lolium* Genome Initiative (ILGI) was consequently founded in 1998 to facilitate and coordinate international collaboration in molecular marker technology for the ryegrasses and their close relatives. The founding concepts of ILGI were:

- The sharing of public domain genetic marker information between partners in order to allow the construction of a common reference genetic map.
- The provision of this information to each affiliated institution for application to trait-specific genetic mapping families.
- The comparison of gene and QTL location information across different crosses to establish consensus views of the genetic architecture of the perennial ryegrass genome.
- Comparative genetic mapping analysis between ryegrasses, fescues and other members of the Poaceae, in order to evaluate conservation of gene/QTL locations across species.

1.2 Reference Genetic Map Development

The key achievement of ILGI to date has been the construction of a comprehensive reference genetic map based on a mapping population with a simplified structure. The p150/112 F_1 progeny set was developed at the Institute for Grassland and Environmental Research (IGER), Aberystwyth, United Kingdom (M.D. Hayward and J.G. Jones, pers. comm.), and is a one-way pseudo-testcross population based on the pair-cross of a multiply heterozygous genotype of complex descent with a doubled haploid genotype (Bert et al. 1999; Jones et al. 2002a). Clonal ramets from each progeny plant were transferred to a number of ILGI-affiliated laboratories, and genotypic data for non-proprietary marker systems was collated from four main sources. The group at Institut National de la Recherche Agronomique (INRA), Clermont-Ferrand, France contributed *Eco*RI/*Mse*I amplified fragment length polymorphism (AFLP) data; the groups at IGER and Department of Primary Industries-Plant Biotechnology Centre (DPI-PBC), Melbourne, Australia contributed wheat, barley and oat cDNA probe-detected RFLP data, and the groups at the Yamanashi Prefectural Dairy Experiment Station (YPDES), Yamanashi, Japan and the National Agricultural Research Centre for Hokkaido Region (NARCH), Sapporo, Japan, contributed rice cDNA probe-detected RFLP data.

Polymorphism data was obtained for 343 markers, of which 192 were AFLP loci and the remainder were primarily heterologous RFLP loci, with the addition of a small number of *L. perenne Pst*I genomic probe RFLP, expressed sequence tag (EST) and isoenzyme markers. A total of 322 loci were assigned to the genetic map, with seven linkage groups (LGs) and a total map distance of 811 cM (Jones et al. 2002a). One hundred and twenty four loci were co-dominant, of which 109 were heterologous RFLP loci. Significant regions of segregation distortion were identified, especially on LG3.

1.3 Comparative Genetic Mapping

The mapping of 109 heterologous RFLP loci on the p150/112 reference genetic map, predominantly detected by the cDNAs from wheat, barley, rice and oat, allowed the evaluation of conserved syntenic relationships with other species of the Poaceae (Gale and Devos, 1998). The syntenic chromosomal regions may be represented either in the form of linkage block ideograms, or as a concentric circle alignment based on circular permutation of chromosome arm orders. At the macrosyntenic level, each of the 7 linkage groups of perennial ryegrass chiefly corresponds to one of the seven basic homeologous chromosome groups of the Triticeae cereals, and they have been numbered accordingly. For instance, LG1 is the syntenic counterpart of wheat chromosomes 1A, 1B and 1D, barley 1H and rye 1R. Perennial ryegrass LGs 1, 3 and 5 showed uninterrupted synteny with their Triticeae counterparts, while LGs 2, 4, 6 and 7 contained non-syntenic regions (Jones et al. 2002a). Overall, 80% of common markers between the perennial ryegrass and Triticeae maps were syntenic, and colinearity was well conserved. The observed relationships between the perennial ryegrass map and the oat and rice maps were also consistent with previous comparisons of these species to the Triticeae (Van Deynze et al. 1995; Devos and Gale, 1997). For instance, LG3 corresponds to rice chromosome 1 and oat groups C and G.

Genetic mapping of perennial ryegrass candidate genes as RFLP loci has provided confirmation of these macrosyntenic relationships. The *LpFT*1 gene is thought to encode a sucrose: fructan 6-fructosyltransferase (6-SFT; Lidgett et al. 2002) and maps to the distal end of LG7 in a region of conserved synteny with the Triticeae group 7 chromosomes, adjacent to the 7S-located marker xpsr119. The putative barley ortholocus maps to the equivalent region on 7HS (Wei et al. 2000). The perennial ryegrass cinnamoyl-CoA reductase (*LpCCR*1) was mapped to a location towards the middle of LG7, in a region of conserved synteny with rice chromosome 8, as defined by the linked marker xr1394A. The *LpCCR*1 gene has a similar structure to a *CCR* orthologue identified in a rice BAC clone mapped to the equivalent region of rice 8 (McInnes et al. 2002).

The development of the comparative genetic map provides the basis for cross-species transfer of genetic information and structured gene location prediction, as described below in Section 2.2 for self-incompatibility, flowering time and cold tolerance genes. The further mapping of perennial ryegrass candidate genes (see Section 3.2) will allow the use of comparative genomics to extend the known relationships.

1.4 SSR Development and Mapping

The RFLP markers mapped in p150/112 are highly reproducible and ideal for comparative mapping, but are not readily transferable across different pedigrees for map alignment. SSR markers are highly reproducible, genetically co-dominant and multiallelic, and suitable for framework mapping (Rafalski et al. 1996). A small number of perennial ryegrass genomic DNA-derived SSR clones were isolated by Kubik et al. (1999), but in numbers insufficient for genetic map development. Enrichment library technology (Edwards et al. 1996) was used to isolate c. 450 unique SSR-containing (LPSSR) clones accessible to primer pair design (Jones et al. 2001, 2002b) and a sub-set of derived markers were tested for polymorphism and ortholocus detection in related species (Jones et al. 2001).

A total of 309 unique LPSSR primer pairs were evaluated for genetic polymorphism in the p150/112 population, with 31% detecting segregating alleles (Figure 1). Ninety three loci were assigned to map positions on the 7 LGs, the majority derived from cloned sequences with $(CA)_n$-type dinucleotide repeats (Jones et al. 2002b). A small proportion of the primer pairs detected multiple loci, and null alleles were also observed at low frequency. The LPSSR locus data was combined with the ILGI map data to produce a composite map of 258 loci covering 814 cM (Figure 2). The SSR loci covered 54% of the genetic map, and show clustering around the putative centromeric regions. A total of 172 markers of the co-dominant type (RFLP and LPSSR) are present on the composite map.

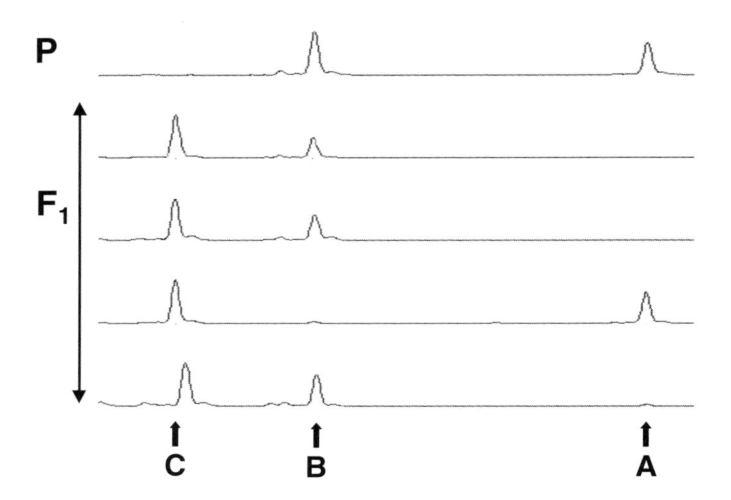

Figure 1. Segregating SSR alleles detected in the p150/112 reference genetic mapping population with primer pairs for locus LPSSRH01H06. PCR products labelled with 6-carboxyfluorescein (FAM) were detected using the ABI 3700 capillary electrophoresis system. P is the heterozygous parent. The xlpssrh01h06 locus shows an AB x CC segregation structure, with the allele sizes A = 167 bp, B = 147 bp, C = 138 bp.

Figure 2. Distribution of 93 loci detected by perennial ryegrass SSR (LPSSR) primer pairs (in bold type) across the seven linkage groups of perennial ryegrass for the cross p150/112. Segregation data is based on LPSSR loci and a selected sub-set of RFLP and AFLP loci from the reference map of Jones et al. (2002a). Locus nomenclature for LPSSR loci is shown in the form xlpssrabbccc, where a indicates the enrichment library of origin (h or k), bb indicates sequencing block number and ccc indicates block coordinate. Multiplicated loci are shown with a numerical extension e.g. lpssrk09f06.1, with .1 indicating the locus was mapped with fragments in the expected size range, and .2 and .3 with fragments not in the expected size range. Markers showing segregation distortion are indicated by * = significant distortion at $P < 0.05$, ** = $P < 0.01$ and *** = $P < 0.001$. The bracketed regions indicate markers that could not be ordered at LOD > 2.

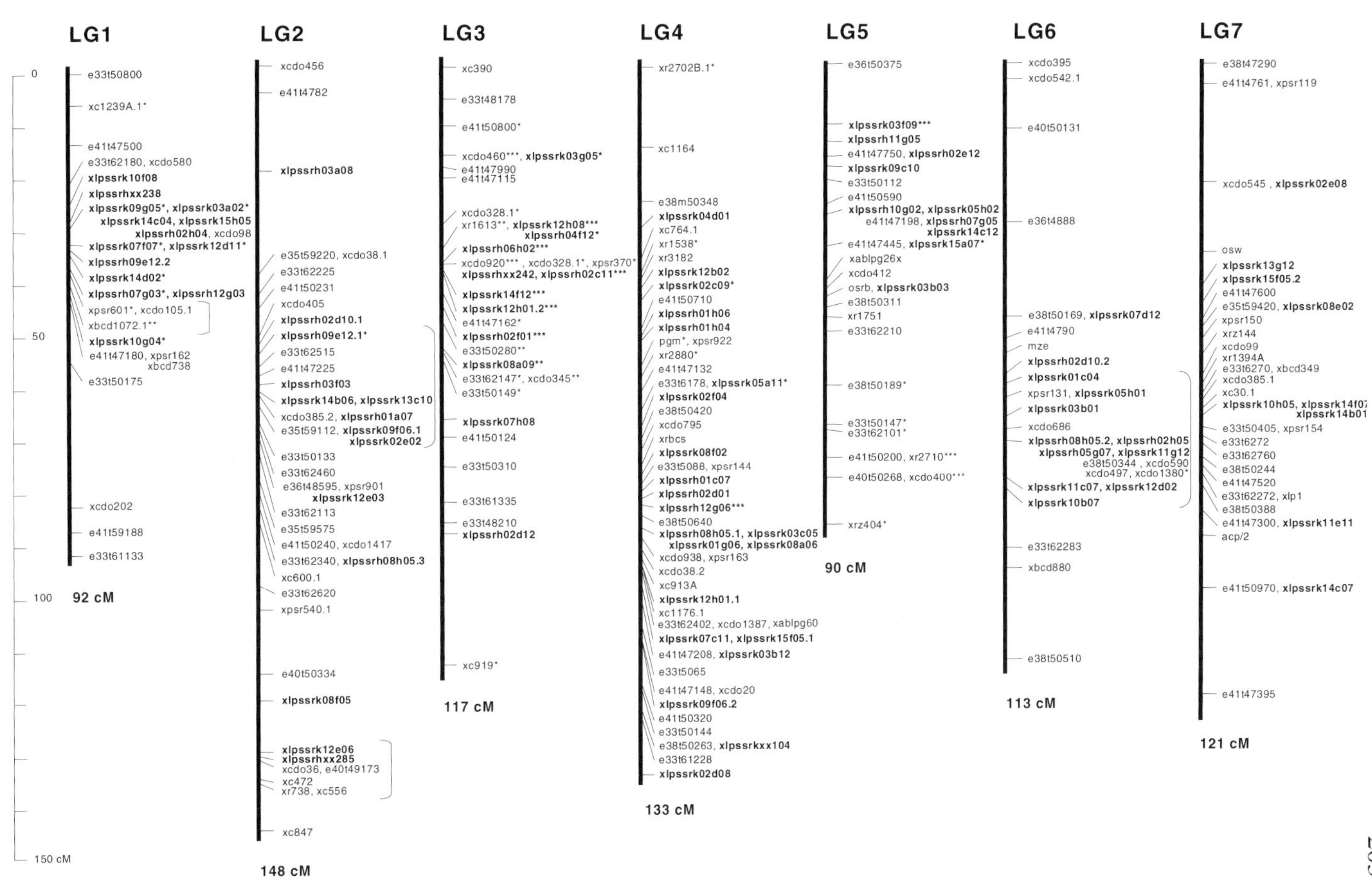
LG1
e33t50800
xc1239A.1*
e41t47500
e33t62180, xcdo580
xlpssrk10f08
xlpssrhxx238
xlpssrk09g05*, xlpssrk03a02*
xlpssrk14c04, xlpssrk15h05
xlpssrh02h04, xcdo98
xlpssrk07f07*, xlpssrk12d11*
xlpssrh09e12.2
xlpssrk14d02*
xlpssrh07g03*, xlpssrh12g03
xpsr601*, xcdo105.1
xbcd1072.1**
xlpssrk10g04*
e41t47180, xpsr162
xbcd738
e33t50175
xcdo202
e41t59188
e33t61133
92 cM
LG2
xcdo456
e41t4782
xlpssrh03a08
e35t59220, xcdo38.1
e33t62225
e41t50231
xcdo405
xlpssrh02d10.1
xlpssrh09e12.1*
e33t62515
e41t47225
xlpssrh03f03
xlpssrk14b06, xlpssrk13c10
xcdo385.2, xlpssrh01a07
e35t59112, xlpssrk09f06.1
xlpssrk02e02
e33t50133
e33t62460
e36t48595, xpsr901
xlpssrk12e03
e33t62113
e35t59575
e41t50240, xcdo1417
e33t62340, xlpssrh08h05.3
xc600.1
e33t62620
xpsr540.1
e40t50334
xlpssrk08f05
xlpssrk12e06
xlpssrhxx285
xcdo36, e40t49173
xc472
xr738, xc556
xc847
148 cM
LG3
xc390
e33t48178
e41t50800*
xcdo460***, xlpssrk03g05*
e41t47990
e41t47115
xcdo328.1*
xr1613**, xlpssrk12h08***
xlpssrh04f12*
xlpssrh06h02***
xcdo920***, xcdo328.1*, xpsr370*
xlpssrhxx242, xlpssrh02c11***
xlpssrk14f12***
xlpssrk12h01.2***
e41t47162*
xlpssrh02f01***
e33t50280**
xlpssrk08a09**
e33t62147*, xcdo345**
e33t50149*
xlpssrk07h08
e41t50124
e33t50310
e33t61335
e33t48210
xlpssrh02d12
xc919*
117 cM
LG4
xr2702B.1*
xc1164
e38m50348
xlpssrk04d01
xc764.1
xr1538*
xr3182
xlpssrk12b02
xlpssrk02c09*
e41t50710
xlpssrh01h06
xlpssrh01h04
pgm*, xpsr922
xr2880*
e41t47132
e33t6178, xlpssrk05a11*
xlpssrk02f04
e38t50420
xcdo795
xrbcs
xlpssrk08f02
e33t5088, xpsr144
xlpssrh01c07
xlpssrh02d01
xlpssrh12g06***
e38t50640
xlpssrh08h05.1, xlpssrk03c05
xlpssrk01g06, xlpssrk08a06
xcdo938, xpsr163
xcdo38.2
xc913A
xlpssrk12h01.1
xc1176.1
e33t62402, xcdo1387, xablpg60
xlpssrk07c11, xlpssrk15f05.1
e41t47208, xlpssrk03b12
e33t5065
e41t47148, xcdo20
xlpssrk09f06.2
e41t50320
e33t50144
e38t50263, xlpssrkxx104
e33t61228
xlpssrk02d08
133 cM
LG5
e36t50375
xlpssrk03f09***
xlpssrh11g05
e41t47750, xlpssrh02e12
xlpssrk09c10
e33t50112
e41t50590
xlpssrh10g02, xlpssrk05h02
e41t47198, xlpssrh07g05
xlpssrk14c12
e41t47445, xlpssrk15a07*
xablpg26x
xcdo412
osrb, xlpssrk03b03
e38t50311
xr1751
e33t62210
e38t50189*
e33t50147*
e33t62101*
e41t50200, xr2710***
e40t50268, xcdo400***
xrz404*
90 cM
LG6
xcdo395
xcdo542.1
e40t50131
e36t4888
e38t50169, xlpssrk07d12
e41t4790
mze
xlpssrh02d10.2
xlpssrk01c04
xpsr131, xlpssrk05h01
xlpssrk03b01
xcdo686
xlpssrh08h05.2, xlpssrh02h05
xlpssrh05g07, xlpssrk11g12
e38t50344, xcdo590
xcdo497, xcdo1380*
xlpssrk11c07, xlpssrk12d02
xlpssrk10b07
e33t62283
xbcd880
e38t50510
113 cM
LG7
e38t47290
e41t4761, xpsr119
xcdo545, xlpssrk02e08
osw
xlpssrk13g12
xlpssrk15f05.2
e41t47600
e35t59420, xlpssrk08e02
xpsr150
xrz144
xcdo99
xr1394A
e33t6270, xbcd349
xcdo385.1
xc30.1
xlpssrk10h05, xlpssrk14f07
xlpssrk14b01
e33t50405, xpsr154
e33t6272
e33t62760
e38t50244
e41t47520
e33t62272, xlp1
e38t50388
e41t47300, xlpssrk11e11
acp/2
e41t50970, xlpssrk14c07
e41t47395
121 cM
0
50
100
150 cM

2. TRAIT-SPECIFIC GENETIC MAPPING IN PASTURE GRASSES

2.1 Structure of Trait-specific Mapping Families

The development of a comprehensive reference genetic map for perennial ryegrass based on the p150/112 population by the members of ILGI has provided the means to anchor and align genetic maps in trait-specific crosses. The c. 200 co-dominant loci on the p150/112 map are accessible for map melding across populations, with the LPSSR markers being of special value.

Trait-specific genetic mapping populations are in general expected to have a more complex structure than the p150/112 population, most commonly showing a two-way pseudo-testcross structure (Ritter et al. 1990; Grattapaglia and Sederoff 1994). Two multiply heterozygous genotypes are pair-crossed, and the F_1 progeny are then genotyped. Co-dominant genetic markers can show a maximum complexity of four segregating alleles in this family type, conforming to an AB x CD structure. Two genetic maps are generated, corresponding to marker segregation from each parent, with dominant markers or co-dominant markers with structures such as AB x BB confined to single maps, and structures such as AB x CD providing bridging loci. Programs such as JOINMAP 2.0 (Stam and van Ooijen 1995) may be used to produce combined maps. Mapping structures involving a second cycle of pair-crossing are also possible, such as pseudo-F_2 families, in which two F_1 genotypes of parallel origin are crossed, or pseudo-backcrosses, in which an F_1 genotype is crossed to a second genotype from the same base population as one of its parents.

2.2 Trait-mapping in the Perennial Ryegrass p150/112 Reference Population

Although the p150/112 population has been primarily used for reference map development, it was originally developed and distributed to ILGI participant laboratories as a tool for the evaluation of genotype x environment (G x E) interactions in a range of geographical locations (M.D. Hayward, pers. comm.). The complex origin of the heterozygous parent, involving ancestry from Romania, northern Italy and the UK, has ensured a high degree of genetic variability in the F_1 progeny. The simplified genetic structure of p150/112 permits the assignment of QTLs for variable traits on a single parental map.

The structure of the reference population makes it highly suitable for the genetic analysis of self-incompatibility (SI). Assuming maximum allelic complexity at the *S* and *Z* loci (Cornish et al. 1979), the structure of the cross was anticipated to be of the form $S_1S_2Z_1Z_2$ x $S_3S_3Z_3Z_3$. In this case, there are four possible genotypic classes in the F_1, which may be readily distinguished given the prior identification of 'tester' genotypes. Using this approach, the *S* and *Z* loci were mapped to linkage groups (LGs) 1 and 2 respectively (Thorogood et al. 2002). The *S* locus is located in the upper part of LG1, close to the heterologous RFLP locus xcdo98. This is in a region of the perennial ryegrass genome showing conserved synteny with the *S* locus region of cereal rye (*Secale cereale* L.: Voylokov et al. 1998; Korzun et al. 2001). In addition, a thioredoxin gene (*Bm*2) is closely linked to the *S* locus of *Phalaris coerulescens* L. (Baumann et al. 2000). The primer pairs for a sequence tagged site (STS) marker designed from the rye ortholocus of *Bm*2 (Hackauf et al. 2000) detected a PCR-based marker with triallelic size polymorphism in perennial ryegrass (*LpBm*2). This marker mapped close to the *S*-linked markers in the expected region (M.P. Dupal, E.S. Jones and J.W. Forster, unpublished data). The *Z* locus mapped to the lower part of LG2, again in the region predicted by conserved synteny with rye (Fuong et al. 1993). Apart from the linked RFLP markers from the ILGI map (Jones et al. 2002a), a number of highly polymorphic SSR loci map close to both *S* and *Z* (Jones et al. 2002b), offering the possibility of associating multiallelic marker loci with the multiallelic SI genes.

A number of more complex phenotypic traits have been analysed in the p150/112 population. Herbage quality traits were analysed using vegetative samples grown at YPDES. Herbage quality data was obtained using NIRS (near infra-red reflectance spectroscopy) calibration to assess target traits such as crude protein (CP), dry matter (DM), estimated metabolisable energy (EstME), *in vivo* dry matter digestibility (IVDMD), neutral detergent fibre (NDF) and water soluble carbohydrate content (WSC). A large genotype x environment component was detected across two years of temporal replication, and no consistent QTLs were detected. Separate analysis for each year detected single QTLs for CP, EstME, IVDMD, NDF and SolCHO and 3 QTLs for DM, with log-of-odds (LOD) values varying from 2.1 to 4.0, accounting for between 9.2 and 21.2% of the total phenotypic variation. A QTL for WSC was assigned to the central region of LG2, close to the SSR locus xlpssrk02e02. A single region on LG3 was associated with 3 different traits (EstME, IVDMD and NDF) suggesting the presence of either a single pleiotropic locus or allelic effects at closely linked loci. Mixed negative and positive signs for the weight components of these 3 QTLs suggest that the marker haplotype associated with an increase in EstME and IVDMD is also associated with a decrease in NDF, both of which are favourable outcomes, providing the basis for an effective marker-based selection strategy (T.Yamada, E.S. Jones, K.F. Smith and J.W. Forster, unpublished). This study has now been extended with further temporal replication and to account for the effects of vegetative growth compared to reproductive maturity.

A number of morphological, developmental and adaptation traits have also been analysed at YPDES and NARCH. The following traits were measured: plant height, tiller number, tiller size, leaf length, leaf width, fresh weight at second harvest, plant type (prostrate or erect), number of spikelets per spike, spike length, heading date, degree of aftermath heading, electrical conductivity (due to ion leakage, as a function of freezing damage, Dexter et al. 1932) and winter survival. Substantial phenotypic variation was observed for the majority of the characters, with evidence for transgressive segregation for the spike length, number of spikelets per spike and winter survival traits. Significant positive correlations were observed between a series of plant size traits, such as plant height with tiller size, leaf length, fresh weight and spike length, as well as heading date with plant type (with late flowering genotypes showing more prostrate growth habits). Negative correlations were seen between plant height and heading date and between plant height and plant type.

Between 1 and 3 QTLs were detected with interval mapping (IM) for the morphological and developmental traits, with coincident groups on LGs 1,3,4 and 5 (Yamada et al. 2003; Yamada et al. submitted). Maximum LOD values varied from 2.1 to 6.5, accounting for between 11.3 and 31.4% of the total phenotypic variation. Coincident QTLs corresponded to significantly correlated traits. LGs 1 and 3 contain clusters of plant size QTLs. This may be due to the pleiotropic effects of a single scale-determining gene, like the *denso* dwarfing gene of barley on chromosome 3H (Bezant et al. 1996) or *cis*-linked alleles at multiple loci. The only traits with non-coincident QTLs were plant type (LG7) and aftermath heading (LG6), providing the potential for selection without correlated effects on other traits. No significant QTLs were detected for winter survival, but a QTL of small effect (maximum LOD value = 2.0) was detected for electrical conductivity, accounting for 11.8% of the total phenotypic variation. This is adjacent to a QTL (maximum LOD value = 4.0) for heading date, accounting for 20.9% of the total phenotypic variation. Although perennial ryegrass LG4 predominantly corresponds to the homeologous group 4 chromosomes of the Triticeae cereals (Jones et al. 2002a), comparative genetic mapping data with meadow fescue and other Poaceae species (Alm, 2001) suggests that the upper region of LG4 may show conserved synteny with the long arms of the group 5 homeologous chromosomes, in which winter-hardiness QTLs (Galiba et al. 1995; Pan et al.

1994) are linked to the vernalisation genes that regulate flowering time. The association of QTLs for homologous traits in the *Lolium* and *Festuca* genera suggests that similar mechanisms may exist.

2.3 Second Generation Perennial Ryegrass Reference Families

Two genetic mapping populations have been independently developed as successors to the p150/112 population, and have been aligned to the reference map using common markers. They share a common base population as the source of one parent in each instance, and are the subject of intensive phenotypic characterisation.

Single genotypes from partially (third generation) inbred lines derived from two agronomically contrasting varieties (cultivar [cv.] Aurora and cv. Perma) were pair-crossed at IGER, UK. A single F_1 hybrid genotype was self-pollinated to derive an F_2 population of 180 genotypes (Turner et al. 2001). This family is here designated F_2 (Aurora x Perma). A total of 74 RFLP and isoenzyme loci were mapped in this population (Armstead et al. 2002), of which 38 were common with p150/112 allowing alignment of the seven linkage groups. A total map length of 515 cM was obtained.

At DPI-Hamilton, Victoria, Australia, a single genotype from a North African (Moroccan) ecotype was pair-crossed with a single genotype from cv. Aurora to derive an F_1 population of 157 genotypes in the first instance. Previous genetic diversity analysis based on AFLP analysis (Forster et al. 2001) demonstrated a large genetic distance between the base populations. This family is here designated $F_1(NA_6 \text{ x } AU_6)$. Repeated cycles of crossing between the clonal replicates of the parents allowed the storage of several thousand F_1 seed which could be used in further experiments. Genotyping has been performed with *Eco*RI/*Mse*I AFLPs and genomic DNA-derived SSR markers (Jones et al. 2001, 2002b) to align the genetic maps of each parent with the p150/112 map. The NA_6 parental map contains 72 markers (50 LPSSR loci and 22 AFLP loci on 7 LGs with a total map length of 372 cM. The AU_6 parental map contains 62 markers (51 LPSSR loci and 11 AFLP loci) with a total map length of 179 cM. The map lengths in each case are much shorter than the p150/112 reference map due to the prevalence of SSR markers, but good alignment to the reference map and between the parental maps was observed (Guthridge et al. 2003). In addition, candidate gene mapping based on EST-RFLPs and EST-SSRs has produced an integrated genetic map for this population (see Section 3.2).

2.4 Analysis of Ryegrass Trait-specific Mapping Families

2.4.1 Disease Resistance

The most important diseases of forage ryegrasses are caused by the fungal pathogens crown rust (*Puccinia coronata* Corda f.sp. *lolii* Brown) and stem rust (*Puccinia graminis* f.sp. *lolii*), the viruses ryegrass mosaic virus (RMV) and barley yellow dwarf virus (BYDV) and the bacterial wilt pathogen (*Xanthomonas campestris* pv. *graminis*). In economic terms, the most damaging foliar disease is crown rust, associated with reductions in herbage quantity, quality and palatability. Consequently, the majority of studies have concentrated on this disease.

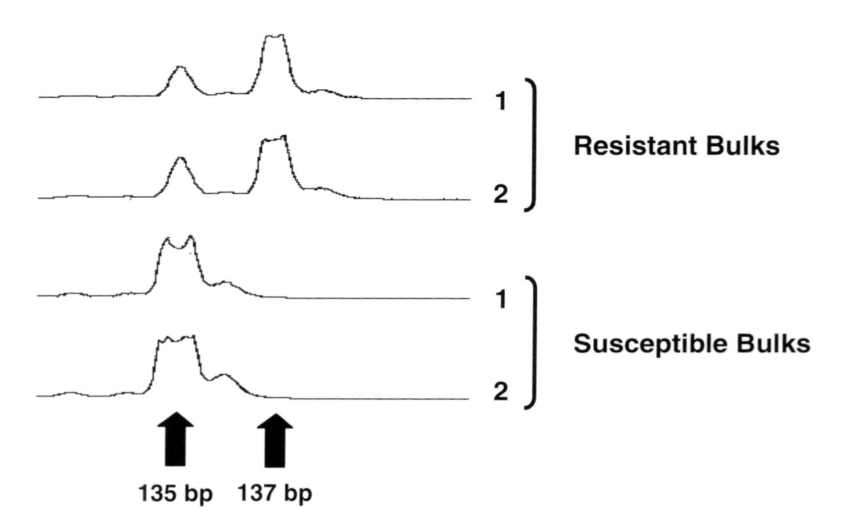

Figure 3. Genetic polymorphism between replicated resistant and susceptible bulks of the $Vedette_6$ x $Victorian_9$ F_1 mapping family detected by the primer pair designed to SSR clone LPSSRK07F08. PCR-products obtained with a primer labelled with the fluorochrome HEX (= hexachloro-6-carboxyfluorescein) were detected using the ABI 3700 automated capillary electrophoresis platform.

A major gene locus for crown rust resistance has been detected in a pair-cross between single genotypes from the resistant base population cv. Vedette and the susceptible base population cv. Victorian. The parents of this F_1($Vedette_6$ x $Victorian_9$) population showed divergent responses to mixed pathogen population from a number of Australian sources, but intermediate responses to a single geographical population from Hamilton, Victoria, Australia. Substantial variation for pathogen resistance was observed in the F_1 progeny, indicating transgressive segregation, with a high degree of correlation between symptom traits. Discontinuous variation in the F_1 distribution suggested the presence of a major gene effect. Bulked segregant analysis (BSA: Michelmore et al. 1991) with LPSSR primer pairs revealed associations with markers assigned to LG2 of the reference map (Jones et al. 2002b, Figure 3). Regression analysis of LG2 SSR marker variation with resistance in the full population identified an interval of c. 5 cM on the reference map between loci xlpssrh03f03 and xlpssrk02e02 as the most likely location for the resistance locus (*LpPc*1). Contributions were observed from both parents, with alleles segregating from $Vedette_6$ and $Victorian_9$ accounting for up to 80% and 26% of the phenotypic variation, respectively (Dumsday et al. 2003a,b,c). This may be due to allelic variation at the same or closely linked loci. Comparative genetics analysis has revealed that the LG2 locus is in a region of conserved synteny with the LGB-located *Pca* gene cluster, which controls resistance to the oat form of crown rust in the diploid species *Avena strigosa* (Yu and Wise, 2000).

QTL analysis of crown rust resistance has been performed in a number of perennial ryegrass populations. In the F_2 (Aurora x Perma) population, four distinct crown rust isolates were used for infection studies (Thorogood et al. 2001; Roderick et al. 2002). Isolates 1 and 2 showed lower levels of virulence, with a quasi-bimodal distribution, suggestive of major gene effects, while the resistance to isolates 3 and 4 was skewed towards susceptibility. Interval QTL mapping located putative QTLs for resistance to isolates 1 and 2 on LGs 2 and 5, with an interval on LG3 solely detected by isolate 2. The LG5 QTL is associated with a recessive allele from cv. Perma (which is typically moderately rust resistant), accounting for c. 60% of the phenotypic variation. A large number of resistance QTLs of low effect were identified to the isolates 3 and 4, with intervals on LGs 2, 5 and 6 associated with significant effects. A number of the minor gene effects were apparently derived from the susceptible cv. Aurora parent.

A single QTL for resistance to a field infection of crown rust in the Hamilton geographical region (maximum LOD value = 2.6, accounting for 7.6% of the phenotypic variation) was identified on the upper part of LG1 in the $Aurora_6$ map from the $F_1(NA_6 \times AU_6)$ population (K.M. Guthridge, pers. comm.).

Resistant and susceptible parental plants were selected from breeding lines at CLO-Gent, Belgium (Muylle et al. 2003) for infection studies and genetic analysis with BSA and QTL mapping. A genetic map containing RFLP, AFLP, SSR and STS markers was developed, with a total length of 833 cM, and was aligned to the ILGI reference map. Four QTLs were detected on 2 different linkage groups, collectively accounting for 45% of the phenotypic variation. AFLP markers associated with the resistance character initially detected by BSA mapped to the QTL-containing regions.

Fujimori et al. (2003) report the linkage analysis of crown rust resistance in Italian ryegrass (*Lolium multiflorum* Lam.). Yamaiku 130, a highly resistant breeding line, was selected as the source of resistant parents for two-way pseudo-testcross development, and may contain both major and minor resistance genes. Of 6 F_1 populations based on crossing Yamaiku 130 genotypes to plants from the susceptible line Yamaiku 131, one showed a 1:1 phenotypic segregation ratio, indicative of a heterozygous major gene. BSA was used to identify 34 putatively linked AFLP markers from 512 *Eco*RI/*Mse*I primer combinations, which were subsequently mapped in the full population of 115 plants. Closely linked markers (< 2 cM) were identified both in coupling and repulsion linkage to the *Pc*1 resistance locus. The Italian ryegrass genetic map has also been enriched with genomic DNA-derived SSR loci, as well as a heterologous RFLP probes. At this level of analysis, *Pc*1 appears to map to a region of conserved synteny with perennial ryegrass LG4 (M. Fujimori, pers. comm.), and consequently *Pc*1 probably does not correspond to the *LpPc*1 resistance gene identified in the F_1 ($Vedette_6$ x $Victorian_9$) population. Other major resistance genes (*Pc*2 and *Pc*3) have been identified in crosses involving the resistant base populations Yamaiku 130 and Harukaze, respectively (Hirata et al. 2003).

These studies have revealed evidence for both major and minor genes for crown rust resistance, consistent with previous genetic studies (Cruickshank 1957; Hayward 1977; Reheul et al. 2001). It is significant that apart from QTLs of large effect segregating from resistant parents, minor resistance QTLs have been identified segregating from susceptible parents, indicating the complexity of the genetics of crown rust resistance. Markers linked to genes may be used to pyramid multiple resistances to increase the durability of resistances. Apart from the further analysis of host resistance genes, a study of genetic variation in the pathogen will be critical for the understanding of resistance variation, and EST-SSR markers are in development for this purpose (Dumsday et al. 2003d). Evidence has been obtained for multiple physiological races of the crown rust pathogen (Wilkins 1978), which may be reflected in patterns of within- and between-population genetic diversity.

Genetic mapping families are also under development to study the mechanisms of resistance to bacterial wilt (R. Kölliker, pers. comm), and genetic diversity analysis based on 16S ribosomal RNA gene sequence variation has been performed to define the molecular pathotypes of *Xanthomonas campestris* pv. *graminis* (Kölliker et al. 2003).

For turf grasses, diseases such as grey leaf spot and dollar spot are also important. Grey leaf spot is caused by the fungal pathogen *Magnaporthe grisea*, which is also responsible for rice blast disease. A genetic map based on a cross between annual and perennial ryegrass genotypes (see Section 2.4.2) was constructed using RFLP, AFLP and RAPD markers, and two genomic regions containing QTLs for grey leaf spot resistance were identified (Curley et al. 2003a). Comparative analysis based on heterologous RFLP loci suggested orthologous QTL locations between rice, barley and perennial ryegrass (Curley et al. 2003b).

2.4.2 Reproductive Development

A pseudo-F_2 mapping family was developed for the analysis of interspecific hybrid characters and floral induction traits associated with perennial and annual growth habit. Parental genotypes were selected from the turf-type perennial ryegrass variety Manhattan and were crossed in two parallel lines of descent to genotypes from the Westerwolds-type Italian ryegrass (*Lolium multiflorum* var. *westerwoldicum*) variety Floregon. Two F_1 genotypes from each interspecific cross were interpollinated to generate the pseudo-F_2 progeny set. Heterologous RFLP, AFLP, isoenzyme and genomic DNA-derived SSR markers were used to construct parental maps, allowing alignment to the perennial ryegrass map as well as inference of comparative relationships (Warnke et al. 2003a,b). A QTL region associated with photoperiod requirement was identified, linked to SSR markers assigned to LG4, along with two QTLs for vernalisation requirement, linked to SSR markers assigned to LGs 2 and 7. Variation for requirement for both photoperiod and vernalisation is a diagnostic feature for annual and perennial ryegrasses. In addition, the first of the vernalisation QTLs is adjacent to the superoxide dismutase (*Sod*-1) isoenzyme locus. Genetic variation at *Sod*-1 provides a useful diagnostic for perennial growth habit (Warnke et al. 2002, Brown et al. 2003).

Substantial genetic variation for characters differentiated between species may also be observed in two-way pseudo-testcross F_1 families. A single genotype from the Westerwolds-type Italian ryegrass variety Andrea ($Andrea_{1246}$) was crossed to a single genotype from the perennial ryegrass variety Lincoln ($Lincoln_{1133}$). The F_1 ($Andrea_{1246}$ x $Lincoln_{1133}$) population varied for traits such as number of inflorescences per plant, length of reproductive tiller and heading date, with high broad sense heritabilities. A series of visual observations were integrated into a scoring system for annuality index. SSR-based framework map construction is in progress for this cross, and preliminary analysis has detected QTLs of large effect (maximum LOD values of 12-13) for annuality index and heading date segregating from the $Andrea_{1246}$ parent (Ponting et al. 2003). Several pseudo-F_2 families have also been generated by crossing individuals from F1($Andrea_{1246}$ x $Lincoln_{1133}$) and a second parallel cross ($Andrea_1$ x $Lincoln_{1120}$). These families show variation for the annual-diagnostic seeedling root exudate fluorescence (SREF) character (Okora et al. 1999), and will be used for linkage analysis of this trait.

2.4.3 Nutrient Use Efficiency

Dolstra et al. (2003) report the development of a one-way pseudo-testcross family based on the pair-cross of a heterozygous parent to a doubled haploid genotype. The outbred parent was bred from parents showing variation for nitrogen-use efficiency (NUE), which is an important breeding objective for environmentally sustainable grassland agriculture. The F_1 progeny were evaluated in hydroponic growth conditions with low nitrogen input for several NUE-associated traits. The genetic map for this cross was aligned with the F_2 (Aurora x Perma) map and 5 genomic regions were detected with QTLs for one or more traits.

2.4.4 Morphogenetic Characters

The F_2 (Aurora x Perma) family was used to measure the morphogenetic characters of plant size, leaf extension and regrowth rate. QTLs accounting for between 23% and 40% of the total phenotypic variation for these morphogenetic traits were located on all 7 linkage groups (Humpheys et al. 2003).

The F_1 (NA_6 x AU_6) family was measured for root, shoot and pseudostem morphological traits in a replicated controlled growth experiment. Single tillers from mature plants were transplanted to free-draining sand pots 1 m long by 10 cm wide, and grown under optimal conditions (fully watered with complete nutrient solution) for seven weeks prior to harvest. The experiment was performed in two successive years with similar design, except that in the second year, the tillers were trimmed prior to transplant and were consequently smaller at establishment. The following characters were measured: root maximum length, root fresh weight, root dry weight and root mass allocation; shoot fresh weight, shoot dry weight and total area; young fully emerged leaf (YFEL) area, YFEL length, YFEL width, YFEL fresh weight; ratios of dry weights and fresh weights for root:shoot; pseudostem fresh weight, pseudostem dry weight and pseudostem number. Limited correlation between the morphological traits was seen between the two temporal replicates, presumably due to the slower growth rate in the second year. Nonetheless, significant variation was detected for the majority of the traits in each case. Strong positive correlations were observed between the morphogenetic traits. For instance, the correlation coefficients between shoot dry weight and pseudostem dry weight were 0.84 and 0.91 respectively. Broad sense heritabilities were generally higher in the year 2 experiment than year 1, with maximum values of 0.66 and 0.84 respectively. A total of 25 QTLs were located on the NA_6 parental map, with maximum LOD values between 2.0 and 5.5, accounting for between 6.5% and 16.7% of the total phenotypic variation. The majority of the QTLs were clustered on LGs 4 (years 1 and 2) and 7 (year 2). A number of traits (root length, shoot dry weight, pseudostem number) reveal conserved QTL location on LG4 across the two years, although the positions are not completely coincident. A total of 22 QTLs for morphological traits were located on the AU_6 parental map, with maximum LOD values between 2.0 and 7.0, accounting for between 6.2% and 22% of the total phenotypic variation. The majority of the QTLs were clustered in two locations on LGs on LGs 1 (close to xlpssrk10f08) and 7 (close to xlpssrh02h04) and were derived from the analysis of the year 2 data. LG1 was also the location of a cluster of plant morphology QTLs in the p150/112 population (Section 2.2). Apart from the sand pot experiments, measurements of leaf area and leaf length in the field have identified QTLs on LGs 4 of the NA_6 map coincident with the main morphogenetic QTL cluster (Guthridge et al. 2003a,b).

2.4.5 Nutritive Quality

The F_2 (Aurora x Perma) family was used to measure nutritive quality characters such as total WSC, CP and NDF (Humphreys et al. 2002; Humphreys et al. 2003). Significant variation was anticipated for WSC, as the parental lines are divergent for this trait. Cultivar Aurora shows a characteristically high WSC content, while cv. Perma shows a lower content, more comparable to other perennial ryegrass varieties (Turner et al. 2002). QTLs for the three traits were located on 4 LGs, accounting individually for between 20 and 25% of the variation. More detailed analysis was then performed on individual WSC components. QTLs for leaf carbohydrate content were not coincident with tiller base carbohydrate content QTLs, but the total WSC QTLs generally coincided with the locations of QTLs for high molecular weight fructan. The WSC QTLs located on LGs 1 and 2 coincided with QTLs for NDF.

The F_1 (NA_6 x AU_6) family sand pot growth experiments were also used to measure pseudostem WSC content. Significant variation was detected in both year 1 and year 2 between the AU_6 parent (high WSC content) and the

NA_6 parent (lower WSC content), with substantial trangressive segregation. No QTLs for WSC content were detected on the NA_6 parental map, but three QTLs (year 1) were located on LGs 1, 2 and 3 of the AU_6 parental map, with maximum LOD values of 2.2, 2.5 and 2.6, accounting for 7.0%, 7.5% and 7.5% of the total phenotypic variation respectively (Figure 4). The year 2 data detected a single QTL of low effect on LG3 (Guthridge et al. 2003b; McFarlane et al. 2003). Consequently, the QTLs in this study on LG1 and 2 map to the same groups as those in the F_2(Aurora x Perma) study, as well as the LG2-located NIRS-calibrated WSC QTL observed in the p150/112 population.

2.4.6 Photosynthetic Traits

The F_1 (NA_6 x AU_6) family sand pot growth experiments were also used to measure variation for photosynthetic traits (McFarlane et al. 2003) Measurements were made of photosynthetic rate, conductance, internal CO_2 concentration, transpiration rate, vapour pressure deficit and leaf temperature. The photosynthetic traits were well correlated across the two years, but showed limited phenotypic variation. The NA_6 parental map contains coincident QTLs for photosynthetic rate and conductance on LGs 3 and 7, and a single QTL for internal CO_2 concentration on LG5, while the AU_6 parental map contains single QTLs for conductance on LG3 and internal CO_2 concentration on LG5 (Figure 4).

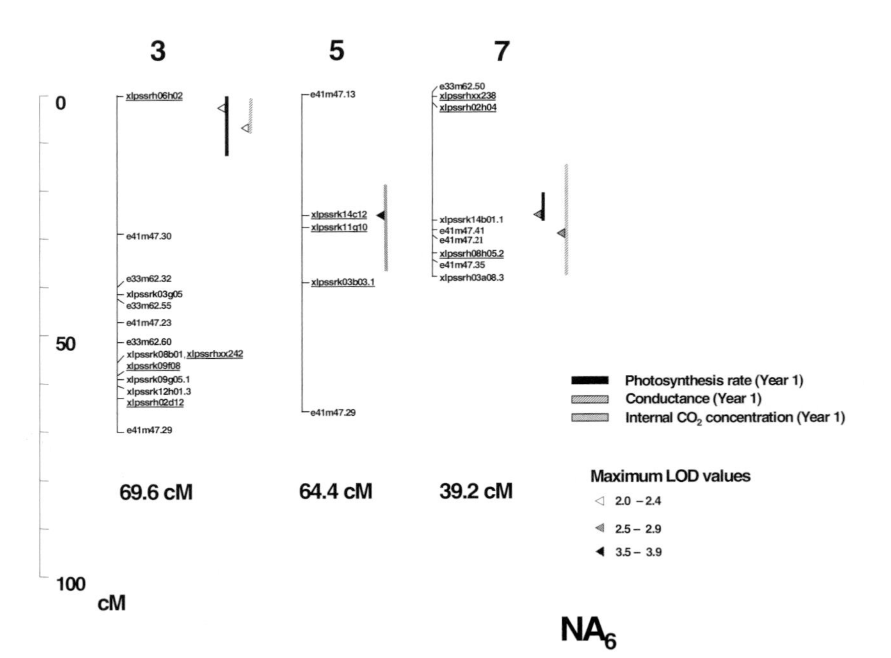

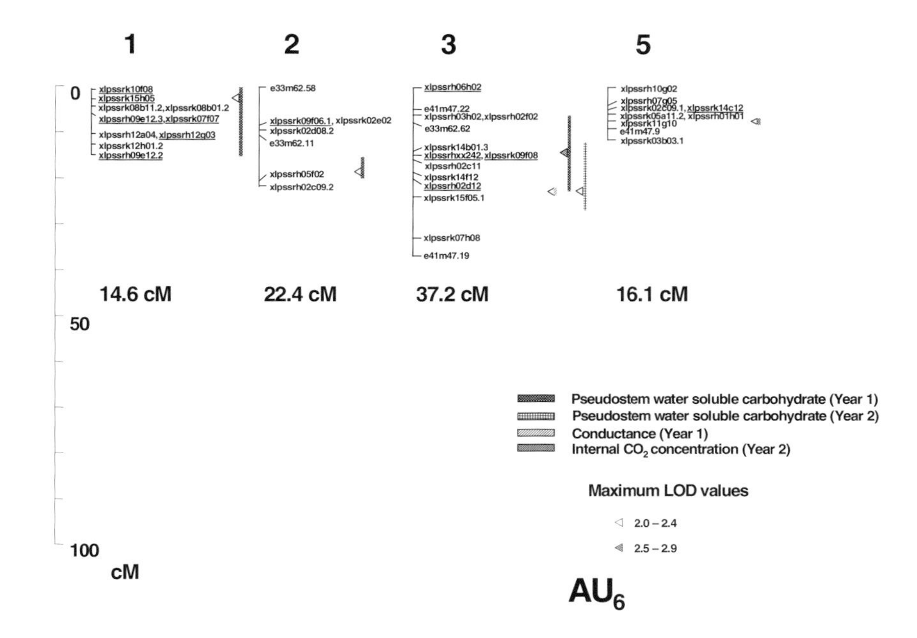

Figure 4. Locations of QTLs for pseudostem water soluble carbohydrate content (WSC) and photosynthetic traits on the framework parental genetic maps of the $F_1(NA_6 \ xAU_6)$ mapping family. Linkage groups (LGs) are labelled in accordance with the p150/112-based reference map. Genomic DNA-derived SSR loci are indicated with the prefix xlpssr. Bridging loci between the equivalent LGs of the parental maps are underlined. AFLP loci are indicated in the format exxmyy.nn, referring to standard primer pair nomenclature.

2.4.7 Abiotic Stress Tolerance

Enhanced drought tolerance is an important breeding objective for perennial ryegrass in the drier regions of temperate climates, such as south-western Victoria in Australia. Conventional breeding programs have so far exploited germplasm from the Mediterranean basin, especially northern Africa, which includes ecotypes showing the property of summer dormancy. This provided the rationale for the selection of the North African parent of the F_1 (NA_6 x AU_6) mapping family. The morphological and physiological traits described in sections 2.4.4-2.4.6 are relevant to adaptation to low rainfall conditions. Large variation for maximum root length has been observed between the NA_6 and AU_6 parents, especially in the year 1 trial, with parental means between in the 600-650 mm interval (NA_6) and the 300-350 mm interval (AU_6). Transgressive segregation, with plants showing root lengths up to 1 m, was observed in the F_1 progeny. Large QTL effects on root length were observed from both parent, especially the LG1-located QTL on the AU_6 map (maximum LOD value = 5.8). Measurement of specific leaf area (SLA) in the field revealed single QTLs on LGs 1 and 4 of the AU_6 map. Variation for SLA has been previously associated with stress tolerance in perennial ryegrass.

The deep-rooting nature of the North African accessions may be important for accessing sub-soil moisture under conditions of extended drought. In addition, the QTLs for WSC content are relevant for drought survival, as fructans have been implicated as osmoprotectants providing tolerance to abiotic stresses (De Roover et al. 2000). Preliminary observations also suggest that the NA_6 parent shows a higher photosynthetic rate under fully watered conditions than AU_6, but also a more efficient rate reduction under drought conditions (McFarlane et al. 2001), possibly related to more efficient water use and the summer dormancy character. For these reasons, the measurement of key traits identified in pre-drought trials of F_1 (NA_6 x AU_6) are being repeated under drought conditions to analyse these characters further.

2.5 Trait Analysis in Fescue Mapping Families

2.5.1 Trait Analysis in Meadow Fescue

A comprehensive genetic map has been constructed for meadow fescue (*Festuca pratensis* Huds.) using a two-way pseudo-testcross F_1 population from the pair-cross of a single genotype from a Norwegian population that has been selected for frost tolerance (HF2) and a genotype from a Yugoslavian variety (B14). The combined data for homologous and heterologous RFLP, AFLP, SSR and isoenzyme markers from the two parental maps defines 466 loci with a total map length of 658 cM (Alm, 2001; Alm et al. submitted). Conserved synteny was analysed through the use of heterologous RFLP anchor probes derived from perennial and Italian ryegrass, wheat, barley, oat, rice, maize and sorghum, demonstrating a high degree of conserved synteny and colinearity with both perennial ryegrass and the Triticeae consensus map. As meadow fescue is a close taxonomic relative of perennial ryegrass, and high levels of genetic recombination are observed in interspecific F_1 hybrids (King et al. 1998), the high level of alignment with the *L. perenne* genetic map was anticipated. The linkage groups of meadow fescue have been designated 1F-7F to correspond with the Triticeae homoeologous chromosomes. Comparative analysis of the structures of chromosomes 4F and 6F reveals differences in syntenic comparisons with different Triticeae genomes, as previously described in Section 2.2, and suggest a more ancestral genome structure for the Poeae grasses than for the Triticeae within the Poaceae family.

The F_1 (B14/16 x HF2/7) progeny set is being developed for multiple trait phenotyping in a similar way to that described for the perennial ryegrass F_2(Aurora x Perma) and the F_1 (NA_6 x AU_6) families. QTL analysis for seed production characteristics detected a total of 57 QTLs for 10 distinct traits, with maximum LOD score values from 2.6 to 13.4 explaining between 5% and 58.5% of the total phenotypic variation (Fang et al. 2003). The relevant traits were plant height, heading date, seed weight/plant, number of fertile tillers, 1000-seed weight, panicle length, seed weight/panicle, panicle fertility, flag leaf length and flag leaf width. Coincident QTLs were identified for a number of traits, including components that are correlated, suggesting, as for the morphogenetic analysis in the p150/112 population, pleiotropic effects of underlying genes.

QTL analysis of herbage quality characteristics in the F_1 (B14/16 x HF2/7) family was based on near infra-red spectroscopy (NIRS) calibrations of traits such as NDF, WSC, CP as well as phosphorus, magnesium, calcium and potassium content (O.-A. Rognli, T.R Solberg, C Fang, V. Alm, Ø. Jørgensen, unpublished data). Measurements were made at the heading time and the subsequent aftermath growth in successive years. A total of 58 QTLs were detected using combined data, the majority of the QTLs being detected from the aftermath analysis. The majority of the QTLs were detected on LGs 3F, 4F, 5F and 6F, with the highest maximum LOD score (12.7, accounting for 39.9% of the total phenotypic variation) associated with a QTL for the phosphorus content calibration on 6F. QTLs of large effect were detected on LG4 for NDF (maximum LOD value = 8.6, accounting for 25.1% of the total phenotypic variation) and acid detergent fibre (ADF) (maximum LOD value = 7.1, accounting for 21.2% of the total phenotypic variation).

Finally, QTL analysis for abiotic stress tolerance (winter survival in the field, freezing tolerance and drought tolerance) was performed in the F_1(B14/16 x HF2/7) family (Alm, 2001). Two major QTLs for frost tolerance (assessed as regrowth after treatment) were identified on LGs 4F and 5F, accounting for 8.4% and 29.5% of the total phenotypic variation, respectively. The 4F-located QTL is an region of conserved synteny with the wheat 5L chromosomes, and is apparently located in a similar chromosomal region to the electrical conductivity QTL reported for perennial ryegrass (Section 2.2). Four QTLs for winter survival were detected on 1F, 2F, 5F and 6F, with maxium LOD score values between 3.24 and 6.48, and accounting for 10.4%, 19.7%, 12.3% and 14.9% of the total phenotypic variation, respectively. A number of dehydrin candidate genes mapped in the vicinity of the winter survival QTLs. Dehydrins are expressed in response to low temperature stress in plants (Close, 1997). Putative syntenic relationships have been proposed for the 1F QTL region with the *Gpi*-2 region on LG1 of perennial ryegrass, which is associated with accumulation of WSC (Humphreys, 1992). The 2F QTL region may correspond to regions of the Triticeae group 2 chromosomes associated with photoperiod response, while the 5F QTL is in the same vicinity as the minor frost tolerance QTL in a region of conserved synteny with a 5H-located winter survival QTL in barley. A single QTL of low effect for tolerance to moderate drought was identified on 5F.

2.5.1 Trait Analysis in Tall Tescue

Although reference genetic map development in tall fescue involved a parental line selected for high *in vitro* dry matter digestibility (Xu et al. 1995), suggesting the potential for QTL mapping of this trait, there have been few reports to date of trait analysis in this species. The large scale development of EST-SSR markers for tall fescue (Saha et al. 2003a) and mapping in a population segregating for digestibility should remedy this situation (Saha et al. 2003b).

2.6 Progress in Marker-Assisted Selection

Several groups have performed preliminary experiments on the consequences of selection for linked markers on target traits. Humphreys et al. (2002, 2003) performed selections based on homozygosity for marker alleles linked to the tiller base fructan content (TBFC) QTLs on LGs 1 and 2 in the F_2(Aurora x Perma) family. The progeny of these marker selections showed small but consistent differences in WSC, with mean differences of 1.4% for LG1 selections and 0.6% for LG2 selections. This suggests that the two QTL effects only represent a small proportion of total variation for WSC.

Dolstra et al. (2003) report the use of markers linked with QTLs for NUE in the progeny derived from self-pollination of the heterozygous mapping population parent. The marker-based selections for NUE were compared at low N supply, demonstrating a strong differential response for regrowth after cutting and tillering. Such differences were not observed at high N supply.

Self-pollination of F_1 genotypes identified as containing favourable QTL alleles is also being performed with Italian ryegrass (Vanderwalle et al. 2003). A group of 24 self-pollinated populations are being screened with markers linked to traits including dry matter yield, WSC, digestibility, disease resistance and heading date. It is proposed that complementary genotypes will be used to compose new varieties.

Strategies for marker-assisted selection (MAS) that involve self-pollination or mating of closely related genotypes are vulnerable to the effects of inbreeding depression, and also reduced fertility due to identity of SI gene alleles. Theoretical strategies to address these problems have been developed (K.F. Smith, E.S. Jones and J.W. Forster, unpublished), based on the use of one or more donor genotypes in a modified polycross scheme. The logistics of recurrent selection for a single marker linked to a major gene or large QTL are relatively simple, but highly complex for multiple marker selection. The main requirement for marker implementation in these schemes is the use of a limited number of parental clones to generate a restricted base variety (Forster et al. 2001; Guthridge et al. 2001). In this context, AFLP-based profiling of the parents of an experimental synthetic variety (KT-2) has been used to determine the genetic distance between genotypes. This data is being used to reconstruct sub-varieties with more restricted genetic bases, to determine the optimal relationship between genetic variability and varietal performance (Bannan et al. 2003).

2.7 Implications of Trait-specific Mapping Analysis

This survey of the status of trait-specific mapping in ryegrasses and fescues demonstrates that important progress has been made, with genes and QTLs for most of the major agronomic traits identified by a number of different groups and using a range of different population types. Evidence for common location of QTLs across different populations has been obtained, and comparative genetics suggests that orthologous QTL locations may be identified between *Lolium* and *Festuca*, and with other Poaceae species. On the other hand, the majority of the QTL effects detected are of low magnitude (maximum LOD values 2.0-4.0) and G x E effects may limit the value of such QTLs. The empirical and theoretical studies of Beavis (1994) also suggest that small mapping population sizes will lead to the overestimation of larger QTL effects and a failure to detect smaller QTLs. Selection for homozygosity of markers associated with target genes may lead to co-selection of *cis*-linked sub-lethal mutations, especially in close-bred pedigrees, reducing genetic fitness. These considerations suggest that a new paradigm based on selection for superior allele content at candidate genes may be of high value for the molecular breeding of pasture grasses.

3. DEVELOPMENT OF CANDIDATE GENE-BASED MARKERS FOR PERENNIAL RYEGRASS

3.1 Limitations of Current Approaches

3.1.1 Genetic Marker Systems

The current systems available for rapid genotypic analysis are highly reliant on anonymous genetic markers (AFLPs, genomic DNA-derived

SSRs). Although the gDNA-derived SSRs show high levels of polymorphism and are consequently ideal for anchoring genetic maps across different mapping populations (Jones et al. 2002b), as selection markers they are in incompletely linked to target genes, and consequently may be separated by genetic recombination during multiple breeding cycles. In addition, the establishment of a marker allele-trait gene linkage for MAS in a new source of germplasm can only be reliably established by a genetic transmission test. The ideal marker system will be reasonably polymorphic but in very close association or linkage disequilibrium (LD) with the target trait, allowing the development of 'perfect markers'.

The technical aspects of high-throughput genotyping with SSRs are also to some extent limiting due to problems with analysis at very high multiplex ratios, and the requirement for allele sizing for discrimination rather than the scoring of digital (presence/absence) data. By contrast, SNP markers are suitable for high-throughput analysis at high multiplex ratios using automated capillary electrophoresis platforms, and are suitable for automatic scoring.

3.1.2 MAS Implementation Strategies

The strategies for MAS (see Section 2.6) that have so far been evaluated require either the pre-introgression of target genes on the basis of linked markers into donor genotypes with compatible genetic backgrounds to other members of the selected polycross parents, or the use of 'exotic' donor genotypes and repeated selection cycles to eliminate undesirable contributions and enrich for the selected character. When a small number of parental genotypes are used, inbreeding depression due to fixation of recessive deleterious alleles may compromise the fitness of the selected plants. In addition, allelic identity at the SI loci may bias the combinations between genotypes and reduce general fertility. For this reason, the ability to pre-select parents on the basis of superior allele content, maximisation of genetic gain due to heterosis and finally the minimisation of self-incompatibility effects are major objectives for the efficient use of markers.

3.2 Candidate Gene-based Markers

3.2.1 Sources of Candidate Gene Markers

The next generation of molecular markers for pasture plants will be derived from expressed sequences, with an emphasis on candidate genes associated with biochemical and physiological processes likely to be correlated with target phenotypic traits. The EST-derived unigene collection from perennial ryegrass developed through the DPI/AgResearch co-funded Plant Genomics Program provides the resource for the development of a comprehensive suite of candidate gene-associated markers. EST analysis based on 5'-single pass sequencing has generated a primary resource of 44,636 sequences (Spangenberg et al. 2001). Cluster analysis has been used to define 'unigenes', consisting of contigs and singletons. A criterion of at least 50 bp of aligned sequence has allowed the definition of 20,633 unigenes, while a more stringent criterion of at least 100 bp of aligned sequence has allowed the definition of 14,767 unigenes (Spangenberg et al. 2001; Sawbridge et al. 2003a,b)

3.2.2 EST-derived Genetic Markers

Selected perennial ryegrass cDNAs from the EST analysis have been mapped as RFLP loci (Vecchies et al. 2003). These clones were selected on the basis of functional annotation by sequence database searches and were classified in terms of core physiological and biochemical processes. In parallel, the same gene set is in use for a number of functional genomic screens including microarray-based expression profiling and transgenic modification. A total of 150 partial or full-length cDNAs were screened for RFLP from the following functional categories: cell wall metabolism (including lignin biosynthesis), carbohydrate metabolism (including fructan biosynthesis), floral development (including homeotic genes), plant defence (including chitinases, proteinase inhibitors, defensins etc.), abiotic stress tolerance (including dehydrins and late embryonic abundant proteins [LEAs]), metal handling enzymes (including metallothioneins), salt stress protection and flavonoid biosynthesis (including chalcone synthase and dihydroflavone reductase genes). A total of 145 probes detected polymorphism in the F_1 (NA_6 x AU_6) cross (97%), with 133 detecting heterozygosity in NA_6 (89%) and 121 detecting heterozygosity in AU_6 (81%). A substantial number of probes detected small multigene families, with several potential polymorphic loci segregating from each parent. Of the 145 polymorphic probes, 101 detected 99 loci segregating from the NA_6 parent and 71 loci segregating from the AU_6 parent. Forty eight of the probes detected polymorphic loci in both parents, which in all cases could be assigned to the same linkage groups and constitute bridging markers (although the possibility of different linked paralogous sequences which are part of multigene families cannot be excluded at this stage). The EST-RFLP data was integrated with the LPSSR and AFLP data for this cross (Section 2.3) to produce a composite map aligned to the reference map (Figure 5).

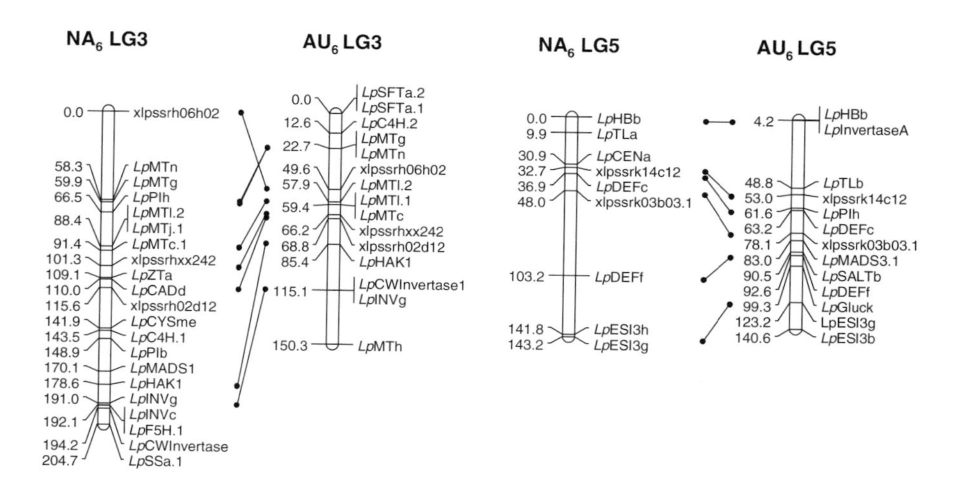

Figure 5. Alignment of the parental genetic maps for LG3 and LG5 from the $F_1(NA_6 \times AU_6)$ mapping family, based on combination of genomic DNA-derived SSR and EST-RFLP locus data. EST-RFLP markers are indicated with an *Lp* prefix and a code related to gene function, e.g. *Lp*CAD is the structural gene for a cinnamyl alcohol dehydrogenase. Bridging loci are indicated by connectors, and SSR loci providing alignment with the p150/112-based reference map are shown with the prefix xlpssr.

In parallel, *in silico* perennial ryegrass EST analysis has identified 1175 EST-SSR loci. A sub-set of 480 loci was evaluated for efficiency of amplification and genetic polymorphism. A total of c. 150 EST-SSR primer pairs detected polymorphism in the F_1 (NA_6 x AU_6) family, generating the two parental linkage maps and a consensus melded map of c. 180 loci (Faville et al. 2003). The EST-RFLP and EST-SSR datasets have now been combined (M.J, Faville and A.C. Vecchies, pers. comm.) to generate parental linkage maps containing c. 200 functionally-associated markers each, aligned to the p150/112 reference map. Candidate gene annotation by sequences corresponding to mapped loci in other Poaceae species will now be used to extend the conserved syntenic relationships described in Section 1.3. This method of *in silico* comparative mapping will permit more accurate assessments of macrosyntenic relationships between taxa.

As RFLPs are not suitable for high-density genetic mapping or high-throughput genotyping, and EST-SSRs are not available for all gene classes, there is a requirement to develop corresponding SNP markers for suites of candidate genes. It would be eventually desirable to identify SNPs and develop the appropriate detection assays for each of the current unigenes, corresponding to c. one-third to one-half of all genes in perennial ryegrass.

These EST-SNPs may then be used to construct high-density linkage maps in selected mapping families such as F_1 (NA_6 x AU_6). The ongoing intensive phenotypic analysis of such families will allow the constant evaluation of new candidate gene loci for co-location with QTLs already detected by whole-genome scans with anonymous markers.

3.2.3 Candidate Gene Selection

The selection of candidate genes for the evaluation of marker-QTL co-location is a major theoretical challenge. Three classes of candidates may be defined. Primary candidates are mainly biosynthetic genes with well-described activities. For instance, the lignin biosynthetic genes are good primary candidates for evaluation against QTLs identified for variation of herbage digestibility traits as assessed by near infra-red reflectance spectroscopy (NIRS). The fructan biosynthetic genes (fructosyltransferases) are good primary candidates for evaluation against QTLs for water soluble carbohydrate content, while plant defence-related candidate genes will be evaluated against QTLs for resistance to pathogens such as crown rust (*Puccinia coronata* Corda f.sp. *lolii*). This approach (Pflieger et al. 2001) has been validated in a number of other plant systems (e.g. Faris et al. 1999; McMullen et al. 1998; Prioul et al. 1999; Thorup et al. 2000). The primary candidate approach assumes that the candidate gene-based marker will constitute a 'perfect marker' for the QTL. QTL variation may be due either to allelic variation in the coding sequence of the gene (producing a functionally altered product) or the control regions of the gene (leading to altered levels of expression).

An intermediate class of secondary candidates may be defined as ESTs that correspond to executor genes in biochemical pathways, but have not as yet been reliably annotated. A large number of such secondary candidates will be available through the concurrent functional genomic analysis, especially microarray-based analysis of transcriptional activity using the perennial ryegrass 15K 'unigene' chip (Sawbridge et al. 2003b; Ong et al. 2003). Reverse genetics methods for functional analysis such as virus induced gene silencing (VIGS: Holzberg et al. 2002), for which systems based on ryegrass mosaic virus (RMV) are currently in development (G. Aldao-Humble, pers. comm.), will be used to further validate these secondary candidates.

The tertiary candidate group will include regulatory genes and represent the most difficult targets for identification. However, a large number of transcriptional control factors have been already been annotated belonging to a range of structural categories (MYB, MYC etc.). A combination of reverse genetics approaches will be required to identify tertiary candidates associated with particular traits, with VIGS and stable gene-silencing transgenic technologies (antisense transcription, RNAi: Hammond et al. 2001, Vance and Vaucheret 2001) used to supplement transcriptional profiling. The development of stable loss-of-function lines based on mutagenised populations may also be of key importance. Transposon mutagenesis is a possibility, but physical mutagenesis based on methylating agents such as ethylmethanesulphonate (EMS: McCallum et al. 2000) or radiation treatments such as fast neutron bombardment (Li et al. 2001) may offer a more attractive alternative and merit careful consideration for parallel programs.

3.3 SNP Discovery in Perennial Ryegrass

3.3.1 *In silico* SNP Discovery

Of 20,633 perennial ryegrass unigene clusters, 3,378 contained three or more independent sequences and are eligible for *in silico* SNP analysis (Sawbridge et al. 2003a,b). Of this group, 1,103 (33%) contained at least one SNP, over an average contig length of 1,153 bp. The total number of SNPs detected was 9,596, corresponding to an incidence of c. 1/132 bp. The *in silico* SNPs were detected in all regions of the transcriptional unit, including 5'-untranslated regions (UTRs), coding sequences (CSs) and 3'-UTRs. The majority of the perennial ryegrass ESTs were obtained from over 20 libraries constructed from plants of the New Zealand-bred variety Grasslands Nui, which is known to have a relatively restricted genetic structure through its origin from 9 parental clones (Armstrong, 1977). This variety has been demonstrated by AFLP-based genetic diversity analysis to show relatively limited genetic variation compared to other perennial ryegrass cultivars and ecotypes (Guthridge et al. 2001). The *in silico* estimate of SNP incidence in perennial ryegrass is consequently likely to be an underestimate of the overall prevalence of SNPs within the gene pool of the species.

3.3.2 *In vitro* SNP Discovery

The process of EST-SNP discovery and validation is being performed with the parental genotypes of the F_1 (NA_6 x AU_6) mapping family (Batley et al. 2003). Genomic amplicons for PCR amplification are obtained, individually from each of the parental genotypes, by primer design from selected full length cDNA sequences to scan intervals of approximately 200-400 bp across the transcriptional unit. The PCR products are visualised using agarose gel electrophoresis and subsequently cloned and sequenced. The presence of introns within amplicons is inferred on the basis of anomalously larger size compared to the predicted values. As the presence of large introns and primer design across intron-exon boundaries will lead to failure of efficient amplification, knowledge of gene organisation is desirable. Access to genomic DNA sequence information for rice *(Oryza sativa* L.) may allow accurate prediction of the location of intron-exon boundaries based on comparative genomics within the Poaceae family.

Two major advantages arise from the sequencing of cloned PCR products compared to direct sequencing. Firstly, genes corresponding to different members of related multiple gene families are discriminated, to allow gene-specific SNP analysis. As a substantial number of perennial ryegrass EST-RFLP profiles reveal multigene family structure (Vecchies et al. 2003), this is an important issue. Secondly, the haplotype structures of each homologous sequence in a heterozygous genotype can readily be determined using this method, with no further analysis required. The sequences are aligned using the Sequencher software package (Genecodes) to generate a map of SNP distribution, either between the parental consensus sequences (corresponding to AA x BB allelic structures), within one parental genotype but not the other (e.g. AB x BB) or within both parental genotypes (e.g. AB x AB). SNP haplotypes are then defined for each parent (Table 1). In the *LpASR* (abiotic stress resistance) candidate gene, this *in vitro* analysis has detected a high SNP frequency, with 15 heterozygous SNPs in 764 bp of amplified sequence for NA_6 and 10 heterozygous SNPs in 756 bp for AU_6. The SNPs were located in 5'- and 3'-UTRs, an intron, but most commonly in the coding sequence. This data suggests that EST-SNP marker development for perennial ryegrass will be relatively straightforward compared to other Poaceae species.

The putative SNPs are then validated using the single nucleotide primer extension (SNuPe) assay followed by capillary electrophoresis (Figure 6) and mapped in the full F_1 population. In those cases where EST-SNPs are derived from cDNAs that have already been mapped as EST-RFLPs, evaluation of map position coincidence will be an important quality control issue, as well as allowing an investigation of the relationships between paralogous members of multigene families.

Table 1. Example of putative SNP haplotypes detected for a perennial ryegrass gene in the AU_6 parent. The first column of the table shows the position of the SNP in the sequence and the rows show the allelic variants at this position in 7 sequenced clones. The first haplotype is shown in the sequences from clones 1, 4, 5 and 7, and the second haplotype is represented by clones 2, 3 and 6.

SNP position (bp)	Clone 1	Clone 2	Clone 3	Clone 4	Clone 5	Clone 6	Clone 7
5	A	T	T	A	A	T	A
26	C	T	T	C	C	T	C
89	G	C	C	G	G	C	G
101	A	T	T	A	A	T	A
125	C	A	A	C	C	A	C
140	G	A	A	G	G	A	G
159	T	G	G	T	T	G	T
240	T	G	G	T	T	G	T
276	A	C	C	A	A	C	A
298	C	G	G	C	C	G	C
344	G	C	C	G	G	C	G
396	A	G	G	A	A	G	A
404	T	A	A	T	T	A	T

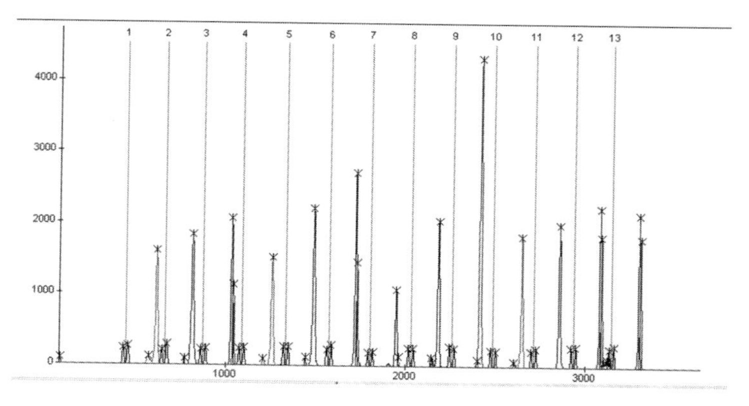

Figure 6. Electropherogram of multiple injections from a SNuPe reaction in a single capillary separation using the MegaBACE 1000 system. The numbered double peaks represent the injection marker and the remaining peaks represent the results of the SNP assay. Heterozygotes are characterised by 2 peaks between an injection marker and homozygotes by a single peak. The MegaBACE has been designed so that up to 12 SNPs can be serially injected into the same matrix within each capillary.

3.4 Population-based Candidate Gene-marker Validation

The major problem with the detection of candidate gene marker – QTL co-location in genetic linkage maps (generally based on the analysis of 100-200 segregants) is uncertainty in the true location of the QTL that may extend over 10-20 cM. For a species such as perennial ryegrass, this map distance could correspond to up to 40 Mb of genomic DNA (assuming that 1 cM = 2 Mb on average), which is sufficient to contain thousands of candidate genes. Consequently, a positive identification of candidate gene status is difficult to obtain. One possible solution is to perform fine genetic linkage mapping in large full-sib segregating families, with c. 1000 individuals. The resolving power for QTL detection has been shown to be considerably enhanced by the use of large segregating populations of this kind by theoretical and empirical studies in maize (Beavis, 1994). Larger populations also permit refined marker ordering and assessment of linkages over shorter distances. However, resolution of linked marker positions over the scale of the transcriptional unit will only be accessible to studies of many thousands (>100,000) of segregants.

The accuracy of detection of QTLs for traits of interest can be refined using LD or association mapping approaches, which exploit multiple recombination over historical time. A newly arising mutation within a population at a QTL is initially in complete LD with all other polymorphic loci (Mackay, 2001), but successive recombination events lead to the decay of LD and restoration of linkage equilibrium with all but closely linked loci. LD most frequently arises due to close physical proximity between genetic loci, but may also arise due to strong epistatic interactions between genes in unlinked positions, population sub-division and population admixture (Hartl and Clark, 1997). Spurious associations between unlinked genes and markers may be removed by a combination of linkage and linkage disequilibrium mapping (Wu et al. 2002). LD mapping is consequently a method for statistical analysis of non-random associations between genetic loci using population samples rather than full-sib mapping populations. This approach is well established for genetic studies of humans (Risch and Merikangas 1996) and some livestock species (Meuwissen et al. 2002) but is still relatively novel for plant species. Studies in maize have defined the extent and distribution of LD across the genome (Remington et al. 2001; Tenaillon et al. 2001), and correlation between candidate genes and markers has been explored for traits such as flowering time and plant height (Thornsberry et al. 2001) and amylose content in the kernel (Prioul et al. 1999).

The level of linkage disequilibrium (LD) between markers and target genes is variable between species, populations and regions of the genome.

The length of the genomic segment surrounding the original mutation in which LD persists depends on the average rate of recombination per generation in that region, the number of elapsed generations and the effective population size (Hill and Robertson 1968). Extensive LD is usually associated with populations derived from recent founder events, or with small population sizes, or inbreeding habit (Hagenblad et al. 2002; Nordborg et al. 2002). In general, we would expect an obligate outbreeding species such as perennial ryegrass to show LD over relatively short molecular distances (Mackay 2001). This is especially true for long established populations derived from a large number of parental individuals, as would be expected for ecotypes and many older varieties, in which many rounds of recombination have occurred. However, short-lived populations of perennial ryegrass that have been subjected to intense selection tend to show higher mean chiasma frequencies than more perennial natural populations (Rees and Dale 1974). Such populations may also prove informative for LD studies.

Highly intensive whole genome scans, currently performed with densely-spaced anonymous genetic markers, are required for LD analysis in complex outbred populations and are logistically demanding (Kruglyak 1999; Rafalski 2002). Several such studies have applied AFLP analysis to correlate genetic variation with ecological diversity for adaptation to low winter temperatures (Skøt et al. 2002) and flowering time (Skøt et al. 2003). However, candidate gene-based markers provide a more focused approach (Wilson et al. 2003), as demonstrated for leaf rust resistance in barley (Brunner et al. 2000) and low temperature tolerance in rice (Abe et al. 2002). Similar principles have been applied to forestry species such as Monterey pine (Wilcox et al. 2002) which show some similarities of population structure to pasture species. The perennial ryegrass candidate gene SNP markers will be evaluated and validated as 'perfect markers' through correlation analysis in stratified population sub-divisions determined on the basis of phenotypic variation.

The design of LD mapping populations is an important theoretical and empirical issue. The key resources for the design of such populations are extensive germplasm collections. A global perennial ryegrass germplasm resource has been established at DPI-Hamilton (N. Bannan and K.F. Smith, unpublished data). The collection contains 477 accessions from all areas of natural distribution and cultivation, with similar numbers of domesticated and non-domesticated sources. The accessions have been established in a spaced-plant nursery with up to 24 plants per accession, and are being subjected to multi-trait phenotyping over recurrent seasons. Hierarchical measurement of genetic variation between and within accessions is being performed using genomic DNA-derived SSR markers (R. Ponting, J. Batley and J.W. Forster, unpublished data), and will provide a comprehensive view of the genetic architecture of the species, supported by previous empirical studies (Guthridge et al. 2001). This data will be used to determine the degree of genetic affinity between different components of the gene pool, and hence to design optimised populations for LD studies in perennial ryegrass.

The feasibility of using LD analysis for pasture plants may be also evaluated using markers which have been demonstrated by linkage analysis to be in very close linkage with gene/QTLs for a known target trait, as demonstrated for the bolting gene (*B*) of sea beet (Hansen et al. 2001). The SREF trait is known to be variable in some perennial ryegrass populations (Nyquist 1963), providing a suitable model system. Empirical estimates of the distribution of LD across genomes are also essential and may be obtained from population-based genetic analysis. Substantial variation of recombinational activity between different genomic regions has been observed in both human (Stumpf 2002) and plant (Dooner et al. 1997) studies, and is consistent with cytogenetic observations.

3.5 Strategies for Implementation of Candidate Gene-based Markers

Given the success of the population-based validation approach, screening of germplasm collections with EST-SNP markers will be performed to identify genotypes with superior alleles at agronomic trait genes as potential parents for varietal development. This method has previously been termed direct allele selection (DAS: Sorrells and Wilson 1997) and gene-assisted selection (GAS: Wilcox et al. 2003). The selected genotypes will be validated by intensive phenotypic analysis, in concert with appropriate pair-crosses to test co-transmission of the selection marker with the relevant trait. The emphasis here is on primary selection at the parental level rather than recurrent selection at the progeny level, although the latter will be clearly necessary to manage the allelic combinations in subsequent generations. The high-throughput capability of SNP genotyping will permit efficient MAS implementation during synthetic variety production.

Apart from selection at the candidate gene loci, that are expected to be 'fixed' in homozygous or close-to-homozygous status in the resulting variety, molecular marker analysis will also be used to identify whole genome genetic distance between the parental clones. The expectation here is that varietal performance will be minimally compromised if variation is maximised at these 'unselected' regions of the genome, so as to minimise inbreeding depression effects. Whole genome genotyping with highly polymorphic genetic markers such as genomic DNA-derived and EST-SSRs will be used to quantify genetic distance between selected genotypes and genetic variability within accessions. The genetic distance between individuals on a whole genome basis will be evaluated by empirical tests as a means of choosing between potential parental genotypes that are otherwise similar in terms of superior allele content at target loci.

4. CONCLUSIONS

The development of specific molecular genetic marker systems and mapping families for perennial ryegrass has permitted rapid progress in genetic map development and trait dissection. QTLs for a majority of primary breeders' traits have been identified in a range of mapping families. However, further work is required to confirm consistent relationships between specific genomic regions and genetic variability for agronomic characters. The stability of QTLs across different environmental conditions is also an important target for study. International collaboration will continue to provide an effective means to achieve these objectives. The implementation of markers for MAS remains a major challenge due to the complexity of the breeding system and the multigenic control of the majority of target traits.

An emphasis on the assessment of molecular diversity associated with gene loci will be the major theme in the next phase of molecular marker technology development for pasture and turf grasses. Candidate gene-derived markers will be validated using population-based analysis and utilised to select parental genotypes for cultivar development. Comparisons of wild and cultivated accessions may allow the identification of genetic loci that have been positively selected during domestication (Brookfield 2001). Substantial empirical work is required to determine the overall levels of genetic diversity in perennial ryegrass and how this variation is partitioned within and between populations, as well as the extent of LD and its variation across the genome. Advances in statistical genetics will be required to exploit these methods effectively.

Problems with the identification of candidate genes for marker development will create the most serious potential obstacles to the successful implementation of this program. However, parallel activities in functional genomics for perennial ryegrass will continuously provide criteria for selection. Comparative genomics with other Poaceae species will also support the choice of candidate genes through orthologous QTL detection and marker co-location. In this context, the establishment of conserved syntenic relationships has been especially beneficial. The whole genome sequence of rice and extensive EST collections from other cereals will provide a critical resource for these activities.

Finally, further advances in high-throughput SNP genotyping such as low-cost automated DNA extraction technology and the use of high density oligonucleotide microarrays (Cho et al. 1999) will increase the scale and precision of marker implementation in grass breeding.

ACKNOWLEDGMENTS

The authors thank members of ILGI, especially Dr. Mervyn Humphreys (IGER, Aberystwyth, UK), Dr. Isabel Roldán-Ruiz (DvP, Gent, Belgium), Dr. Oene Dolstra (PRI, Wageningen, Netherlands), Dr. Odd-Arne Rognli (AUN, Ås, Norway) and Dr. Roland Kölliker (FAL, Zürich, Switzerland) for communicating research results ahead of publication. The authors also thank Prof. German Spangenberg, Prof. Michael Hayward, Dr. Noel Cogan, Kathryn Guthridge and Eline Van Zijll de Jong for critical reading for the manuscript. The contributions and support of past and present research colleagues at DPI-PBC and DPI-Hamilton is gratefully acknowledged. The original research described here was supported by the Victorian Department of Primary Industries and the Cooperative Research Centre for Molecular Plant Breeding.

REFERENCES

Abe F, Saito K, Miura K, Toriyama K (2002) A single nucleotide polymorphism in the alternative oxidase gene among rice varieties differing in low temperature tolerance. FEBS Lett. 527: 181-185.

Alm V (2001) Comparative genome analyses of meadow fescue (*Festuca pratensis* Huds.): Genetic linkage mapping and QTL analyses of frost and drought tolerance. Ph.D. thesis, Agricultural University of Norway.

Alm V, Cheng F, Busso C, Devos K, Vollan K, Grieg Z, Rognli O-A (2003) A linkage map of meadow fescue (*Festuca pratensis* Huds.) and comparative mapping with the Triticeae species, *Lolium*, oat, rice, maize and sorghum. Theor. Appl. Genet. (in press).

Armstead IP, Turner LB, King IP, Cairns AJ, Humphreys MO (2002) Comparison and integration of genetic maps generated from F_2 and BC_1-type mapping populations in perennial ryegrass. Plant Breed. 121: 501-507.

Armstrong CS (1977) 'Grasslands Nui' perennial ryegrass. N.Z. J. Exp. Agr. 5: 381-384.

Bannan NR, Smith KF, Forster JW (2003) Efficient use of marker assisted selection in the breeding of perennial ryegrass (*Lolium perenne* L.). In: Abstracts of Molecular Breeding of Forage and Turf, Third International Symposium, p.27. May 18-22, Dallas, Texas and Ardmore, Oklahoma, USA.

Batley J, Vecchies AC, Drayton MC, Dumsday JL, Edwards D, Emmerling M, Sawbridge T, Spangenberg GC, Forster JW (2003) Development and application of molecular technologies in forage and turf improvement. In: Abstracts of Molecular Breeding of Forage and Turf, Third International Symposium, p.30. May 18-22, Dallas, Texas and Ardmore, Oklahoma, USA.

Baumann U, Juttner J, Bian X, Langridge P (2000) Self-incompatibility in the grasses. Ann. Bot. 85 (Supplement A): 203-209.

Beavis WD (1994) The power and deceit of QTL experiments: lessons from comparative QTL studies. In: 49th Annu. Corn Sorghum Res. Conf. pp. 252-268. Am. Seed Trade Assoc., Washington DC.

Bert PF, Charmet G, Sourdille P, Hayward MD, Balfourier F (1999) A high-density molecular map for ryegrass (*Lolium perenne*) using AFLP markers. Theor. Appl. Genet. 99: 445-452.

Bezant J, Laurie D, Pratchett N, Chojecki J, Kearsey M (1996) Marker regression mapping of QTL controlling flowering time and plant height in a spring barley (*Hordeum vulgare* L.) cross. Heredity 77: 64-73.

Bhattramakki D, Dong J, Chhabra AK, Hart GE (2000) An integrated SSR and RFLP map of *Sorghum bicolor* (L.) Moench. Genome 43: 988-1002.

Brookfield JFY (2001) Population genetics: the signature of selection. Curr. Biol. 11: R388-R390.

Brown R, Warnke S, Barker R (2003) Toward a DNA-based screening test for annual ryegrass (*Lolium multiflorum*) contamination of perennial ryegrass (*L. perenne*) seed. In: Abstracts of Molecular Breeding of Forage and Turf, Third International Symposium, p.80. May 18-22, Dallas, Texas and Ardmore, Oklahoma, USA.

Brunner S, Keller B, Feuillet C (2000) Molecular mapping of the *Rph7.g* leaf rust resistance gene in barley (*Hordeum vulgare* L.). Theor. Appl. Genet. 101: 783-788.

Cho RJ, Mindrinos M, Richards DR, Sapolski RJ, Anderson M, Drenkard E, Dewdney J, Reuber TL, Stammers M, Federspiel N, Theologis A, Yang W, Hubbell E, Au M, Chung EY, Lashkari D, Lemieux B, Dean C, Lipshutz RJ, Ausubel FM, Davis RW, Oefner PJ (1999) Genome wide mapping with biallelic markers in *Arabidopsis thaliana*. Nat. Genet. 23: 203-207.

Close TJ (1997) Dehydrins: a commonality in the response of plants to dehydration and low temperature. Physiol. Plant 100: 291-296.

Cornish MA, Hayward MD, Lawrence MJ (1979) Self-incompatibility in ryegrass. I. Genetic control in diploid *Lolium perenne* L. Heredity 43: 95-106.

Cruickshank IAM (1957) Crown rust of ryegrass. N.Z. J. Sci. Tech. 38: 539-43.

Curley J, Sim S, Leong S, Warnke S, Barker R, Jung G (2003a) Comparative mapping of *Magnaporthe grisea* resistance genes and QTL in rice, barley and perennial ryegrass. In: Abstracts of Plant & Animal Genome XI Conference, P162. January 11-15, San Diego, CA, USA.

Curley J, Sim S, Leong S, Warnke S, Barker R, Jung G (2003b) Comparison of putative orthologous *Magnaporthe grisea* resistance QTL in rice, barley and perennial ryegrass. In: Abstracts of Molecular Breeding of Forage and Turf, Third International Symposium, p.1. May 18-22, Dallas, Texas and Ardmore, Oklahoma, USA.

De Roover J, Vandenbranden K, Van Laere A, Van den Ende W (2000) Drought induces fructan synthesis and 1-SST (sucrose:sucrose fructosyltransferase) in roots and leaves of chicory seedlings (*Chicorium intybus* L.). Planta 210: 808-814.

Devos KM, Gale MD (1997) Comparative genetics in the grasses. Plant Mol. Biol. 35: 3-15

Dexter ST, Tottingham WE, Graber LF (1932) Investigations of the hardiness of plants by measurement of electrical conductivity. Plant Physiol. 7: 63-78.

Dolstra O, Boucoiran CFS, Dees D, Denneboom C, Groot PJ, De Vos ALF, Van Loo EN (2003) Genome-wide search of genes relevant to nitrogen-use-efficiency in perennial ryegrass. In: Abstracts of Plant & Animal Genome XI Conference, W208. January 11-15, San Diego, CA, USA.

Dooner HK, Martinez-Ferez IM (1997) Recombination occurs uniformly within the bronze gene, a meiotic recombination hotspot in the maize genome. Plant Cell 9: 1633-1646.

Dumsday JL, Trigg PJ, Jones ES, Batley J, Smith KF, Forster JW (2003a) SSR-based genetic linkage analysis of resistance to crown rust (*Puccinia coronata* Corda f.sp. *lolii*) in perennial ryegrass (*Lolium perenne* L.). In: Abstracts of Plant & Animal Genome XI Conference, P709. January 11-15, San Diego, CA, USA.

Dumsday JL, Smith KF, Forster JW, Jones ES (2003b) SSR-based genetic linkage analysis of resistance to crown rust (*Puccinia coronata* Corda f. sp. *lolii*) in perennial ryegrass (*Lolium perenne* L.). Plant Path. (in press).

Dumsday JL, Smith KF, Forster JW, Jones ES (2003c) SSR-based genetic linkage analysis of resistance to crown rust (*Puccinia coronata* Corda f. sp. *lolii*) in perennial ryegrass (*Lolium perenne* L.). In: Abstracts of Molecular Breeding of Forage and Turf, Third International Symposium, p.15. May 18-22, Dallas, Texas and Ardmore, Oklahoma, USA.

Dumsday JL, Batley J, McFarlane NM, Croft V, Smith KF, Forster JW (2003d) Genetic diversity in the perennial ryegrass crown rust pathogen *Puccinia coronata* Corda f.sp. *lolii* Brown. In: Abstracts of Molecular Breeding of Forage and Turf, Third International Symposium, p.6. May 18-22, Dallas, Texas and Ardmore, Oklahoma, USA.

Edwards KJ, Barker JHA, Daly A, Jones C, Karp A (1996) Microsatellite libraries enriched for several microsatellite sequences in plants. Biotechniques 20: 758-759.

Fang C, Alm V, Aamlid TS, Rognli O-A (2003) Comparative mapping of QTLs for seed production and related traits in meadow fescue (*Festuca pratensis* Huds.). In: Abstracts of Molecular Breeding of Forage and Turf, Third International Symposium, p.2. May 18-22, Dallas, Texas and Ardmore, Oklahoma, USA.

Faris JD, Li WH, Liu DJ, Chen PD, Gill BS (1999) Candidate gene analysis of quantitative disease resistance in wheat. Theor. Appl. Genet. 98: 219-225.

Faville MJ, Schreiber M, Bryan GT, Forster JW, Sawbridge T, Spangenberg GC (2003) An EST-SSR map for the perennial ryegrass genome. In: Abstracts of Molecular Breeding of Forage and Turf, Third International Symposium, p.6. May 18-22, Dallas, Texas and Ardmore, Oklahoma, USA.

Forster JW, Jones ES, Kölliker R, Drayton MC, Dupal MP, Guthridge KM, Smith KF (2001) DNA profiling in outbreeding forage species. In: Plant genotyping – the DNA fingerprinting of plants, Henry R (ed.), pp. 299-320. CABI Press.

Fujimori M, Hayashi K, Hirata M, Mizuno K, Fujiwara T, Akiyama F, Mano Y, Komatsu T, Takamizo T (2003) Linkage analysis of crown rust resistance in Italian ryegrass (*Lolium multiflorum Lam.*). In: Abstracts of Plant & Animal Genome XI Conference, W203. January 11-15, San Diego, CA, USA.

Fuong FT, Voylokov AV, Smirnov VG (1993) Genetic studies of self-fertility in rye (*Secale cereale* L.). 2. The search for molecular marker genes linked to self-incompatibility loci. Theor. Appl. Genet. 87: 619-623.

Gale MD, Devos KM (1998) Plant comparative genetics after 10 years. Science 282: 656-659.

Galiba G, Quarrie SA, Sutka J, Morgounov A, Snape JW (1995) RFLP mapping of the vernalisation (*Vrn-1*) and frost resistance (*Fr-1*) genes on chromosome 5A of wheat. Theor. Appl. Genet. 90: 1174-1179.

Grattapaglia D, Sederoff R (1994) Genetic linkage maps of *Eucalyptus grandis* and *Eucalyptus urophylla* using a pseudo-testcross: mapping strategy and RAPD markers. Genetics 137: 1121-1137.

Guthridge KM, Dupal MD, Kölliker R, Jones ES, Smith KF, Forster JW (2001) AFLP analysis of genetic diversity within and between populations of perennial ryegrass (*Lolium perenne* L.). Euphytica 122: 191-201.

Guthridge KM, Ciavarella T, McFarlane N, Batley J, Jones ES, Smith KF, Forster JW (2003a) Molecular marker-based analysis of drought-associated phenotypic traits in perennial ryegrass (*Lolium perenne* L.). In: Abstracts of Plant & Animal Genome XI Conference, P710. January 11-15, San Diego, CA, USA.

Guthridge KM, McFarlane NM, Ciavarella TA, Batley J, Jones ES, Smith KF, Forster JW (2003b) Molecular marker-based analysis of morphological traits in perennial ryegrass (*Lolium perenne* L.). In: Abstracts of Molecular Breeding of Forage and Turf, Third International Symposium, p.58. May 18-22, Dallas, Texas and Ardmore, Oklahoma, USA.

Hackauf B, Makarova N, Wehling P (2000) Development of STS markers for the self-incompatibility loci in rye. In: Abstracts of Plant, Animal & Microbe Genome VIII Conference, P239. January 9-12, San Diego, CA, USA.

Hagenblad J, Nordborg M (2002) Sequence variation and haplotype structure surrounding the flowering time locus *FRI* in *Arabidopsis thaliana*. Genetics 161: 289-298.

Hammond SM, Caudy AA, Hannon GJ (2001) Post-transcriptional gene silencing by double-stranded DNA. Nature Rev. Gen. 2: 1110-1119.

Hansen M, Kraft T, Ganestam S, Säll T, Nilsson N-O (2001) Linkage disequilibrium mapping of the bolting gene in sea beet using AFLP markers. Genet. Res. (Cambs.) 77: 61-66.

Hartl DL, Clark AG (1997) Principles of population genetics. 3rd edition. pp. 542. Sinauer Sunderland, MA, USA.

Hayward MD (1977) Genetic control of resistance to crown rust (*Puccinia coronata* Corda) in ryegrass (*Lolium perenne* L.) and its implication in plant breeding. Theor. Appl. Genet. 51: 49-53.

Hayward MD, McAdam NJ, Jones JG, Evans C, Evans GM, Forster JW, Ustin A, Hossain KG, Quader B, Stammers M, Will JAK (1994) Genetic markers and the selection of quantitative traits in forage grasses. Euphytica 77: 269 – 275.

Hayward MD, Forster JW, Jones JG, Dolstra O, Evans C, McAdam NJ, Hossain KG, Stammers M, Will JAK, Humphreys MO, Evans GM (1998) Genetic analysis of *Lolium*. I. Identification of linkage groups and the establishment of a genetic map. Plant Breed. 117: 451-455.

Hill WG, Robertson A (1968) Linkage disequilibrium in finite populations. Theor. Appl. Genet. 38: 226-231.

Hirata M, Fujimori M, Inoue M, Miura Y, Cai H, Satoh H, Mano Y, Takamizo T (2003) Mapping of a new crown rust resistance gene, *Pc*2, in Italian ryegrass cultivar 'Harukaze'. In: Abstracts of Molecular Breeding of Forage and Turf, Third International Symposium, p.11. May 18-22, Dallas, Texas and Ardmore, Oklahoma, USA.

Holzberg S, Brosio P, Gross C, Pogue GP (2002) Barley stripe mosaic virus-induced gene silencing in a monocot plant. Plant J. 30: 315-327.

Humphreys MO (1992) Association of agronomic traits with isozyme loci in perennial ryegrass (*Lolium perenne* L.). Euphytica 59: 141-150.

Humphreys MO, King IP, Humphreys MW, Thomas HM (1998). Genome Research in *Festuca* and *Lolium*. In: Proc. Intl. Workshop Utilization of Transgenic Plants and Genome Analysis in Forage Crops, Nakagawa H (ed.), pp 31-39. National Grasslands Research Institute, Nishnasuno, Tochigi, Japan.

Humphreys MO, Turner LB, Armstead IP (2002) QTL identification and marker-assisted selection for feeding value in ryegrass. In: Genotype-Phenotype: Narrowing the Gaps. December 16-18, 2002, Royal Agricultural College, Cirencester, UK.

Humphreys M, Turner L, Armstead L (2003) QTL mapping in *Lolium perenne*. In: Abstracts of Plant & Animal Genome XI Conference, W207. January 11-15, San Diego, CA, USA.

Jefferies SP, Barr AR, Karakousis A, Kretschmer JM, Manning S, Chalmers KJ, Nelson JC, Islam AKMR, Langridge P (1999) Mapping of chromosome regions conferring boron toxicity tolerance in barley. Theor.Appl. Genet. 98: 1293-1303.

Jones ES, Dupal MP, Kölliker R, Drayton MC, Forster JW (2001) Development and characterisation of simple sequence repeat (SSR) markers for perennial ryegrass (*Lolium perenne* L.). Theor. Appl. Genet. 102: 405-415.

Jones ES, Mahoney NL, Hayward MD, Armstead IP, Jones JG, Humphreys MO, King IP, Kishida T, Yamada T, Balfourier F, Charmet C, Forster JW (2002a) An enhanced molecular marker-based map of perennial ryegrass (*Lolium perenne* L.) reveals comparative relationships with other Poaceae species. Genome 45: 282-295.

Jones ES, Dupal MD, Dumsday JL, Hughes LJ, Forster JW (2002b) An SSR-based genetic linkage map for perennial ryegrass (*Lolium perenne* L.). Theor. Appl. Genet. 105: 577-584.

King IP, Morgan WG, Armstead IP, Harper JA, Hayward MD, Bollard A, Nash JV, Forster JW, Thomas HM (1998) Introgression mapping in the grasses. I. Introgression of *Festuca pratensis* chromosomes and chromosome segments into *Lolium perenne*. Heredity 81: 462-467.

Kruglyak L (1999) Prospects for whole-genome linkage disequilibrium mapping of common disease genes. Nat. Genet. 22: 139-144.

Kölliker R, Krähenbühl R, Schubiger F, Widmer F (2003) Genetic diversity and pathogenicity of the grass pathogen *Xanthomonas campestris* pv. *graminis*. In: Abstracts of Molecular Breeding of Forage and Turf, Third International Symposium, p.5. May 18-22, Dallas, Texas and Ardmore, Oklahoma, USA.

Korzun V, Malyshev S, Voylokov AV, Börner A (2001) A genetic map of rye (*Secale cereale* L.) combining RFLP, isozyme, protein, microsatellite and gene loci. Theor. Appl. Genet. 102: 709-717.

Kubik C, Meyer WA, Gaut BS (1999) Assessing the abundance and polymorphism of simple sequence repeats in perennial ryegrass. Crop Sci 39: 1136-1141.

Li X, Song Y, Century K, Straight S, Ronald P, Dong X, Lassner M, Zhang Y. (2001) A fast neutron deletion mutagenesis-based reverse genetics systems for plants. Plant J. 27: 235-242.

Lidgett A, Jennings K, Johnson X, Guthridge K, Jones E, Spangenberg G. (2002) Isolation and characterisation of fructosyltransferase gene from perennial ryegrass (*Lolium perenne*). J. Plant Physiol. 159: 1037-1043.

Mackay TFC (2001) The genetic architecture of quantitative traits. Ann. Rev. Genet. 35: 303-309.

McCallum CM, Comai L, Greene EA, Henikoff S. (2000) Targeting Induced Local Lesions IN Genomes (TILLING) for plant functional genomics. Plant Physiol. 123: 439-442.

McFarlane NM, Guthridge KM, Smith KF, Jones ES, Forster JW (2001) Photosynthetic variation in genotypes of perennial ryegrass (*Lolium perenne* L) selected to map drought tolerance. Proceedings of the 10th Australian Agronomy Conference. www.regional.org.au/au/asa/2001.

McFarlane NM, Guthridge KM, Ciavarella TA, Batley J, Jones ES, Smith KF, Forster JW (2003) Molecular marker-based analysis of drought-tolerance associated phenotypic traits in perennial ryegrass (*Lolium perenne* L.). In: Abstracts of Molecular Breeding of Forage and Turf, Third International Symposium, p.17. May 18-22, Dallas, Texas and Ardmore, Oklahoma, USA.

McInnes R, Lidgett A, Lynch D, Huxley H, Jones E, Mahoney N, Spangenberg G (2002) Isolation and characterisation of a cinnamoyl-CoA reductase gene from perennial ryegrass (*Lolium perenne*). J. Plant Physiol. 159: 415-422.

McMullen MD, Byrne PF, Snook P, Wiseman BR, Lee EA, Widstrom NW, Coe EH (1998) Quantitative trait loci and metabolic pathways. Proc. Natl. Acad. Sci. USA 97: 11192-11197.

Meuwissen THE, Karlsen A, Lien S, Olsaker I, Goddard ME (2002) Fine mapping of a quantitative trait locus for twinning rate using combined linkage and linkage disequilibrium mapping. Genetics 161: 373-379.

Michelmore RW, Paran I, Kesseli RV (1991) Identification of markers linked to disease-resistance genes by bulked segregant analysis: a rapid method to detect markers in specific genomic regions by using segregating populations. Proc. Natl. Acad. Sci. USA 88: 9828-32.

Muylle H, Van Bockstaele E, Roldán-Ruiz I (2003) Identification of four genomic regions involved in crown rust resistance in a *L. perenne* population. In: Abstracts of Molecular Breeding of Forage and Turf, Third International Symposium, p.9. May 18-22, Dallas, Texas and Ardmore, Oklahoma, USA.

Nordborg M, Borevitz JO, Bergelson J, Berry CC, Chory J, Hagenblad J, Kreitman M, Maloof JN, Noyes T, Oefner PJ, Stahl EA, Weigel D (2002) The extent of linkage disequilibrium in *Arabidopsis thaliana*. Nat. Genet. 30: 190-193.

Nyquist WE (1963) Fluorescent perennial ryegrass. Crop Sci. 3: 223-226.

Okora JO, Watson CE, Gourley LM, Keith BC, Vaughn CE (1999) Comparison of botanical characters and seedling root fluorescence for distinguishing Italian and perennial ryegrass. Seed Sci. Tech. 27: 721-730.

Ong EK, Webster T, Sawbridge T, Nguyen N, Rhodes C, Tian P, Winkworth A, Spangenberg GC (2003) Ryegrass and clover 15K unigene microarrays: generation, validation and applications in genome-wide gene expression profiling and promoter activity. In: Abstracts of Molecular Breeding of Forage and Turf, Third International Symposium, p.73. May 18-22, Dallas, Texas and Ardmore, Oklahoma, USA.

Pan A, Hayes PM, Chen F, Chen THH, Blake T, Wright S, Karsai I, Bedö Z (1994) Genetic analysis of the components of winter-hardiness in barley (*Hordeum vugare* L.). Theor. Appl. Genet. 89: 900-910.

Pflieger S, Lefebvre V, Causse M (2001) The candidate gene approach in plant genetics: a review. Mol. Breed. 7: 275-291.

Ponting RC, McFarlane NM, Dupal MP, Batley J, Jones ES, Smith KF, Forster JW (2003) Molecular marker-based analysis of perennial/annual growth habit in ryegrasses (*Lolium* spp.). In: Abstracts of Molecular Breeding of Forage and Turf, Third International Symposium, p.59. May 18-22, Dallas, Texas and Ardmore, Oklahoma, USA.

Prioul JL, Pelleschi S, Séne M, Thévenot C, Causse M, de Vienne D, Leonard A. (1999) From QTLs for enzyme activity to candidate genes in maize. J. Exp. Bot. 50: 1281-1288.

Rafalski JA, Vogel JM, Morgante M, Powell W, Andre C, Tingey SV (1996) Generating and using DNA markers in plants. In: Non-mammalian genomic analysis: a practical guide, Birren B, Lai E (eds.), pp 75-135. Academic Press Inc., San Diego.

Rafalski A (2002) Applications of single nucleotide polymorphisms in crop genetics. Curr. Op. Plant Biol. 5: 94-1000.

Rees H, Dale P (1974) Chiasmata and variability in *Lolium* and *Festuca* populations. Chromosoma 47: 335-351.

Reheul D, Baert J, Boller B, Bourdan O, Cagas B (2001) Crown rust *Puccinia coronata* Corda: recent developments. Proc. 3rd Intl Conf. Harmful and Beneficial Microorganisms in Grassland, Pastures and Turf. Paul VH, Dapprich PD (eds.). pp. 17-28. September 26, 2000. Soest, Germany.

Remington DL, Thornsberry JM, Matsuoka Y, Wilson LM, Whitt SR, Doebley J, Kresovich S, Goodman MM, Buckler IV ES (2001) Structure of linkage disequilibrium and phenotypic associations in the maize genome. Proc. Natl. Acad. Sci. USA 98: 11479-11484.

Risch N, Merikangas K (1996) The future of genetic studies of complex human diseases. Science 273: 1516-1517.

Ritter E, Gebhardt C, Salamini F (1990) Estimation of recombination frequencies and construction of RFLP linkage maps in plants from crosses between heterozygous parents. Genetics 135: 645-654.

Roderick HW, Humphreys MO, Turner L, Armstead I, Thorogood D (2003) Isolate specific trait loci for resistance to crown rust in perennial ryegrass. Proc. 24th EUCARPIA Fodder Crops and Amenity Grasses Section Meeting, September 22-26 2002. Braunschweig, Germany (in press).

Saha MC, Chekhovksi K, Zwonitzer JC, Eujayl I, Mian MAR (2003a) Development of microsatellite markers for forage grass and cereal species. In: Abstracts of Plant & Animal Genome XI Conference, P201. January 11-15, San Diego, CA, USA.

Saha MC, Zwonitzer JC, Hopkins AA, Mian MAR (2003b) A molecular linkage map of a tall fescue population segregating for forage quality traits. In: Abstracts of Molecular Breeding of Forage and Turf, Third International Symposium, p.3. May 18-22, Dallas, Texas and Ardmore, Oklahoma, USA.

Sawbridge T, Ong E-K, Emmerling M, Lewis B, Wearne K, Simmonds J, Nunan K, O'Neill M, O'Toole F, Winkworth A, Tian P, Rhodes C, Meath K, Spangenberg G (2003a) Generation and analysis of a collection of expressed sequence tags in perennial ryegrass. In: Abstracts of Molecular Breeding of Forage and Turf, Third International Symposium, p.35. May 18-22, Dallas, Texas and Ardmore, Oklahoma, USA.

Sawbridge T, Ong E-K, Binnion C, Emmerling M, McInnes R, Meath K, Nguyen N, Nunan K, O'Neill M, O'Toole F, Rhodes C, Simmonds J, Tian P, Wearne K, Webster T, Winkworth A, Spangenberg G (2003b) Generation and analysis of expressed sequence tags in perennial ryegrass (*Lolium perenne* L.). Plant Sci. (in press).

Skøt L, Sackville-Hamilton NR, Mizen S, Chorlton KH, Thomas ID (2002) Molecular genecology of temperature response in *Lolium perenne*: 2. Association of AFLP markers with ecogeography. Mol. Ecol. 11: 1865-1876.

Skøt L, Humphreys M, Mizen S, Armstead I, Dhanoa D, Sackville-Hamilton NR, Thomas I, Chorlton K (2003) Association mapping of molecular markers for flowering time using natural populations of *Lolium perenne*. In: Abstracts of Molecular Breeding of Forage and Turf, Third International Symposium, p.2. May 18-22, Dallas, Texas and Ardmore, Oklahoma, USA.

Sorrells ME, Wilson WA (1997) Direct classification and selection of superior alleles for crop improvement. Crop Sci. 37: 691-697.

Spangenberg G, Kalla R, Lidgett A, Sawbridge T, Ong E-K, John U (2001) Breeding forage plants in the genome era. In: Molecular breeding of forage crops. Spangenberg, G. (ed.), pp. 1-39, Kluwer academic press, Dordrecht.

Stam P, van Ooijen JW (1995) Joinmap™ version 2.0: software for the calculation of genetic linkage maps. CPRO-DLO, Wageningen, The Netherlands.

Stumpf MPH (2002) Haplotype diversity and the block structure of linkage disequilibrium. Trends Genet. 18: 226-228.

Tenaillon MI, Sawkins MC, Long AD, Gaut RL, Doebley JF, Gaut BS (2001) Patterns of DNA sequence polymorphism along chromosome 1 of maize (*Zea mays* spp. *mays* L.). Proc. Natl. Acad. Sci USA 98: 9161-9166.

Thomas H, Evans C, Thomas HM, Humphreys MW, Morgan WG, Huack B, Donnison I (1997) Introgression, tagging and expression of a leaf senescence gene in *Festulolium*. New Phytol. 137: 29-34.

Thornsberry JM, Goodman MM, Doebley J, Kresovich S, Nielsen D, Buckler IV ES (2001) *Dwarf8* polymorphisms associate with variation in flowering time. Nature Gen. 28: 286-289.

Thorogood D, Paget M, Humphreys M, Turner L, Armstead I, Roderick H (2001) QTL analysis of crown rust resistance infection in perennial ryegrass. Intl. Turfgrass Soc. Res. J. 9: 218-223.

Thorogood D., Kaiser WJ, Jones JG, Armstead I (2002) Self-incompatibility in ryegrass 12. Genotyping and mapping the *S* and *Z* loci of *Lolium perenne* L. Heredity 88: 385-390

Thorup TA, Tanyolac B, Livingstone KD, Popovsky S, Paran L, Jahn M (2000) Candidate gene analysis in the Solanaceae. Proc. Natl. Acad. Sci. USA 97: 11192-11197.

Turner LB, Humphreys MO, Cairns AJ, Pollock CJ (2001) Comparison of growth and carbohydrate accumulation in seedlings of two varieties of *Lolium perenne*. J. Plant Physiol. 158: 891-897.

Turner LB, Humphreys MO, Cairns AJ, Pollock CJ (2002) Carbon assimilation and partitioning into non-structural carbohydrate in contrasting varieties of *Lolium perenne*. J. Plant Physiol. 159: 257-263.

Van Deynze AE, Nelson JC, O'Donoghue LS, Ahn S, Siriponnwiwat W, Harrington SE, Yglesias ES, Braga DP, McCouch SR, Sorrells ME (1995) Comparative mapping in grasses. Oat relationships. Mol. Gen. Genet. 249: 349-356.

Vance V, Vaucheret H (2001) RNA silencing in plants – defence and counterdefence. Science 292: 2277-2280.

Vandewalle M, Caslyn E, Van Bockstaele E, Baert J, De Riek J (2003) DNA marker-assisted selection in the breeding of Italian ryegrass. In: Abstracts of Molecular Breeding of Forage and Turf, Third International Symposium, p.26. May 18-22, Dallas, Texas and Ardmore, Oklahoma, USA.

Vecchies AC, Drayton MC, Hughes LJ, Batley J, Guthridge KM, Jones ES, Smith KF, Sawbridge T, Spangenberg GC, Forster JW (2003) Candidate gene mapping in perennial ryegrass (*Lolium perenne* L.). In: Abstracts of Molecular Breeding of Forage and Turf, Third International Symposium, p.13. May 18-22, Dallas, Texas and Ardmore, Oklahoma, USA.

Voylokov AV, Korzun V, Börner A (1998) Mapping of three self-fertility mutations in rye (*Secale cereale* L.) using RFLP, isozyme and morphological markers. Theor. Appl. Genet. 97: 147-153.

Warnke SE, Barker RE, Brilman LA, Young III, Cook RL (2002) Inheritance of superoxide dismutase (*Sod*-1) in a perennial x annual ryegrass cross and its allelic distribution among cultivars. Theor. Appl. Genet. 105: 1146-1150.

Warnke SE, Barker RE, Sung-Chu S, Jung G, Forster JW (2003a) Genetic map development and syntenic relationships of an annual x perennial ryegrass mapping population. In: Abstracts of Plant & Animal Genome XI Conference, W206. January 11-15, San Diego, CA, USA.

Warnke S, Barker R, Sim S, Jung G, Mian MAR (2003b) Identification of flowering time QTLs in an annual x perennial ryegrass mapping population. In: Abstracts of Molecular Breeding of Forage and Turf, Third International Symposium, p.43. May 18-22, Dallas, Texas and Ardmore, Oklahoma, USA.

Wei J-Z, Chatteron NJ, Larson SR, Wang R R-C (2000) Linkage mapping and nucleotide polymorphisms of the 6-SFT gene in cool-season grasses. Genome 43: 931-938.

Wilcox PL, Cato S, Ball RD, Kumar S, Lee JR, Kent J, Richardson TE, Echt CE (2002) Genetic architecture of juvenile wood density in *Pinus radiata* and implications for design of linkage disequilibrium studies. PAG X, San Diego, California.

Wilcox PL, Echt CS, Cato S, McMillan L, Kumar S, Ball RD, Burdon RD, Pot D (2003) Gene assisted selection – a new paradigm in forest tree species? In: Abstracts of Plant, Animal & Microbe Genome X Conference, W143. January 12 – 16, San Diego, CA, USA.

Wilkins PW (1978) Specialisation of crown rust on highly and moderately resistant plants of perennial ryegrass. Ann. Appl. Biol. 88: 179-184.

Wilson ID, Barker GL, Edwards KJ (2003) Genotype to phenotype: a technological challenge. Ann. Appl. Bot. 142: 33-39.

Wu R, Ma C-X, Casella G (2002) Joint linkage and linkage disequilibrium mapping of quantitative trait loci in natural populations. Genetics 160: 779-792.

Xu WW, Sleper DA, Chao S (1995) Genome mapping of polyploid tall fescue (*Festuca arundinacea* Schreb.) with RFLP markers. Theor. Appl. Genet. 91: 947-955.

Yamada T, Jones ES, Nomura T, Hisano H, Shimamoto Y, Smith KF, Forster JW (2003) QTL analysis of morphological, developmental and winter hardiness-associated traits in perennial ryegrass (*Lolium perenne* L.). In: Abstracts of Molecular Breeding of Forage and Turf, Third International Symposium, p.16. May 18-22, Dallas, Texas and Ardmore, Oklahoma, USA.

Yamada T, Jones ES, Cogan NOI, Vecchies AC, Nomura T, Hisano H, Shimamoto Y, Smith KF, Hayward MD, Forster JW QTL analysis of morphological, developmental and winter-hardiness associated traits in perennial ryegrass (*Lolium perenne* L.). Crop Sci. (submitted).

Yu G-X, Wise RP (2000) An anchored AFLP- and retrotransposon-based map of diploid *Avena*. Genome 43: 736-749.

EST-SSRs for Genetic Mapping in Alfalfa

Mary Sledge[1], Ian Ray[2] and M. A. Rouf Mian[1]
[1]*Forage Improvement Division, Samuel Roberts Noble Foundation, 2510 Sam Noble Parkway, Ardmore, OK 73401, USA.* [2]*Agronomy and Horticulture Department, New Mexico State University, PO Box 30003, Las Cruces, NM 88003-8003, USA. (Email:mksledge@noble.org).*

Keywords: alfalfa, EST-SSR, *Medicago*, SDRF

Abstract:

Several genetic maps of alfalfa have been published, both at the diploid (*Medicago sativa* subsp. *coerulea*) and tetraploid (*Medicago sativa* L.) levels. These maps have been constructed primarily with RFLP and RAPD markers, and have not been integrated into a single, reference map. The development of reliable, publicly available PCR-based markers would provide a basis for unifying the variously available alfalfa genetic maps. The goal of the present research is to construct a molecular map of tetraploid alfalfa using EST-derived SSR markers. Two alfalfa backcross populations have been constructed from a cross between a water-use efficient, *M. falcata* genotype and a low water-use efficient *M. sativa* genotype of Chilean origin. These populations are also segregating for yield, fall dormancy, and winter hardiness. We have screened 333 primer pairs amplifying *M. truncatula* EST-SSRs, approximately 50% of which are polymorphic between the Chilean and *M. falcata* parents. We have currently identified a total of 216 single-dose alleles from 125 polymorphic EST-SSRs, and are constructing an EST-SSR map that will be used for identifying QTLs for drought tolerance in autotetraploid alfalfa.

A. Hopkins, Z.Y. Wang, R. Mian, M. Sledge and R.E. Barker (eds.), Molecular Breeding of Forage and Turf, 239-243.

1. INTRODUCTION

Alfalfa (*Medicago sativa* L.), is the most widely grown forage legume in the world. In terms of acreage, alfalfa is the fourth largest crop grown in the USA, behind, maize (*Zea mays* L.), soybean [*Glycine max* (L.) Merr.], and wheat. Cultivated alfalfa is an autotetraploid species (2n=4x=32). Due to the complexity of tetrasomic inheritance, most genetic maps of alfalfa have been constructed in diploids (Brummer et al. 1993; Kiss et al. 1993; Echt et al. 1994; Tavoletti et al. 1996; Kalo et al. 2000). The difficulties of mapping in autotetraploids can be overcome by mapping single dose restriction fragments (Wu et al. 1992). The drawback to this method is the large number of markers needed to construct a map. A linkage group must be constructed for each homologue of each chromosome, and the homologues aligned into a single chromosome linkage group based on the common presence of SDRFs generated by the same DNA marker. Despite these difficulties, one autotetraploid map has been produced using single-dose restriction fragments generated from RFLP markers (Brouwer and Osborn 1999).

While the various alfalfa maps have some markers in common, there is no single map that unifies the currently available alfalfa genetic maps. EST-derived SSR markers are well suited for mapping SDRFs in autotetraploids such as alfalfa, and could provide a set of easily shared markers that could be used unify and cross reference the established alfalfa and *M. truncatula* genetic maps. EST-SSRs are PCR-based markers, which makes them are more efficient than hybridization-based RFLP markers. In contrast to SSR markers derived from genomic libraries, EST-SSRs are both inexpensive and efficient to generate, since no library construction or sequencing of clones is required. Since the EST-SSRs are derived from gene sequences, they should be more informative than genomic SSRs, which may be derived from intergenic DNA sequences. Finally, the growing number of available EST sequences in public databases makes the EST-SSRs both abundant and easy to identify.

2. MATERIALS AND METHODS

2.1 Plant Materials

Two alfalfa backcross populations were constructed from a cross between a water-use efficient, fall-dormant, winter hardy *M. falcata* genotype (MF) and a low water-use efficient, non-fall dormant, winter-sensitive *M. sativa* genotype (CH) of Chilean origin. A single F1 from the cross of MFxCH was used as the female parent for each of the two backcross populations (MFBC and CHBC). For SSR analysis, 93 individuals from each population were analyzed.

2.2 SSR Identification and Detection

The PERL program, Simple Sequence Repeat Identification Tool (SSRIT) was used to search approximately 150,000 *Medicago truncatula* ESTs, from NCBI's dbEST, for the presence of SSRs (Eujayl et al. 2003). Forward and reverse primers were synthesized by Qiagen/Operon Technologies (Alameda, CA, USA) with the additional 21 nucleotides from the M13 universal primer appended to the 5' end of the forward primer. PCR reactions were prepared according to the protocol of Schuelke (2000) with the following modifications. The total reaction volume of 10µl contained 20ng of template DNA, 2.5mM MgCl2, 1X PCR buffer II (Perkin-Elmer), 0.15mM dNTPs, 2.5 pmol of each reverse and M13 (-21) universal primer, 0.5pmol of the forward primer, 0.5 U Ampli Taq Gold DNA polymerase (Perkin Elmer). The M13 (-21) universal primer was labeled either with blue (6-FAM), green (HEX), or yellow (NED) fluorescent tags. PCR products with different fluorescent labels and with different fragment sizes were pooled for detection. PCR products (1.6µl) were combined with 12µl of deionized formamide and 0.5µl of GeneScan-500 ROX internal size standard and analyzed on the ABI3100 Capillary Genetic Analyzer (PE Applied Biosystems). The SSR fragments were visualized with GeneScan 3.7 software, and manually scored.

The Chilean and *M. falcata* parents, and the F_1 were initially screened for polymorphisms. The polymorphic EST-SSR markers were then used to amplify DNA from 10 BC_1 individuals from each backcross population, in order to identify markers that potentially segregated 1:1. Finally, the potential single dose EST-SSR markers were used to amplify DNA from 93 BC1 individuals from each population, and 1:1 segregation was verified by χ^2 analysis.

2.3 SSR Linkage Analysis

Linkage maps were constructed for each of the backcross populations using JoinMap 3.0 software (Van Ooijen 2001). Within each population, data were divided into two groups according to the parent from which each single dose fragment was derived. Linkage analysis was performed separately for each of four mapping groups: the MFBC population with alleles derived from the *M. falcata* parent (MFBCfal), the MFBC population with alleles derived from the Chilean parent (CHBCch), the CHBC population with alleles derived from the *M. falcata* parent (CHBCfal), and the CHBC population with alleles derived from the Chilean parent (CHBCch). The groups were analyzed as population type BC_1, using the Kosambi mapping function. SSR fragments derived from the same SSR primer pairs were used to identify and join homologous linkage groups into composite linkage groups.

3. RESULTS AND DISCUSSION

Currently, we have performed linkage analysis with 216 single-dose alleles from 125 polymorphic loci. These results are summarized in Table 1.

Table 1. Summary of single-dose alleles segregating in two tetraploid alfalfa backcross populations constructed from a cross between *M. sativa* of Chilean origin and *M. falcata*

	Mapping Groups*				
	CHBCch	CHBCfal	MFBCch	MFBCfal	Total
Loci	33	70	49	31	125
Single dose alleles	38	87	60	31	216
Unlinked alleles	14	11	31	9	65
Linkage groups	9	12	6	8	35

*ChBCch – CHBC population, alleles derived from the CH parent; CHBCfal – CHBC population, alleles derived from the MF parent; MFBCch – MFBC population, alleles derived from the CH parent; MFBCfal – MFBC population, alleles derived from the MF parent.

Single dose alleles that are generated by the same EST-SSR primer pair can be used to align homologous linkage groups. Using Joinmap 3.0, at least two common markers are required to join linkage groups. Linkage groups within the same backcross population were joined prior to joining groups from both populations. Numbers of single-dose alleles segregating in the four mapping groups are summarized in Table 2.

Table 2. Number of single-dose alleles amplified by the same EST-SSR locus

Single-dose alleles per locus	CHBCch	CHBCfal	MFBCch	MFBCfal	Total
1 allele	28	53	38	31	63
2 alleles	5	17	11	0	37
3 alleles	--	--	--	--	21
4 alleles	--	--	--	--	4

4. CONCLUSIONS

M. truncatula derived SSRs appear to be an ideal source of genetic markers for mapping in autotetraploid alfalfa. They are plentiful, and have a high incidence of single-dose alleles, with multiple single-dose alleles per locus that can be used to align homologs. With further development, this map will be used to identify drought tolerance QTL in alfalfa.

REFERENCES

Brouwer DJ, Osborn TC (1999) A molecular linkage map of tetraploid alfalfa (*Medicago sativa* L.). Theor. Appl. Genet. 1999: 1194-1200.

Brummer EC, Bouton JH, Kochert G (1993) Development of an RFLP map in diploid alfalfa. Theor. Appl. Genet. 86: 329-332.

Echt CS, Kidwell KK, Knapp SJ, Osborn TC, McCoy MJ (1994) Linkage mapping in diploid alfalfa (*Medicago sativa*). Genome 37: 61-71.

Eujayl I, Sledge MK, Wang L, May GD, Chekhovskiy K, Zwonitzer JC, Mian MAR (2003) *Medicago truncatula* EST-SSRs reveal cross-specific genetic markers for *Medicago* spp. Theor. Appl. Genet. (in Press).

Kalo P, Endre G, Zimanyi L, Csanadi G, Kiss GB (2000) Construction of an improved linkage map of diploid alfalfa (*Medicago sativa*). Theor. Appl. Genet. 100: 641-657.

Kiss GB, Csanadi G, Kalman K, Kalo P, Kresz LO (1993) Construction of a basic genetic map for alfalfa using RFLP, RAPD, isozyme, and morphological markers. Mol. Gen. Genet. 238: 129-137.

Schuelke M (2000) An economic method for the fluorescent labeling of PCR fragments. Nature Biotech. 18: 233-234.

Tavoletti S, Veronesi F, Osborn TC (1996) RFLP linkage map of an alfalfa meiotic mutant based on an F1 population. J. Hered. 87: 167-170.

Van Ooijen JW, Voorrips RE (2001) JoinMap® 3.0, Software for the calculation of genetic linkage maps. Plant Research International, Wageningen, the Netherlands.

Wu KK, Burnquist W, Sorrells ME, Tew TL, Moore PH, Tanksley SD (1992) The detection and estimation of linkage in polyploids using single dose restriction fragments. Theor. Appl. Genet. 83: 294-300.

Controlling Transgene Escape in Genetically Modified Grasses

Hong Luo, Qian Hu, Kimberly Nelson, Chip Longo and Albert P. Kausch
HybriGene Inc., 530 Liberty Lane, West Kingston, RI 02892, USA. (Email: hongluo@hybrigene.com).

Keywords: antisense, *barnase*, gene flow, male sterility, perennial grasses, tapetum-specific promoter, transgene escape, turfgrass

Abstract:

Trait improvement of turfgrass through genetic engineering is important to the turfgrass industry and the environment. However, the possibility of transgene escape to wild and non-transformed species raises commercial and ecological concerns. Male sterility provides an effective way for interrupting gene flow. We have designed and synthesized two chimeric gene constructs consisting of a rice tapetum-specific promoter (TAP) fused to either a ribonuclease gene *barnase*, or the antisense of a rice tapetum-specific gene *rts*. Both constructs were linked to the *bar* gene for selection by resistance to the herbicide glufosinate. Using *Agrobacterium*-mediated transformation, we have successfully introduced those gene constructs into creeping bentgrass (*cv* Penn-A-4), producing a total of 219 stably transformed individual events. Tapetum-specific expression of *barnase* or antisense *rts* gene did not affect the vegetative phenotype compared with the control plants, and male-sterile flowers were obtained with both constructs. Microscopic studies confirmed the failure of mature pollen formation in male-sterile transgenics. Mendelian segregation of herbicide tolerance and male sterility has been observed in T_1 progeny derived from crosses with wild-type plants. Male sterility in transgenic grasses provides the best tool to evaluate gene flow in genetically modified perennial plants and should facilitate the application of genetic engineering in producing environmentally responsible grasses with enhanced traits.

A. Hopkins, Z.Y. Wang, R. Mian, M. Sledge and R.E. Barker (eds.), Molecular Breeding of Forage and Turf, 245-254.

1. INTRODUCTION

Trait improvement of turfgrass through genetic engineering is important to the turfgrass industry and the environment. Beneficial traits such as herbicide and fungus resistance to reduce chemical use, drought and stress tolerance that will reduce water usage, insect and pest resistance that will cut pesticide applications, phyto-remediation of soil contaminants, and horticultural qualities such as aluminum tolerance, stay-green appearance, pigmentation and growth habit among a long list of others, can be improved in turfgrass. However, the possibility of transgene escape to wild and non-transformed species raises commercial and ecological concerns. Although numerous risk assessment studies have been conducted on transgenic plants of annual and/or self-pollinating crops (Altieri 2000; Dale 1992; Dale et al. 2002; Eastham and Sweet 2002; Ellstrand et al. 1999; Ellstrand and Hoffman 1990; Hoffman 1990; Rogers and Parkes 1995), there is a lack of information on the potential risks from the commercialization and large-scale seed production of perennial transgenic grasses. However, in a three-year field study on gene flow of transgenic bentgrass, it was observed that pollen from the transgenic nursery traveled at least 411.5 feet (Wipff and Friker 2001). Therefore, there is a need to develop methods that decrease, or even prevent transgene escape in perennial plants. In flowering plants, gene flow occurs through movement of pollen grains and seeds, with pollen flow often contributing the major component. With the availability of current molecular technologies, various gene-containment strategies have been developed to alter gene flow by interfering with flower pollination, fertilization, and/or fruit development (reviewed by Daniell 2002).

Interfering with the development of male reproductive structures through genetic engineering has been widely used as an effective strategy for the development of male sterility in plants. The tapetum is the innermost layer of the anther wall that surrounds the pollen sac and is essential for the successful development of pollen (Mariani et al. 1990; Moffatt and Somerville 1988; Tsuchiya et al. 1995; Xu et al. 1995a). It has already been shown that tapetum produces a number of highly expressed messenger RNAs (Scott et al. 1991). Genes expressed exclusively in the anther are most likely to include those that control male fertility. Indeed, a variety of anther and tapetum-specific genes have been identified that are involved in normal pollen development in many plant species, including maize (Hanson et al. 1989), rice (Xu et al. 1995b; Zou et al. 1994), tomato (Twell et al. 1989), *Brassica campestris* (Theerakulpisut et al. 1991), and *Arabidopsis* (Xu et al. 1995a). Selective ablation of tapetal cells by cell-specific expression of cytotoxic molecules (De Block et al. 1997; Jagannath et al. 2001; Mariani et al. 1990; Moffatt and Somerville 1988; Tsuchiya et al. 1995) or an antisense gene essential for pollen development (Goetz et al. 2001; Luo et al. 2000; Xu et al. 1995a) blocks pollen development, giving rise to male sterility. Therefore, male sterility, resulting in the lack of significant numbers of viable pollen grains, when linked to the genes of interest provides an effective way for interrupting gene flow.

Here we have studied the feasibility of using cell-specific expression of cytotoxic molecules and an antisense gene that controls male fertility to block pollen development in transgenic turfgrass. Using this strategy, we have successfully engineered male sterility in transgenic bentgrass for the purpose of analyzing the control of transgene escape in genetically modified perennials.

2. MATERIALS AND METHODS

2.1 Production of Transgenic Plants

Following the procedure described in Luo et al. (2003b), friable embryogenic callus was induced from mature seeds of creeping bentgrass (*Agrostis stolonifera* L.), *cv* Penn-A-4 in callus induction medium MMSG containing MS basal medium (Murashige and Skoog 1962), 3% (w/v) sucrose, 0.05% (w/v) casein hydrolysate, 6.6 mg/l 3,6-dichloro-*o*-anisic acid (dicamba), 0.5 mg/l 6-benzylaminopurine (BAP) and 0.2% (w/v) Phytagel. Embryogenic calli were visually selected, divided into small pieces (1-2 mm) and placed on fresh MMSG medium incubating for 1 week before co-cultivation with *Agrobacterium tumefaciens* for genetic transformation.

Agrobacterium strains LBA4404 (Hoekema et al. 1983) containing derivatives of the binary vector pSB11 (Hiei et al. 1994) were used in all experiments. The T-DNA regions of these pSB11 derivatives are shown in Figure 1. pTAP:*barnase*-Ubi:*bar* (Figure 1A) and pTAP:a*rts*-35S:*bar* (Figure 1B) are vectors containing a rice tapetum-specific promoter TAP driving either the antisense of a rice tapetum-specific gene *rts* (Lee et al. 1996), or a ribonuclease gene *barnase* (Hartley 1988). Both chimeric gene constructs were linked to the *bar* gene driving either by a rice ubiquitin (*ubi*) promoter or the CaMV35S promoter. A control plasmid pUbi-*gus*/Act-*hyg* (Figure 1C) containing the maize *ubi* promoter driving the reporter *gusA* gene and the rice actin promoter driving the *hygromycin* resistance gene was kindly provided by Dr. Barbara Zilinskas (Rutgers University).

One day before agro-infection the embryogenic callus was subcultured into 1-2 mm pieces and placed on the MMSG medium supplemented with 100 μM acetosyringone. Following 3 days of co-cultivation with

Agrobacterium in the dark at 25°C, the callus was transferred and cultured for 2 weeks on MMSG medium plus 125 mg/l cefotaxime and 250 mg/l carbenicillin to suppress bacterial growth, and then moved to MMSG medium containing 250 mg/l cefotaxime and 10 mg/l phosphinothricin (PPT) or 200 mg/l hygromycin for 8 weeks at 3-week intervals at room temperature in the dark. Plantlets were then regenerated in the regeneration medium containing MS basal medium, 3% (w/v) sucrose, 0.01% (w/v) myo-inositol, 1 mg/l BAP and 0.2% (w/v) Phytagel supplemented with cefotaxime, PPT or hygromycin. After rooting in the hormone-free regeneration medium containing PPT or hygromycin and cefotaxime, the regenerated plantlets were transferred to soil and grown either in the greenhouse or in the field.

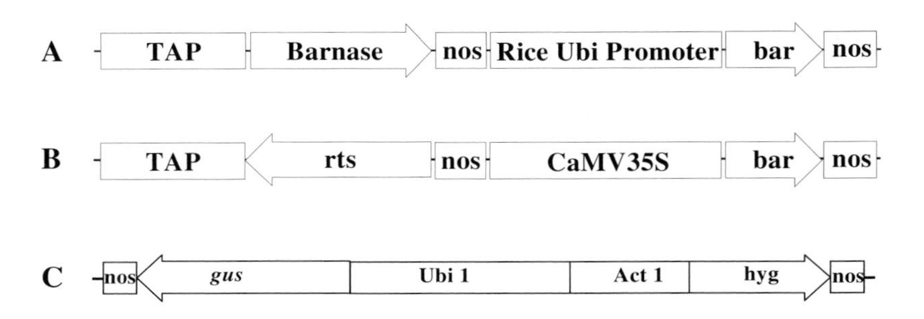

Figure 1. Diagram of the plasmid constructs used for turfgrass transformation. pTAP:barnase-Ubi:bar (**A**) and pTAP:arts-35S:bar (**B**) are vectors containing a rice tapetum-specific promoter TAP driving either the antisense of a rice tapetum-specific gene rts, or a ribonuclease gene barnase. Both chimeric gene constructs were linked to the bar gene driving either by a rice ubi promoter or the CaMV35S promoter. Plasmid pUbi-gus/Act-hyg (**C**) containing the maize ubi promoter driving the reporter gusA gene and the rice actin promoter driving a hygromycin resistance gene was used as a control for transformation.

To examine the foreign gene delivery in regenerated plants, the genomic DNA from leaf tissue of the representative plants of each independent transformation event in greenhouse was isolated as previously described (Luo et al. 1995) and analyzed by PCR and Southern blot hybridization as described by Sambrook et al. (1989).

2.2 Test of Transgenic Plants and Their Progenies for Tolerance to the Herbicide Glufosinate

Transgenic plants and their progenies were evaluated for tolerance to glufosinate for screening seedlings containing expressed foreign gene. The seedlings were sprayed twice with herbicide Finale® (AgrEvo USA,

Montvale, NJ) containing 11% glufosinate as the active ingredient. Tolerant and sensitive seedlings were clearly distinguishable one week after the application of various concentrations (1-10%) of Finale®.

2.3 Vernalization, Pollen Viability Analysis and Out-crossing of Transgenic Plants

Transgenic plants were maintained outside in a containment nursery (3-6 months) until the winter solstice in December. The vernalized plants were then shifted back to the greenhouse at 25°C in artificial light under a 16/8 h (day/light) photoperiod and surrounded by non-transgenic wild-type plants, physically isolated from other pollen sources. Plants started flowering three to four weeks after being moved back into the greenhouse. The viability of pollen taken one day before anthesis was determined by IKI staining (Johansen, 1940) for starch accumulation. The male sterility/fertility status was documented by photomicrography. The flowering transgenic plants were then out-crossed with the pollen from the surrounding wild type plants. The seeds collected from each individual transgenic plant were germinated in soil at 25°C and grown in the greenhouse for further analysis.

3. RESULTS AND DISCUSSION

3.1 Production of Transgenic Plants

To induce male sterility in turfgrass, we fused the 1.2-kb rice *rts* gene regulatory fragment TAP with two different genes. One was the antisense of rice *rts* gene that is predominantly expressed in the anther's tapetum during vigorous meiosis (Lee et al. 1996). Another was a natural ribonuclease gene from *Bacillus amyloliquefaciens* called *barnase* (Hartley 1988). Both chimeric gene constructs were linked to the *bar* gene driving either by a rice *ubi* promoter or the CaMV35S promoter for selection by resistance to the herbicide PPT. These two constructs, pTAP:*barnase*-Ubi:*bar* (Figure 1A) and pTAP:a*rts*-35S:*bar* (Figure 1B) were introduced separately into bentgrass (*Agrostis stolonifera* L.), *cv* Penn-A-4 using *Agrobacterium tumefaciens*-mediated transformation. Transgenic plants were screened from two independent transformation events by PPT selection. A total of 219 primary transgenic callus lines (23 from pTAP:*barnase*-Ubi:*bar* transformation and 196 from pTAP:a*rts*-35S:*bar* transformation) were recovered and regenerated into plants. Under greenhouse conditions, the insertion and expression of the two gene constructs did not affect the vegetative phenotype. The transgenic plants were vigorous and morphologically indistinguishable from untransformed control plants.

PCR assays and Southern blot analysis on genomic DNA from the leaf samples of independent transgenic plants were carried out to assess the stable integration of the transgenes in the host genomes. The *bar* gene was present in all the transformants, and the *barnase* or the antisense *rts* gene was also detected in the respective transgenic plants. All the transformation events had less than three copies of transgene insertion, and a majority of them (60-65%) contained only a single copy of foreign gene integration with no apparent rearrangement.

3.2 Herbicide Tolerance in Regenerated Transgenic Plants

Herbicide Finale® containing 11% glufosinate as the active ingredient was applied on plants regenerated from all independent transformation events to check herbicide tolerance conferred by the *bar* gene expression. The untransformed control plants died within one week after 1-10% (v/v) of Finale® spray, while the transgenic plants showed no damage at these concentrations, indicating that the level of the *bar* gene expression in the transgenic plants was sufficient enough to be resistant to as high as 10% of Finale®. It should also be noted that some transgenic lines have survived 100% of Finale® application.

3.3 Expression of Barnase or Antisense *rts* Gene Causes Male Sterility

To check the sterility/fertility status of pollen from transgenic plants expressing *barnase* or antisense *rts*, the vernalized transgenic and non-transgenic control plants were grown in the greenhouse and flowered at 25°C in artificial light under a 16/8 h (day/light) photoperiod.

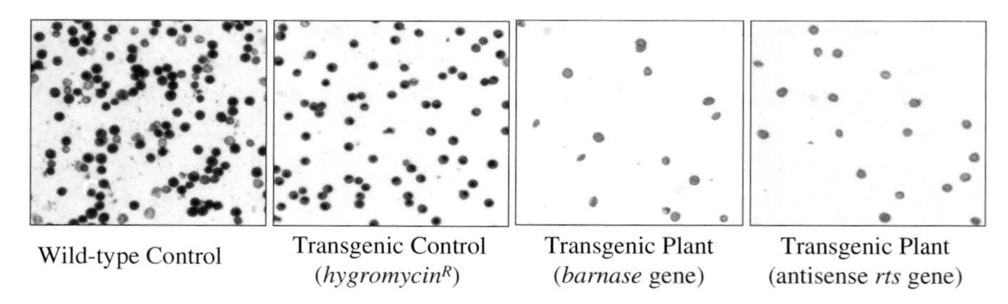

Figure 2. T_0 pollen viability using IKI staining. The viability of pollen taken one day before anthesis from wild-type, non-transgenic control plant; hygromycin-resistant transgenic control plant; transgenic plant expressing barnase or antisense rts, respectively, was determined by IKI staining for starch accumulation. The viable pollen is stained darkly. The male sterility/fertility status was documented by photomicrography.

Pollen was taken one day before anthesis for viability analysis using IKI staining. More than 90% of the plants containing *barnase* and around 50% of the plants containing the antisense *rts* gene were completely male sterile, without viable pollens that are normally stained darkly by IKI as observed in the wild-type control plants and *hygromycin*-resistant transgenic plants that do not contain the *barnase* or the antisense *rts* gene (Figure 2), indicating that cell-specific expression of the *barnase* or the antisense *rts* gene in transgenic plants blocks pollen development, giving rise to male sterility.

3.4 Male Sterility Is Inherited in the Progeny

To assess whether the transgenes are able to be inherited by the progeny in a normal Mendelian fashion, a subset of independent transgenic T_0 plants that have a single copy of the intact integrated transgene (both from the *barnase*-containing and the antisense *rts*-containing plants) were pollinated (out-crossed) with pollen from wild type plants. The T_1 progeny showed an expected 1:1 segregation for glufosinate tolerance. PCR assays confirmed the presence of the transgene *bar* as well as the *barnase* or the antisense *rts* gene in all the glufosinate-tolerant progeny, but not in the sensitive ones. A small portion of the T_1 plants flowered after vernalization and was examined for pollen viability. Male sterility was observed in the progeny. Although more samples still need to be analyzed to determine the inheritance and expression of male sterility, our preliminary data suggests that *barnase* and antisense *rts* were transmitted to and expressed in the progeny.

Male sterility in transgenic grasses provides an effective way of controlling transgene escape to wild and non-transformed species through movement of pollen grains. This will adequately control contamination in commercial production fields. In consumer products (golf courses, recreational lands, etc.) out-crossing with wild type *Agrostis stolonifera* will result in 50% *Agrostis stolonifera* progeny that are herbicide-tolerant and male-sterile. Out-crosses resulting from inter-specific (*Agrostis* spp.) pollination is very low (frequency less than 0.01%, J. Wipff, personal communication). Since the probability of inter-specific crossing is very low, if not impossible (Belanger et al. 2003), even though some hybrid seeds could be produced through out-crossing, the non-selected transgene in a male-sterile inter-specific background should be progressively diluted and eventually extinct in the populations (Figure 3).

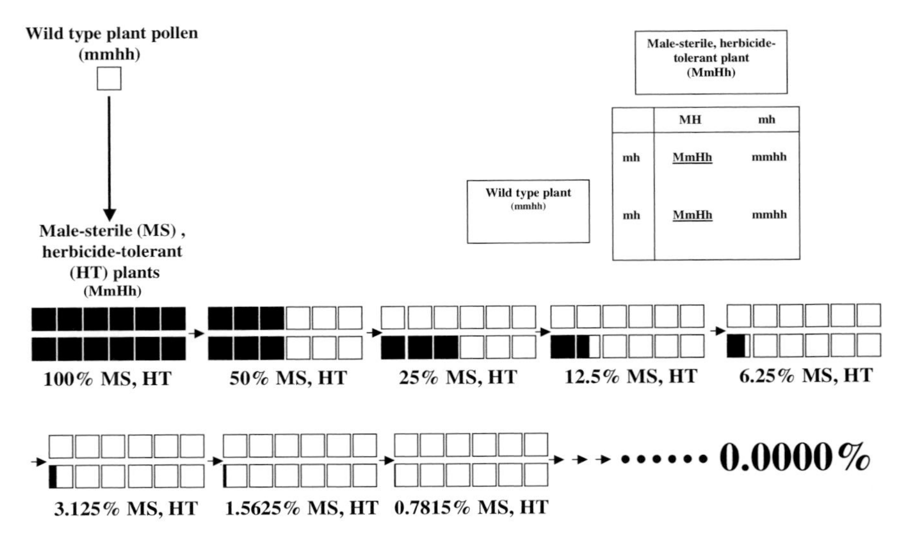

Figure 3. Assuming a low frequency of inter-specific out-crossing, progressive dilution of a non-selected transgene will occur in a male-sterile background. Here male sterility (MS) is linked to herbicide-tolerant gene (HT) in the transgenic *Agrostis stolonifera* plants. The population of hemizygous male-sterile transgenic plants (in black), when pollinated with pollen from wild-type inter-specific plants will gradually diminish and eventually extinguish.

Although seeds produced by pollination with pollen from wild, non-transgenic plants consist of 50% of male-sterile transgenic and 50% of wild genotypes, when linked with an herbicide-tolerant gene, plants with wild genotype can be readily eliminated by herbicide spray in the field. In addition, if an inducible promoter (Gatz 1997; Gatz and Lenk 1998; Shimizu-Sato et al. 2002; Zuo et al. 2000) and/or a site-specific DNA recombination system (Luo and Kausch 2002; Luo et al. 2003a; Ow and Medberry 1995) are used to control the male sterility in transgenic plants, seed production from transgenic plants with male-sterile background should be greatly facilitated. Similarly, this strategy can also be used to generate female-sterile or totally sterile transgenic plants to fully control gene flow in transgenic perennial grasses.

REFERENCES

Altieri MA (2000) The ecological impacts of transgenic crops on agroecosystem health. Ecosystem Health 6: 13-23.

Belanger FC, Meagher TR, Day PR, Plumley K, Meyer WA (2003) Interspecific hybridization between *Agrostis stolonifera* and related *Agrostis* species under field conditions. Crop Sci. 43: 240-246.

Dale PJ (1992) Spread of engineered genes to wild relatives. Plant Physiol. 100: 13-15.

Dale PJ, Clarke B, Fontes EMG (2002) Potential for the environmental impact of transgenic crops. Nat. Biotechnol. 20: 567-574.

Daniell H (2002) Molecular strategies for gene containment in transgenic crops. Nat. Biotechnol. 20: 581-586.

De Block M, Debrouwer D, Moens T (1997) The development of nuclear male sterility system in wheat: expression of the *barnase* gene under the control of tapetum specific promoters. Theor. Appl. Genet. 95: 125-131.

Eastham K, Sweet J (2002) Genetically modified organisms (GMOs): the significance of gene flow through pollen transfer. European Environment Agency, Denmark.

Ellatrand NC, Hoffman CA (1990) Hybridization as an avenue of escape for engineered genes. Bioscience 40: 438-442.

Ellstrand NC, Prentice HC, Hancock JF. (1999). Gene flow and introgression from domesticated plants into their wild relatives. Annu. Rev. Ecol. Syst. 30: 539–563.

Gatz C (1997) Chemical control of gene expression. Annu. Rev. Plant Physiol. Plant Mol. Biol. 48: 89-108.

Gatz C, Lenk I (1998) Promoters that respond to chemical inducers. Trend. Plant Sci. 3: 322-328.

Goetz M, Godt DE, Guivarch A, Kahmann U, Chriqui D, Roitsch T (2001) Induction of male sterility in plants by metabolic engineering of the carbohydrate supply. Proc. Natl. Acad. Sci. USA 98: 6522-6527.

Hanson DD, Hamilton DA, Travis JL, Bashe DM, Mascarenhas JP (1989) Characterization of a pollen-specific cDNA clone from *Zea mays* and its expression. Plant Cell 1: 173-179.

Hartley RW (1988) Barnase and Barstar: expression of its cloned inhibitor permits expression of a cloned ribonuclease. J Mol Biol 202: 913-915.

Hiei Y, Ohta S, Komari T, Kumashiro T (1994) Efficient transformation of rice (*Oryza sativa* L.) mediated by *Agrobacterium* and sequence analysis of the boundaries of the T-DNA. Plant J. 6: 271-282.

Hoekema A, Hirsch PR, Hooykaas PJJ, Schilperoort RA (1983) A binary vector strategy based on separation of *vir*- and T-region of the *Agrobacterium tumefaciens* Ti-plasmid. Nature 303: 179-180.

Hoffman CA (1990) Ecological risks of genetic engineering of crop plants. BioScience 40: 434-437.

Jagannath A, Bandyopadhyay P, Arumugam N, Gupta V, Kumar P, Pental D (2001) The use of spacer DNA fragment insulates the tissue-specific expression of a cytotoxic gene (*barnase*) and allows high-frequency generation of transgenic male sterile lines in *Brassica juncea* L. Mol. Breed. 8: 11-23.

Johansen DA (1940) Plant microtechnique. McGraw-Hill, New York.

Lee J-Y, Aldemita RR, Hodges TK (1996) Isolation of a tapetum-specific gene and promoter from rice. Int. Rice Res. Newsl. 21: 2-3.

Luo H, Hu Q, Nelson K, Kausch AP (2003a) FLP-mediated site-specific DNA recombination in the cells of grasses. Plant Sci. (submitted).

Luo H, Hu Q, Nelson K, Longo C, Kausch AP, Chandlee JM, Wipff JK, Fricker CR (2003) *Agrobacterium tumefaciens*-mediated creeping bentgrass transformation using phosphinothricin selection results in high frequency of single-copy transgene integration. Plant Cell Rep. (in press).

Luo H, Kausch AP (2002) Application of FLP/*FRT* site-specific DNA recombination system in plants. In: Genetic Engineering, Principles and Methods, Setlow JK (ed.), Vol 24, pp.1-16. Kluwer Academic/Pleum Publishers, New York.

Luo H, Lyznik LA, Gidoni D, Hodges TK (2000) FLP-mediated recombination for use in hybrid plant production. Plant J. 23: 423-430.

Luo H, Van Coppenolle B, Seguin M, Boutry M (1995) Mitochondrial DNA polymorphism and phylogenetic relationship in *Hevea brasiliensis*. Mol Breed 1: 51-63.

Mariani C, De Beuckeleer M, Truettner J, Leemans J, Goldberg RB (1990) Induction of male sterility in plants by a chimeric ribonuclease gene. Nature 347: 737-741.

Moffatt B, Somerville C (1988) Positive selection for male-sterile mutants of *Arabidopsis* lacking adenine phosphoribosyl transferase activity. Plant Physiol. 86: 1150-1154.

Murashige T, Skoog F (1962) A revised medium for rapid growth and bioassays with tobacco tissue cultures. Physiol. Plantarum 15: 473-497.

Ow DW, Medberry SL (1995) Genome manipulation through site-specific recombination. Crit. Rev. Plant Sci. 14: 239-261.

Rogers HJ, Parkes HC (1995) Transgenic plants and the environment. J. Exp. Bot. 46: 467-488.

Sambrook J, Fritsch EF, Maniatis T (1989) Molecular Cloning: A Laboratory Manual, 2nd ed., Cold Spring Harbor Laboratory Press, New York.

Scott R, Hodge R, Paul W, Draper J (1991) The Molecular biology of anther differentiation. Plant Sci. 80: 167-191.

Shimizu-Sato S, Huq Enamul, Tepperman JM, Quail PH (2002) A light-switchable gene promoter system. Nat. Biotechnol. 20: 1041-1044.

Theerakulpisut P, Xu H, Sipgh MB, Pettitt JM, Knox RB (1991) Isolation and developmental expression of *Bcpl,* an anther-specific cDNA clone in *Brassica campestris.* Plant Cell 3: 1073-1084.

Tsuchiya T, Toriyama K, Yoshikawa M, Ejiri S, Hinata K (1995) Tapetum-specific expression of the gene for an endo-β-1,3-glucanase causes male sterility in transgenic tobacco. Plant Cell Physiol. 36: 487-494.

Twell D, Wing R, Yamaguchi J, McCormick S (1989) Isolation and expression of an anther-specific gene from tomato. Mol. Gen. Genet. 217: 240-245.

Wiff JK, Fricker C (2000) Determining gene flow of ransgenic creeping bentgrass and gene transfer to other bentgrass species. Diversity 16: 36-39.

Wipff JK, Fricker C (2001) Gene flow from transgenic creeping bentgrass (*Agrostis stolonifera* L.) in the Willamette valley, Oregon. Int. Turfgrass Soc. Res. J. 9: 224-242.

Xu H, Knox RB, Taylor PE, Singh MB (1995a) *Bcp1*, a gene required for male fertility in *Arabidopsis*. Proc. Nat. Acad. Sci. USA 92: 2106-2110.

Xu H, Theerakulpisut P, Taylor PE, Knox RB, Singh MB, Bhalla PL (1995b) Isolation of a gene preferentially expressed in mature anthers of rice (*Oryza sativa* L.). Protoplasma 187: 27-131.

Zou JT, Zhan XY, Wu HM, Wang H, Cheung AY (1994) Characterization of a rice pollen-specific gene and its expression. Am. J. Bot. 81: 552-561.

Zuo J, Niu Q-W, Chua N-H (2000) An estrogen receptor-based transactivator XVE mediates highly inducible gene expression in transgenic plants. Plant J. 24: 265-273.

Comparison of Transgene Expression Stability after *Agrobacterium*-mediated or Biolistic Gene Transfer into Perennial Ryegrass (*Lolium perenne* L.)

Fredy Altpeter[1*], Yu-Da Fang[2], Jianping Xu[3] and Xinrong Ma[4]
[1]University of Florida, IFAS, Agronomy Department, 2191 McCarty Hall A, Gainesville FL 32611-0300, USA. [2]Cold Spring Harbor Laboratory, One Bungtown Road, Cold Spring Harbor, NY 11724, USA. [3]University of Nebraska Lincoln, E249 Beadle Center, Lincoln, NE 68588-0666, USA. [4]Sichuan University, Chendu Institute of Biology, Chinese Academy of Sciences, Chengdu, 610064, P.R. CHINA. (Email: FAltpeter@mail.ifas.ufl.edu).

Key words: transgene integration, transgene expression stability, transgene rearrangements, *Agrobacterium tumefaciens*, biolistic gene transfer, perennial ryegrass.

Abstract:

Perennial ryegrass (*Lolium perenne* L.) is the most widely distributed grass species in areas with temperate climate. Genetic engineering of perennial ryegrass is complementing traditional breeding in the development of improved germplasm. The generation of large numbers of transgenic perennial ryegrass plants from turf type- or forage type cultivars following biolistic or *Agrobacterium-mediated* gene transfer was recently reported by us. Transgenic plants generated by the two alternative transformation systems were compared regarding transgene integration pattern, fertility and stability of transgene expression after vegetative and sexual reproduction. Advantages and limitations of the two perennial ryegrass transformation systems are discussed.

A. Hopkins, Z.Y. Wang, R. Mian, M. Sledge and R.E. Barker (eds.), Molecular Breeding of Forage and Turf, 255-260.

1. INTRODUCTION

Perennial ryegrass is one of the most widely cultivated grasses in the temperate regions (Watschke and Schmidt, 1992). Genetic engineering of grasses is complementing traditional breeding in the development of improved germplasm. Recently we reported the development of RNA-mediated virus resistance in perennial ryegrass (Xu et al. 2001). A range of gene transfer protocols have been described for the production of transgenic perennial ryegrass plants including biolistic gene transfer using DNA coated microprojectiles (Spangenberg et al., 1995; Dalton et al., 1999) or silicon carbide fibre-mediated gene transfer (Dalton et al., 1998) or direct gene transfer into protoplast (Wang et al., 1997). These protocols required a long tissue culture period, which is more likely to result in undesirable somaclonal variation (Creemers-Molenaar and Loeffen, 1991). We recently presented an accelerated biolistic transformation and selection protocol for the production of large numbers of fertile transgenic perennial ryegrass plants and demonstrated its applicability to commercial turf and forage type cultivars (Altpeter et al. 2000). Meanwhile an *Agrobacterium*-mediated perennial ryegrass transformation protocol has been developed in our laboratory (Altpeter et al., in preparation).

Agrobacterium-mediated gene transfer offers potential advantages over biolistic gene transfer, including preferential integration of single T-DNA copies (Tingay et al. 1997) into transcriptionally active regions of the chromosome (Czernilofsky et al. 1986), unlinked integration of co-transformed T-DNAs that facilitates the elimination of selectable marker genes (McKnight et al. 1987; Komari et al. 1996), exclusion of vector DNA following T-DNA transfer (Hiei et al. 1997, Fang et al. 2002) that allows the integration of defined transgene expression cassettes, and transfer of large DNA fragments (Hamilton et al. 1996) that will support positional cloning and pathway engineering.

This study describes differences in transgene integration patterns and transgene expression stability after vegetative and sexual reproduction, following biolistic or *Agrobacterium*-mediated gene transfer.

2. RESULTS AND DISCUSSION

Both biolistic (Bio-Rad PDS 1000) or *Agrobacterium*-mediated (AGL1) transfer of a constitutive *npt II* expression cassette (pCAMBIA vector) in freshly established embryogenic calli, followed by selection with paromomycin (Altpeter et al. 2000) resulted in transgenic perennial ryegrass plants within four to six months after excision of explants. Between 1.3 and 4.0 % of the bombarded calli or 8 to 16 % of the *Agrobacterium*-inoculated calli regenerated independent transgenic plants. Southern blot analysis confirmed the independent nature of the transgenic plants (data not shown). Reproducibility and efficiency in these perennial ryegrass transformation protocols were controlled by multiple factors including genotype dependent tissue culture response, a short tissue culture- and selection period and the efficient suppression of *Agrobacterium* growth following *Agrobacterium*-mediated gene transfer. The majority of transgenic lines from both biolistic and *Agrobacterium*-mediated gene transfer had a simple transgene integration pattern with one to four transgene copies, were fertile and the transgene was stably expressed in sexual progenies (Table 1). Approximately 20 % of the ryegrass lines generated with biolistic gene transfer had very complex integration patterns with more than 5 and up to twenty transgene copies, while none of the lines generated by *Agrobacterium*-mediated gene transfer had more than five T-DNA inserts (Table 1). The integration pattern usually observed after biolistic gene transfer into grasses and cereals is multiple transgene copy inserts (Hartman et al., 1994; Spangenberg et al., 1995; Altpeter et al., 1996; Stöger et al. 1998). However Dalton et al. (1999) described that the transgene expressing ryegrass plants following biolistic gene transfer carried only one to two transgene copy inserts. Their observation that transgene expression might be negatively affected by a higher copy number is in agreement with an earlier report by Matzke and Matzke (1995).

Our data also suggest that selection protocols maintaining antibiotic selection pressure during regeneration of transgenic plantlets from tissue cultures might eliminate most of the unstable multiple copy events and thus should result in recovery of events with more simple integration patterns. Multicopy inserts after biolistic gene transfer into perennial ryegrass were commonly inserted at the same locus, whereas the majority of transgenic lines after *Agrobacterium*-mediated gene transfer showed two transgene inserts at independent loci, segregating into single locus events in the following progeny. This contributed to a large number of independent single copy events following sexual reproduction after *Agrobacterium*-mediated gene transfer into perennial ryegrass and would be expected to support a higher transgene expression stability in these sexual progenies. All of the thirty six fertile transgenic ryegrass lines generated with *Agrobacterium*-mediated gene transfer stably expressed the transgene in sexual progenies (Table 1).

Table 1. Transgene integration pattern, fertility and expression stability of transgenic perennial ryegrass lines after biolistic or Agrobacterium-mediated gene transfer and sexual reproduction

Copy number	**Biolistic** Transgenic lines / Fertile lines / Lines with transgene expression in sexual progenies	***Agrobacterium*** Transgenic lines / Fertile lines / Lines with transgene expression in sexual progenies
1	8 / 6 / 5	5 / 4 / 4
2	8 / 7 / 5	20 / 14 / 14
3	6 / 4 / 3	16 / 11 / 11
4	5 / 3 / 3	4 / 4 / 4
5	2 / 1 / 1	4 / 3 / 3
>5	6 / 4 / 2	0 / 0 / 0
Total	35 / 25 / 19	49 / 36 / 36

Gene silencing after sexual (Table 1) or one year vegetative reproduction (Table 2) was most frequently observed in lines with 5 or more transgene copies, generated by biolistic gene transfer. Our data suggest that both gene transfer systems have a high potential to produce fertile and stably expressing transgenic perennial ryegrass lines. The described biolistic gene transfer protocol is applicable to a wide range of turf- and forage type genotypes, while *Agrobacterium*-mediated ryegrass transformation is so far limited to a few responsive ryegrass genotypes.

Table 2: Transgene integration pattern, fertility and expression stability of transgenic perennial ryegrass lines after biolistic or Agrobacterium-mediated gene transfer and one year vegetative reproduction

Copy number	**Biolistic** Transgenic lines / Lines with transgene expression after 1 year vegetative propagation	***Agrobacterium*** Transgenic lines / Lines with transgene expression after 1 year vegetative propagation
1	8 / 6	5 / 5
2	8 / 7	20 / 16
3	6 / 6	16 / 14
4	5 / 3	4 / 3
5	2 / 1	4 / 4
>5	6 / 3	0 / 0
Total	35 / 26	49 / 42

Agrobacterium-mediated gene transfer might be the preferred method to transfer large DNA fragments (Hamilton et al. 1996). This will support positional cloning and pathway engineering. An additional advantage of *Agrobacterium*-mediated gene transfer is the simple elimination of selectable marker genes by unlinked integration of co-transformed T-DNA followed by their segregation in sexual progenies (McKnight et al. 1987; Komari et al. 1996).

ACKNOWLEDGEMENTS

The authors thank R. Jefferson for pCAMBIA2300 vectors, U.K. Posselt for providing seeds of perennial ryegrass genotypes.

REFERENCES

Altpeter F, Vasil V, Srivastava V, Vasil IK (1996) Integration and expression of the high-molecular-weight glutenin subunit 1Ax1 gene into wheat. Nature Biotechnol. 14: 1155-1159.

Altpeter F, Xu J, Ahmed S (2000) Generation of large numbers of independently transformed fertile perennial ryegrass (*Lolium perenne* L.) plants of forage- and turf type cultivars. Mol. Breed. 6: 519-528.

Czernilofsky AP, Hain R, Baker B, Wirtz U (1986) Studies of the structure and functional organization of foreign DNA integrated into the genome of *Nicotiana tabacum*. DNA 5: 473-482.

Creemers-Molenaar J, Loeffen JPM (1991) Regeneration from protoplasts of perennial ryegrass; progress and applications. In: Fodder crops breeding: achievements, novel strategies and biotechnology, Proceedings of the 16th meeting of the Fodder Crops Section of Eucarpia, Den Nijs, A.P.M., and A. Elgersma (eds.), pp 123-128.

Dalton SJ, Bettany AJE, Timms E, Morris P (1998) Transgenic plants of *Lolium multiflorum*, *Lolium perenne*, *Festuca arundinacea* and *Agrostis stolonifera* by silicon carbide fibre-mediated transformation of cell suspension cultures. Plant Sci. 132: 31-43.

Dalton SJ, Bettany AJE, Timms E, Morris P (1999) Co-transformed, diploid *Lolium perenne* (perennial ryegrass), *Lolium multiflorum* (Italian ryegrass) and *Lolium temulentum* (darnel) plants produced by microprojectile bombardment. Plant Cell Rep. 18: 721-726.

Fang YD, Akula C, Altpeter F (2002) *Agrobacterium*-mediated barley (*Hordeum vulgare* L.) transformation using green fluorescent protein as a visual marker and sequence analysis of the T-DNA::barley genomic DNA junctions. J. Plant Physiol. 159: 1131-1138.

Hamilton CM, Frary A, Lewis C, Tanksley SD (1996) Stable transfer of intact high molecular weight DNA into plant chromosomes. Proc. Natl. Acad. Sci. U S A 93: 9975-9979.

Hartman, CL, Lee L, Day PR, Nilgun ET (1994) Herbicide resistant turfgrass (*Agrostis palustris* Huds.) by biolistic transformation. Bio/Technology 12: 919-923.

Hiei Y, Komari T, Kubo T (1997) Transformation of rice mediated by *Agrobacterium tumefaciens*. Plant Mol. Biol. 35: 205-218.

Komari T, Hiei Y, Saito Y, Murai N, Kumashiro T (1996) Vector carrying two separate T-DNAs for co-transformation of higher plants mediated by *Agrobacterium tumefaciens* and segregation of transformants free from selection markers. Plant J. 10: 165-174.

Matzke MA, Matzke AJM (1995) How and why do plants inactivate homologous (trans)genes? Plant Physiol. 107: 679-685.

McKnight TD, Lillis MT, Simpson RB (1987) Segregation of genes transferred to one plant cell from two separate *Agrobacterium* strains. Plant Mol. Biol. 8: 439-445.

Spangenberg G, Wang ZY, Wu XL, Nagel J, Potrykus I (1995) Transgenic perennial ryegrass (*Lolium perenne*) plants from microprojectile bombardment of embryogenic suspension cells. Plant Sci. 108: 209-217.

Stöger E, Williams S, Keen D, Christou P (1998) Molecular characteristics of transgenic wheat and the effect on transgene expression. Transgenic Res. 7: 463-471.

Tingay S, McElroy D, Kalla R, Fieg S, Wang M, Thornton S, Brettell R 1997. *Agrobacterium tumefaciens*-mediated barley transformation. Plant J. 11: 1369-1376.

Wang GR, Binding H, Posselt UK (1997) Fertile transgenic plants from direct gene transfer to protoplasts of *Lolium perenne* L. and *Lolium multiflorum* Lam. J. Plant Physiol. 151: 83-90.

Watschke TL, Schmidt RE (1992) Ecological aspects of turfgrass communities. In: Turfgrass Agron. Monogr. 32, Waddington DV et al. (eds.), pp 129-174. ASA-CSSA-SSSA, Madison.

Xu J, Schubert J, Altpeter F (2001) Dissection of RNA mediated virus resistance in fertile transgenic perennial ryegrass (*Lolium perenne* L.). Plant J. 26: 265-274.

Bioinformatics: Bringing Data to a Usable Form for Breeders

H. J. Ougham and L. S. Huang
Institute of Grassland and Environmental Research, Plas Gogerddan, Aberystwyth, Ceredigion, SY23 3EB, Wales, UK. (Email: helen.ougham@bbsrc.ac.uk).

Key words: genetic maps, genome databases, marker-assisted selection, metabolomics, proteomics, transcriptomics

Abstract:

Over the past two decades, the trickle of data emerging from DNA sequencing projects has increased to a flood. For example, by early 2003 there were over 36 billion bases of DNA sequence publicly available. New molecular marker technologies have similarly resulted in a huge increase in the availability of detailed genetic maps of model organisms and crop species. These advances have necessitated the development of bioinformatics tools for curation, dissemination, analysis and comparison of genome data, facilitated by the pace of improvement in computing power and storage capacity. Many genome databases use the object-oriented ACEDB database management system. ACEDB offers powerful search tools and sophisticated cross-linking between data items, and has tools to facilitate comparative map and sequence displays. Because many crops, including most forages and turf, have large genomes they are unlikely to be sequenced in the near future. However, comparative genomics offers breeders many opportunities for targeted molecular breeding approaches, and there are now several major of comparative plant genome bioinformatics projects worldwide. Bioinformatics can assist in development or selection of markers associated with quantitative traits, and identification and cloning of candidate genes for control of important characters. Post-genomic techniques such as transcriptomics and metabolomics, which can aid characterisation of mapping populations and varieties, also require bioinformatics tools for the curation, analysis and displaying of the very large data sets which they can generate.

A. Hopkins, Z.Y. Wang, R. Mian, M. Sledge and R.E. Barker (eds.), Molecular Breeding of Forage and Turf, 261-274.

1. INTRODUCTION

What is bioinformatics? There are almost as many definitions of the term as there are people working in the field. Some of these definitions restrict the term to work on classes of biological macromolecule - usually DNA or protein - and/or computer algorithm development. For example, Luscombe et al. (2001) proposed, and submitted to the Oxford English Dictionary for consideration, the definition: "Bioinformatics is conceptualising biology in terms of molecules (in the sense of Physical chemistry) and applying 'informatics techniques' (derived from disciplines such as applied maths, computer science and statistics) to understand and organise the information associated with these molecules, on a large scale. In short, bioinformatics is a management information system for molecular biology...". Such a description excludes many of the types of data which are of most interest to breeders, such as genetic maps and quantitative trait data. However, most modern definitions of bioinformatics are much more inclusive. For example, the US National Center for Biotechnology Information, NCBI (http://www.ncbi.nlm.nih.gov/Education/), calls it "the field of science in which biology, computer science, and information technology merge into a single discipline ... to enable the discovery of new biological insights". Using such a definition, bioinformatics becomes relevant to everyone whose work involves biology in the broadest sense. Nevertheless, its origins can certainly be attributed to needs associated with a single molecule: DNA.

Over the past two decades, the trickle of data emerging from DNA sequencing projects has increased to a flood. When researchers began to develop publicly-accessible databases to store sequence data in the early 1980s, they initially contained only a few thousand DNA bases. By early 2003, over 36 billion bases of DNA, from tens of thousands of organisms, were available from the European EMBL and US GenBank databases. New molecular marker technologies have similarly resulted in a huge increase in the availability of detailed genetic maps of model organisms and agriculturally-important species. More recently, other "omics" technologies such as transcriptomics, proteomics and metabolomics have begun to add to the diversity and volume of data. These advances have necessitated the development of bioinformatics tools for curation, dissemination, analysis and comparison of genome data. Fortunately, the pace of improvement in computing power and storage capacity has kept pace with progress in the "wet" science. However, there are many challenges facing the rapidly-evolving field of bioinformatics, and among the greatest of these is to make data easily available in a user-friendly and validated form to end-users who are neither computer specialists nor molecular biologists. For a plant breeder wishing to capitalise on bioinformatics resources to facilitate a marker-assisted selection or transgenic crop programme, it is not always easy to locate the necessary information or extract it in a usable form. This paper reviews some of the existing and developing public bioinformatics resources which may be of special relevance to breeders now or in the future. In several cases URLs have been used to indicate appropriate web sites; since any web site may be removed or undergo a change of address, the long-term validity of these URLs unfortunately cannot be guaranteed

2. GENOME DATABASES

2.1 History

One of the earliest genome sequencing projects focused on the worm *Caenorhabditis elegans*, a model species for studies on development. The (essentially) complete genome sequence was released in 1998 (C. elegans Sequencing Consortium, 1998). By the end of the 1980s, researchers had already concluded that their data could not adequately be accommodated by conventional database management systems, which had been developed predominantly for commercial organisations. The complexity of biological data, and the extent of cross-referencing required between different data types, necessitated an alternative approach. Jean Thierry-Mieg and Richard Durbin therefore designed and implemented a new database management system, AceDB (short for "A Caenorhabditis elegans Database). Background, documentation and software downloads for AceDB, which is free software, can be found at http://www.acedb.org/. AceDB has since become a standard for many genome databases worldwide, though it is also used for non-genome biological data. A large number of the databases for single crop species or genera use this system, and several of them will be described in section 2.5.

2.2 Principles of AceDB Databases

The first requirement for the developers of the AceDB system was that it must be capable of containing genetic maps, together with all relevant associated information - loci, marker types, DNA sequences, mapping data, publications and other types. Maps, which may be physical or genetic linkage maps, are the focus for the majority of AceDB databases, and Figure 1 shows a typical map display in which the linkage group is represented by a bar on the left, with a scale showing genetic distances in centimorgans and a series of loci. Each of these loci can be clicked to provide more information about the marker type and associated data.

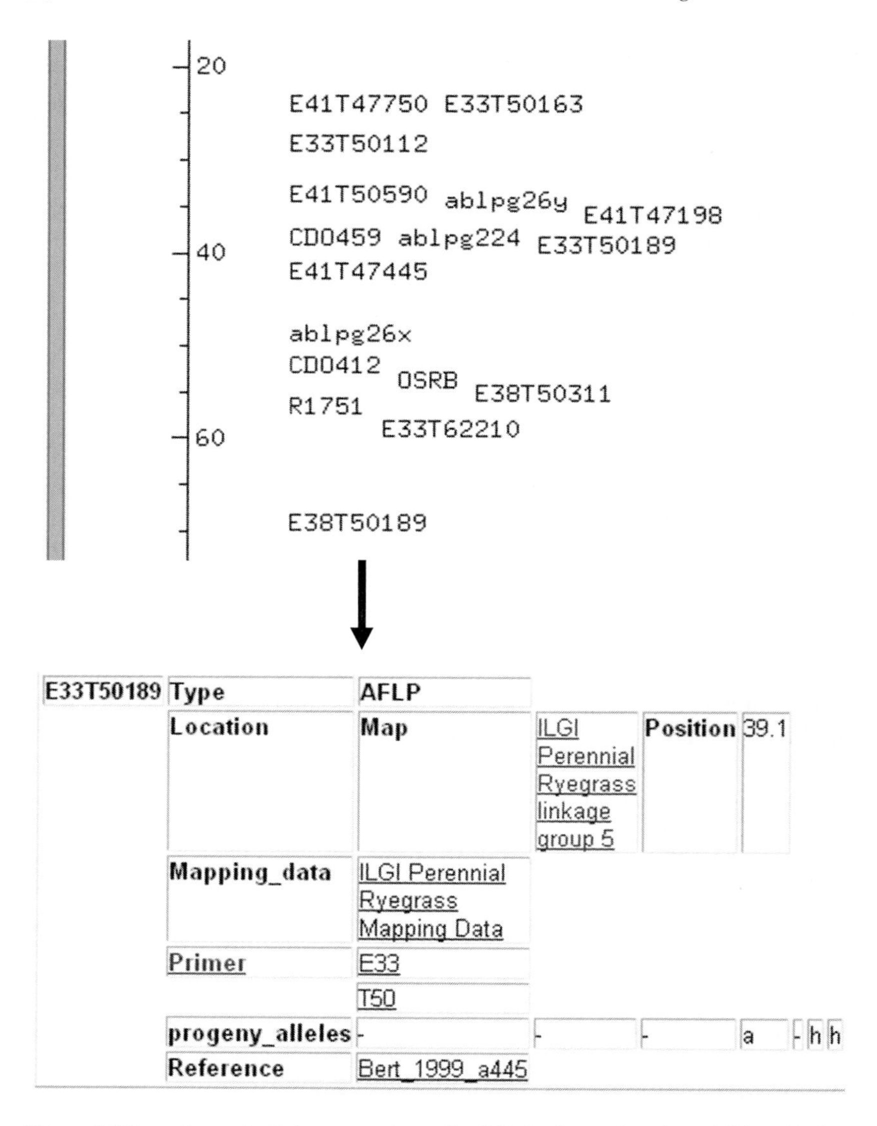

Figure 1. View of genetic linkage map in an AceDB database, showing additional information available by selecting an individual locus.

From this display, many items can be selected and clicked to provide further information - in this instance, about mapping data, AFLP primers used, and the relevant publication - each of which in turn may lead on to other related data through clickable links, eventually returning to the original map. This extensive cross-referencing is an important feature of AceDB genome databases.

Many other data types can be displayed alongside the map, for example quantitative trait loci or the regions of the linkage group containing coding sequences; the developer can customise the database according to the characteristics of the organism(s) represented and the data types available. This is advantageous for the database curator and its principal users, but the variations in design can occasionally cause problems for those who need frequently to move between different AceDB databases, particularly when constructing queries (see section 2.3).

2.3 Querying AceDB Databases

AceDB offers a number of options for interrogating a database, ranging from simple text queries to a powerful query language which allows sophisticated questions to be posed. They are:

- Text query (fast or in-depth) - this searches the whole database for a word or phrase. The wildcard character * can be included, so a text search for Jon* would identify all entries in the database beginning with Jon, such as authors Jones and Jonsson and a marker named Jon3
- Class browser - lists all items in the database in a particular class, for example maps, sequences or images. They can then be examined individually.
- BLAST search - a DNA sequence can be submitted and its sequence compared with all those in the database using the BLAST algorithm.

The two remaining query methods require a little more explanation.

- TableMaker queries allow a table of information to be constructed a column at a time, with the option of applying a condition at each stage. For example, Table 1 represents the first 4 lines returned by a TableMaker search of the Forage Grass Genome Database FoggDB. The search requested a column of all sequences where the species of origin was any *Lolium* (*Lolium**), followed by a column listing the relevant species in each case.

Table 1. Results of a TableMaker search for *Lolium* DNA sequences in the FoggDB database

EMBL:A31057	Lolium perenne
EMBL:A31058	Lolium perenne
EMBL:A31060	Lolium perenne
EMBL:AB016122	Lolium multiflorum

- For the final method, a knowledge of the Ace query language is required. This language is not highly intuitive, and this is the likely reason why statistics show it to be the least-frequently used query method in the plant genome databases. Nevertheless, it allows some extremely complex queries to be made and, for the serious user, acquiring some skill in the language greatly improves access to the information in the databases. In order to construct an Ace query, the user needs some knowledge of the way data is structured in the specific database he or she is using, since, as already indicated, each genome database structures its data in a slightly different way according to its developers' and users' needs. For each class of data - map, sequence, probe, publication and so on - there is a *model* which shows how the data is organised, and these models are themselves a class of data in an Ace database, which the user can consult while setting up a query.

A simple Ace query might be

*find author Bark**

which would list all authors in the database whose names begin with Bark.

find author COUNT paper > 2

returns all authors with more than two publications in the database;

find Map COUNT locus > 40; follow locus type = "RFLP"*

finds all the maps on which more than 40 loci have been mapped, then lists all the loci on those maps which are of marker type RFLP.

As a final example, supposing a grass breeder wishes to identify all those loci which have been mapped, using RFLP markers, on linkage maps from the International Lolium Genome Initiative programme (Jones et al. 2002) and also maps showing the S and Z self-incompatibility loci (Thorogood et al. 2002). Here the query to the Forage Grass Genome Database, FoggDB, would be

find locus map = "ILGI" & map = "S and Z*"*

2.4 The New Ace Query Language, AQL

The majority of plant genome databases still use only the query language described above. However, the AceDB developers have in recent years implemented a new query language, AQL, which is similar to the SQL language used extensively in querying other database types. The GrainGenes database (project home page http://wheat.pw.usda.gov/index.shtml; database entry point http://www.graingenes.org/) offers both query languages and, in acknowledgement of the fact that AQL is even less intuitive and more difficult to learn than the older Ace query language, they have also included a number of pre-prepared AQL queries for common requests, to act as examples. At the time of writing the majority of plant genome databases have not yet made AQL querying available to end-users.

2.5 AceDB Plant Genome Databases

Since the early 1990s, genome databases using the AceDB system have been developed for more than 30 plant species, genera or families. The groups most active in this area have been funded by the USDA in the US, and by the UK's Biotechnology and Biological Sciences Research Council under the UK CropNet program (Dicks et al. 2000). Development of several of the US databases was suspended towards the end of the 1990s, and the USDA site which formerly housed them was closed down, but many of them are still available through sites maintained by UK CropNet (http://ukcrop.net), which hosts over 30 databases, and the INRA Genome Database Mirror in France (http//grain.jouy.inra.fr/gendatabasemirror.html) which maintains a large subset, including the AlfaGenes database of information relevant to *Medicago* species. Several of the US AceDB databases have continued development independently; these include TreeGenes at UC Davis (http://dendrome.ucdavis.edu/Treegenes/), covering more than 70 tree species, SoyBase, curated at the National Center for Genome Research (http://soybase.ncgr.org/ace/), which focuses on soybean and its pathogens, and GrainGenes (http://wheat.pw.usda.gov/index.shtml), a compilation of molecular and phenotypic information on wheat, barley, rye, triticale, and oats. The UK CropNet consortium curates databases on Arabidopsis, barley, brassicas and millet, and the Forage Grass Genome database which includes data on over 70 grass species, predominantly temperate, with the majority of map and sequence information derived from *Lolium perenne* and its relatives. CropNet also develops a resource for comparative sequence analysis, which will be described in section 3.2.

2.6 Non-AceDB Genome Databases

Not all public genome databases use the AceDB system. Some groups have opted instead for a commercial database management system, either because they wish to include specialised data types for which AceDB is not ideal, or because they may require less computational knowledge on the part of the database curator/developer, or in some cases because of technical support available from the manufacturer. The systems chosen are usually relational rather than - like AceDB - object-oriented; relational databases use the principle of tables of data between which relationships are established on the basis of common data items. Examples of major non-AceDB plant genome databases include

- MaizeDB (http://www.agron.missouri.edu/index.html; Polacco et al. 2002) which includes information on genetic stocks and maize genetics as well as maps and other genome data.

- ZmDB (http://www.zmdb.iastate.edu/), another maize database concentrating on genomic data. MaizeDB and ZmDB are being combined into the new maize genetics and genomics database Maize GDB (http://www.maizegdb.org/)

- The US-based Arabidopsis Information Resource TAIR (Garcia-Hernandez et al. 2002; http://www.arabidopsis.org/home.html). TAIR uses a relational database system, and covers germplasm and genetic stock data as well as genome data about this model species. In contrast its UK complement, the Arabidopsis Genome Resource AGR (http://ukcrop.net/agr/), uses AceDB to house genome data and provides links to the Nottingham Arabidopsis Stock Centre catalogue (http://nasc.nott.ac.uk/) so that users can order seed stocks. MATDB (http://www.mips.biochem.mpg.de/proj/thal/; Schoof et al. 2002), the other major international Arabidopsis genome database, is another non-AceDB system.

- Of special relevance to breeders of leguminous fodder crops, the *Medicago truncatula* Consortium's database MtDB (http://www.medicago.org/MtDB/; Lamblin et al. 2003), makes available information on ESTs from this model legume using a relational database system,

3. COMPARATIVE GENOMICS

For many plant breeders, the potential to exploit genomics and bioinformatics may be limited by a paucity of information on their crop of interest. Although sequencing technology becomes ever more efficient, cost considerations mean that it is still unlikely that the complete genome sequence will be available in the near future for most large-genome species. However, where a crop is closely related to a model species which has been the subject of intensive research, it may be possible to exploit syntenic relationships for some regions of the genome to identify genes of interest or appropriate markers.

3.1 Gramene

A number of bioinformatics projects are devoted to comparative plant genomics; one of the largest is Gramene (http://www.gramene.org), which is developing a comparative genomics resources for grasses and cereals (Ware et al. 2002). Gramene has incorporated the physical and genetic map data which were contained in the single-genus AceDB database RiceGenes, and as well as continuing to develop the rice dataset, maps have so far been added for barley, wheat, oat, maize and sorghum. Wherever two or more maps bear a relationship based on common features, they can be displayed in a comparative map view, making it easier to identify regions of synteny. Because many of the markers used to generate these maps are based on anchor probes (http://greengenes.cit.cornell.edu/anchors/; van Deynze et al. 1998) which have also been used by the International Lolium Genome Initiative in mapping *Lolium perenne* (Jones et al. 2002; maps and loci available from FoggDB), it is possible to identify corresponding regions in the Lolium and rice genomes and - where synteny holds good - exploit this information for map-based cloning.

3.2 CropSeqDB

While most of the databases being developed by the UK CropNet consortium are for single or closely-related species, one is designed primarily for comparative genomics work. CropSeqDB (http://ukcrop.net/perl/ace /search/CropSeqDB) extracts all the DNA sequence data from the EMBL database which is relevant to crops of importance in UK agriculture and combines it for ease of sequence homology searching. In early 2003 it contained over half a million sequences from nearly 200 species.

3.3 Sputnik

For many crop plants, there is not only little likelihood of a complete genome sequence in the near future; even full-length gene sequences may be scarce. However, there is often a wealth of Expressed Sequence Tag (EST) data, and the Sputnik project (http://mips.gsf.de/proj/sputnik/; Rudd et al. 2003) is designed to assist researchers in exploiting this for comparative plant genomics. The database provides the results of extensive computational analyses of individual ESTs, EST clusters, and derived peptide sequences, for over 25 species including cereals - barley, maize, rice, rye, sorghum and wheat - and the legumes *Medicago truncatula* and *Lotus corniculatus*.

3.4 PlantGDB

PlantGDB (http://www.plantgdb.org/) is a database of plant DNA sequences from over 20 monocot and dicot species. The sequences are predominantly Expressed Sequence Tags (ESTs) organized into annotated contigs that represent tentative unique genes.

4. BEYOND THE GENOME: TRANSCRIPTOMICS, PROTEOMICS, METABOLOMICS

In dissecting complex traits, it is often necessary to know where, when and in response to which stimuli genes are expressed. This is not only of importance in applying transgenic methodologies to crop improvement; selection of appropriate candidate genes to use as the basis for marker-assisted selection strategies can also benefit from an understanding of gene expression and gene products. In the wake of genomic techniques and their associated bioinformatics, other areas of science have developed to address these post-transcriptional processes. Transcriptomics is the study of all gene transcripts (mRNAs) present in a given cell or tissue under specified conditions, and usually involves a comparison of two or more treatments, tissue types or genotypes (Dunwell et al. 2001). Proteomics concerns the complement of proteins, usually excluding insoluble and very small polypeptides (Guo et al. 2002), and metabolomics is the term used for studying the set of small organic molecules - sugars, amino acids, lipids etc. - in the tissue at a given time (Fiehn 2002; Sumner et al. 2003). The methodologies used in these fields generate large amounts of data, and a specialised area of bioinformatics is developing to meet the needs in each case (Kanehisa & Bork 2003). Transcriptomics is probably the most advanced in this respect.

4.1 Transcriptomics

Transcriptomics projects generally employ some variant of array hybridisation technology, in which a series of DNA sequences (for example, ESTs, cDNAs or genomic sequences) is dotted onto a support matrix to generate a two-dimensional array of spots; this array is challenged, simultaneously or sequentially, with labelled probes representing the gene transcripts present in two contrasting tissues or treatments. The probes may be RNA or, more commonly, DNA derived from RNA by reverse transcription; the labels used are most frequently fluorescent dyes. The primary results of a transcriptome analysis experiment are generally in the form of an image showing relative fluorescence intensities of the different spots in the array. Bioinformatics tools being developed to serve the needs of transcriptomics projects include databases that can accommodate primary image data, methods for normalising data and eliminating artefacts, and machine-learning methodologies for extracting useful information from the large datasets that result - for example, identifying clusters of genes showing similar patterns of transcriptional response to particular environmental factors. In comparison with databases for sequence and map data, there are as yet few publicly-accessible plant transcriptome databases; the following three contain data from Arabidopsis, tomato and soybean respectively, but others will emerge in the near future, some integrated with existing genome databases.

4.1.1 CATMA

The Complete Arabidopsis Transcriptome Microarray project CATMA (http://www.catma.org/) is intended to design high-quality Gene Sequence Tags (GSTs) covering most Arabidopsis genes; they will be used primarily for creating and transcription profiling of arrays to determine the expression patterns of the corresponding genes. In early 2003 the CATMA database already contained information on over 20000 Arabidopsis GSTs. CATMA is one of several plant genome bioinformatics projects within the Génoplante programme (http://genoplante-info.infobiogen.fr/), which is developing databases and analysis tools for work on five major European crop species together with Arabidopsis (Samson et al. 2003).

4.1.2 TED

The Tomato Expression Database TED (http://ted.bti.cornell.edu/) includes information about over 12000 Expressed Sequence Tags from tomato, with figures for the relative expression levels of their transcripts in different tissues (for example, leaf compared with ripe fruit) and links to the corresponding sequence and other data for each clone in GenBank and the Solanaceae Genomics Network database (http://www.sgn.cornell.edu/index.html).

4.1.3 SGMD

For soybean, sequence and microarray data have been collected together in the Soybean Genomics and Microarray Database (http://psi081.ba.ars.usda.gov/SGMD/Default.htm), which includes extensive information about the arrays themselves, the probes and hybridisation conditions used, and the results. It is possible, for example, to ask the question "which genes are upregulated in response to high compared with low potassium?"

5. INTEGRATING GENETIC RESOURCES DATABASES AND GENOME BIOINFORMATICS

Many curators of plant genetic resources collections have made stock information publicly available using Web access to databases containing their catalogues. Plant breeders have often been able to capitalise on this to gain access to valuable germplasm for crop improvement. Hitherto, these databases have focused on essential information about the origins and holders of the germplasm, often using the descriptor standards defined by the International Plant Genetic Resources Institute (http://www.ipgri.cgiar.org/). In some cases, they have also included characterisation and evaluation information based on plant morphology, field performance and other well-established properties. However, availability of modern molecular marker systems means that many germplasm collections are now undergoing molecular characterisation, so that the worlds of plant genome analysis and plant genetic resources are converging.

5.1 The GENE-MINE Project

New bioinformatics approaches are needed to accommodate this more integrated approach, and one project addressing this need is GENE-MINE (http://www.gene-mine.org/), an EU-funded project involving nine European partners and the US National Center for Genome Resources. Its aims include developing a web-based system for storing and accessing large amounts of molecular genetic data, trait data and passport data from genebanks; linking this information to genomic and other biological databases; developing a querying system allowing biologically-meaningful questions to be asked of the data.. Such approaches should ultimately allow plant breeders to access the information they need through the genetic resources interfaces which are often already a part of their working practices, but in this area there remains a huge amount of work to be done, in the lab and at the computer, before practical implementation becomes a reality.

6. CONCLUSIONS

Bioinformatics, which had its origins in the needs of molecular biologists carrying out fundamental research on model organisms, is developing to the point where it can potentially be of direct benefit to those engaged in crop improvement programmes. The volume of data now produced by molecular marker, transcriptome and metabolome methodologies requires data management and analysis tools on a scale which were unnecessary for traditional methods of genotype analysis. The next decade is likely to see an increase in integration of disparate data types, from sequence and map through transcriptome and proteome to germplasm characterisation; and the development of more sophisticated data mining approaches to draw strands of useful information from the tangled mass of data. For breeders to gain the maximum benefit from these advances, there is also a need for more user-friendly interfaces and querying methods, and for continuing dialogue between bioinformaticists and breeders so that each can improve their understanding of the others' requirements and limitations. Finally, the international community of publicly-funded plant genome scientists and bioinformaticists has hitherto demonstrated a commitment to making their data available for the common good and to the widest possible community. This ethos must continue if the true potential of bioinformatics is to be realised.

REFERENCES

C. elegans Sequencing Consortium (1998) Genome Sequence of the Nematode *C. elegans*: A Platform for Investigating Biology. Science 1998: 2012-2018. (Note: the full author listing for this publication, consisting of over 350 names, is available from http://www.sanger.ac.uk/Projects/C_elegans/)

Dicks J, Anderson M, Cardle L, Cartinhour S, Couchman M, Davenport G, Dickson J, Gale MD, Marshall D, May S, McWilliam H, O'Malia A, Ougham HJ, Trick M, Walsh S, Waugh R (2000) UK CropNet: a collection of databases and bioinformatics resources for crop plant genomics. Nucleic Acids Res. 28: 104-107.

Dunwell JM, Moya-Leon MA, Herrera R (2001) Transcriptome analysis and crop improvement. Biol. Res. 34: 153-164.

Fiehn O (2002) Metabolomics - the link between genotypes and phenotypes. Plant Mol. Biol. 48: 155-171.

Garcia-Hernandez M, Berardini TZ, Chen G, Crist D, Doyle A, Huala E, Knee E, Lambrecht M, Miller N, Mueller LA, Mundodi S, Reiser L, Rhee SY, Scholl R, Tacklind J, Weems DC, Wu Y, Xu I, Yoo D, Yoon J, Zhang P (2002) TAIR: a resource for integrated Arabidopsis data. Funct. Integr. Genomics 2: 239–253.

Guo YM, Shen SH, Jing YX, Kuang TY (2002) Plant proteomics in the post-genomic era. Acta Bot. Sin. 44: 631-641.

Jones ES, Mahoney NL, Hayward MD, Armstead IP, Jones JG, Humphreys MO, King IP, Kishida T, Yamada T, Balfourier F, Charmet G, Forster JW (2002) An enhanced molecular marker based genetic map of perennial ryegrass (*Lolium perenne*) reveals comparative relationships with other Poaceae genomes. Genome 45: 282-295.

Kanehisa M, Bork P (2003) Bioinformatics in the post-sequence era. Nat. Genet. 33: 305-310 Suppl.

Lamblin AJ, Crow JA, Johnson JE, Silverstein KAT, Kunau TM, Kilian A, Benz D, Stromvik M, Endré G, VandenBosch KA, Cook DR, Young ND, Retzel EF (2003) MtDB: a database for personalized data mining of the model legume Medicago truncatula transcriptome. Nucleic Acids Res. 31: 196-201.

Luscombe NM, Greenbaum D, Gerstein M (2001) What is bioinformatics? A proposed definition and overview of the field. Methods Inf. Med. 40: 346-58.

Polacco ML, Coe E, Fang, Z, Hancock DC, Sanchez-Villeda H, Schroeder S (2002) MaizeDB - a functional genomics perspective. Comp. Funct. Genom. 3:128-131.

Rudd S, Mewes HW, Mayer KFX (2003) Sputnik: a database platform for comparative plant genomics. Nucleic Acids Res. 31: 128-132.

Samson D, Legeai F, Karsenty E, Reboux S, Veyrieras JB, Just J, Barillot E (2003) GenoPlante-Info (GPI): a collection of databases and bioinformatics resources for plant genomics. Nucleic Acids Res. 31: 179-182.

Schoof H, Zaccaria P, Gundlach H, Lemcke K, Rudd S, Kolesov G, Arnold R, Mewes HW, Mayer KF (2002) MIPS Arabidopsis thaliana Database (MAtDB): an integrated biological knowledge resource based on the first complete plant genome. Nucleic Acids Res. 30: 91-93.

Sumner LW, Mendes P, Dixon RA (2003) Plant metabolomics: large-scale phytochemistry in the functional genomics era. Phyrtochem. 62: 817-836.

Thorogood D, Kaiser WJ, Jones JG, Armstead IP (2002) Self-incompatibility in ryegrass 12 Genotyping and mapping the S and Z loci of Lolium perenne L Heredity 88: 385-390.

van Deynze AE, Sorrells ME, Park WD, Ayres NM, Fu H, Cartinhour SW, Paul E, McCouch SR (1998) Anchor probes for comparative mapping of grass genera. Theor. Appl. Genet. 97: 356-369.

Ware D, Jaiswal P, Ni JJ, Pan XK, Chang K, Clark K, Teytelman L, Schmidt S, Zhao W, Cartinhour S, McCouch S, Stein L (2002) Gramene: a resource for comparative grass genomics. Nucleic Acids Res. 30: 103-105.

Data Integration and Target Selection for *Medicago* Genomics

L. Wang and Y. Zhang
Plant Biology Division, The Samuel Roberts Noble Foundation, Ardmore, OK 73401, USA. (Email: lwang@noble.org).

Key words: genomics, target selection, data integration, database, bioinformatics

Abstract:

A large collection of expressed sequence tags (ESTs) have been generated for *Medicago truncatula*. With the progress of genome sequencing, *M. truncatula* has become a model legume for genomics. However, the sequence and related information is distributed across several sites, and thus not organized in an optimal way for genomic studies. We developed computational methods to transform, clean and load the sequence data into a local warehouse, and to integrate the data with other information from major public databases (*e.g.* GenBank, Pfam and PDB). The integrated database system allows us to build web-based tools for genomics target selection. For our own research, the system permits a genome-wide search for genes that may be involved in legume-specific biological processes. The database and related bioinformatic tools are available at http://bioinfo.noble.org/.

A. Hopkins, Z.Y. Wang, R. Mian, M. Sledge and R.E. Barker (eds.), Molecular Breeding of Forage and Turf, 275-288.

1. INTRODUCTION

Medicago truncatula, a close relative of alfalfa, has been used as a model legume for genomic studies, including large-scale EST and genome sequencing (Cook 1999; Frugoli and Harris 2001). By the end of year 2002, NCBI's dbEST had over 170,000 *M. truncatula* EST records. With the progress of genome sequencing at the University of Oklahoma (http://www.genome.ou.edu/medicago.html) and advancement of genetic transformation (Boisson-Dernier et al. 2001; Somers et al. 2003), *M. truncatula* can serve as an excellent model organism for genomic studies of plant-microbe interactions, natural product biosynthetic pathways and stress responses. Unlike the other model plants such as *Arabidopsis* and rice, *M. truncatula* establishes symbiotic relationships with the nitrogen-fixing *Rhizobia* and beneficial arbuscular mycorrhizal fungi (Harrison 1999; Stougaard 2000, 2001), and has a variety of legume-specific metabolic pathways (*e.g.*, the isoflavonoid pathway) (Dixon and Steele 1999; Dixon and Sumner 2003).

Large-scale single pass sequencing of cDNA clones, or EST sequencing, has been used for rapid gene discovery in many plant species. Currently, the public EST dataset of *M. truncatula* has been obtained from over 35 cDNA libraries representing different tissue types and/or various experimental treatments. These EST sequences have been assembled into contigs by The Institute for Genomic Research (TIGR) to generate a non-redundant dataset, called *M. truncatula* Gene Index or MtGI (Quackenbush et al. 2000). MtGI version 6.0 released on 12/31/02 has a total of 36,262 sequences, which may represent most of the transcriptome. Furthermore, it is possible to roughly estimate the gene expression pattern of a contig by counting the frequency of its tags in different cDNA libraries (Ewing et al. 1999). This method, called EST counting or 'electronic northern', has been used to select interesting targets for studies of rhizobial and arbuscular mycorrhizal symbioses in *M. truncatula* (Journet et al. 2002; Fedorova et al. 2002). Although TIGR has a website for public access to individual MtGI records and EST information (http://www.tigr.org/tdb/tgi/mtgi/), no bioinformatic tool is provided for genome-wide analyses based on user-specified queries. In addition, the protein sequences and domain information of most *M. truncatula* genes are still not available in major public domain databases, partly due to the relatively low quality of EST sequence data.

The Center for *Medicago* Genomics Research (http://www.mtruncatula.org/) was established at the Noble Foundation in 1999. Since then, various research projects have been initiated to understand the biological events and environmental interactions of the model legume from a global view. To facilitate these genomics activities, we have developed a local warehouse and associated software tools to process and integrate data from various sources. The integrated system supports efficient selection of gene targets for *Medicago* genomics.

2. GENOMICS DATA INTEGRATION

The public resources for *Medicago* genomics are currently located in several sites, including the public domain ESTs in dbEST (http://www.ncbi.nlm.nih.gov/dbEST/), non-redundant MtGI datasets from TIGR (http://www.tigr.org/tdb/tgi/mtgi/), and available genomic sequences at http://www.genome.ou.edu/medicago.html. Integration of data from these heterogeneous sources is critically important for genome-wide analysis and target selection.

Data integration may be achieved using WWW links, database federation or data warehousing. The method of WWW links, although commonly used in the scientific community to provide a single entry point of access to multiple databases, does not support large-scale data analysis. It is rather designed for human navigation of documents through hyperlinks.

The database federation approach is to construct a central query system for heterogeneous source databases. The central system maintains only a global schema, while the source databases manage all the data. In other words, upon a user query, the central system calls the appropriate source databases to process the query, and then combine the results to generate the final output. Thus, the main disadvantage of federation is the system response time to execute queries, which may become intractable for genome-wide analyses. In addition, this approach may entail a major standardization effort within genomics databases for interoperability.

In contrast to database federation, the warehousing approach loads all the source data into a single location using an integrated schema. Since the source data are often in heterogeneous formats, transformation and cleaning are required prior to loading. The warehousing approach has a number of properties that are advantageous for integration of genomics databases (Leser et al. 1998; Paton et al. 2000; Shoop et al. 2001). First, the source data can be integrated locally based on semantics. Genomics databases have rich object and complex relationship types, but lack standards in nomenclature and term definitions. The source databases may also contain inconsistent or noisy data. Transformation and cleaning can be applied to data without affecting the source databases. Second, datasets that are generated locally and not contained in any of the source databases can be integrated into a warehouse. These datasets can stem from analysis of the source data or local genomics experiments. Third, data access to a single warehouse is faster than to several source databases. Furthermore, materialized views can be implemented in the warehouse to further enhance query performance and thus system response time (Shepherd et al. 2002). The major disadvantage of warehousing is to keep the data up to date. When the source databases release new data, the warehouse needs to be updated accordingly.

3. *MTGENES*: A DATA WAREHOUSE FOR *MEDICAGO* GENOMICS

We are using the data warehousing approach to integrate data for *Medicago* genomics. As shown in Figure 1, the local warehouse, *MtGenes*, integrates data from GenBank (ESTs, NR and taxonomy data), TIGR (MtGI), Pfam (protein domain models) and PDB (protein structure information). The available genomic sequences of *M. truncatula* are currently being integrated into the warehouse. We also anticipate that experimental results from genomics research such as microarray profiling, proteomics and genetic mapping will be integrated into *MtGenes* in the future.

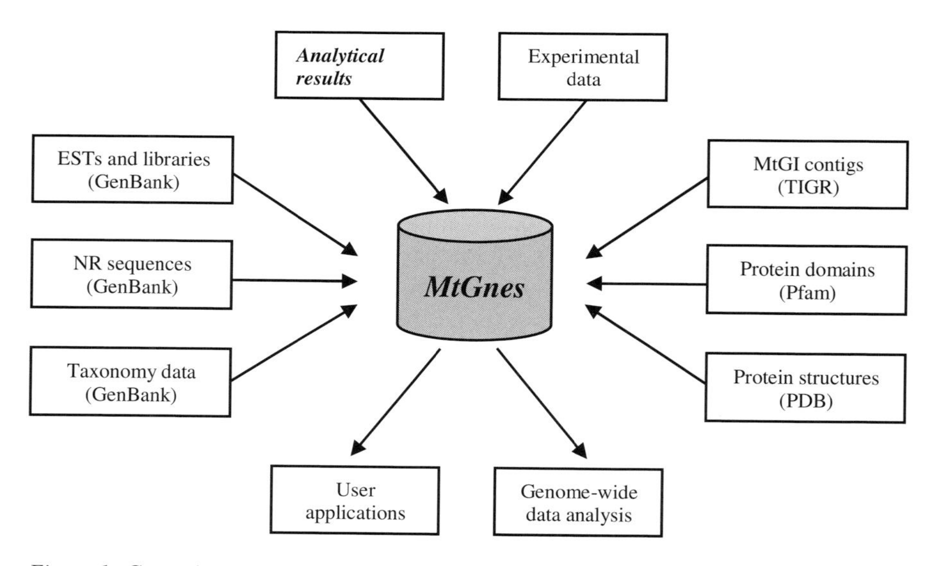

Figure 1. Genomics data integration in the local warehouse *MtGenes*.

MtGenes is a relational database currently with 25 tables to store transformed source data and meta-data. Meta-data is defined as the data about data, and in this project includes the description of source data (*e.g.*, source URL, sequence identifiers) and results from our own data analysis. All the sequence data downloaded from source databases are analyzed using our local tools to provide more information about the record and in some cases to establish the relationships between heterogeneous data entries. For example, protein coding information of MtGI sequences is not yet provided by TIGR. We developed and used the *EST-Analyzer* tool (http://bioinfo.noble.org/estanalyzer.htm) to derive the *M. truncatula* protein set, and searched the protein sequences against Pfam domain models. The results from these analyses were then imported into *MtGenes* to provide the protein information about MtGI records and to establish the relationships between MtGI sequences and Pfam domains (Figure 2). Keeping analytical results in the warehouse greatly improves system response to execute complex queries (*e.g.*, find all the *Medicago* genes that have the Myb DNA-binding domain).

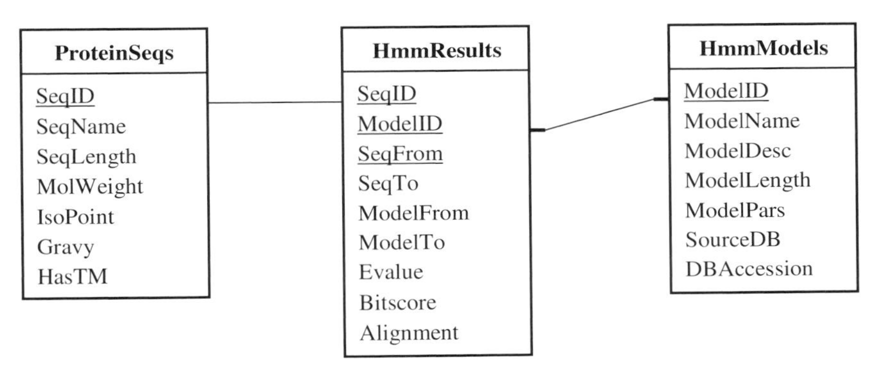

Figure 2. Part of the *MtGenes* database schema designed for protein domain information of *M. truncatula* genes (table keys underlined).

4. TARGET SELECTION TOOLS FOR FUNCTIONAL GENOMICS

MtGenes provides an integrated platform for *Medicago* genomics research. The web-based interface of *MtGenes* (Figure 3) supports three types of queries by using sequence identifiers (TIGR's TC numbers or NCBI's GB identifiers), gene annotations (text search) and Pfam domain names. When queried using a sequence identifier, *MtGenes* provides four categories of gene information as shown in Figure 4, including functional annotations and sequence homologues; protein domain and structural information; gene expression patterns (based on EST counts); and cDNA clone information. The different categories of information may help understand gene functions. For example, the *M. truncatula* gene TC51862 is annotated as hypothetical or unknown protein based on sequence similarity or BLAST search (Figure 4). Nevertheless, protein domain search reveals that TC51862 contains a Myb-like DNA-binding domain, and EST expression analysis suggests that it may be highly expressed during drought and development of leaves. Thus, TC51862 appears to be a Myb-like transcription factor, and may be involved in leaf development and/or drought responses. In addition, WWW links to source databases are included on the query output page for further references. The other search options for target selection are described below.

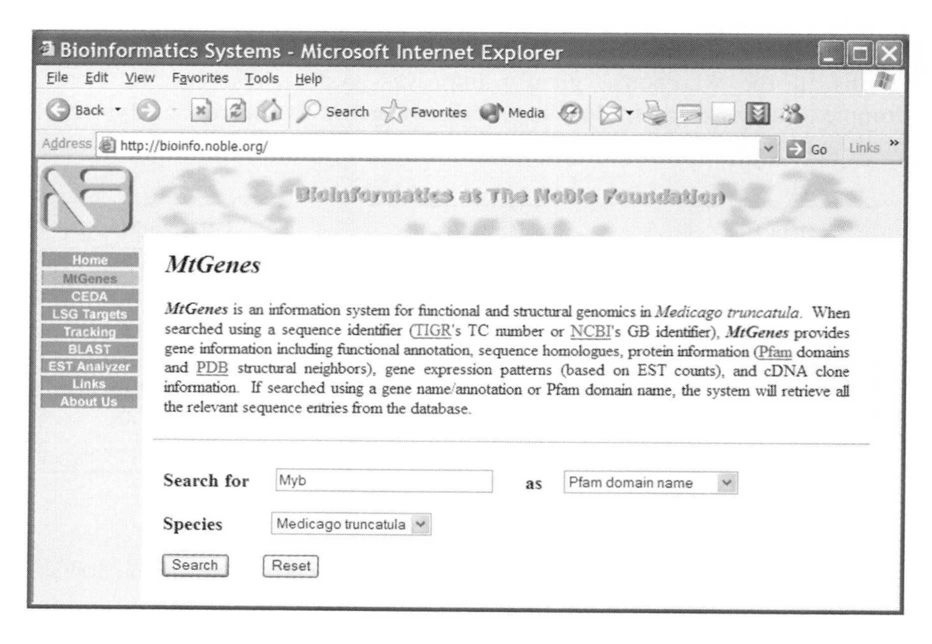

Figure 3. Web interface of *MtGenes*.

4.1 Text Search

The *MtGenes* interface supports queries using key words from gene names or annotations. The system searches the database for sequence records that contain the key words, and then displays a list of all the relevant entries, including sequence identifiers and annotations. The user can click on the sequence identifiers to retrieve all the gene information as shown in Figure 4. However, since the current gene annotations still lack standard nomenclatures and Gene Ontology (GO) terms, text searches could be neither accurate nor complete. We encourage users to explore the protein domain or BLAST search options.

TC51862

Our Annotation:	similar to ref|NP_173269.1| (NM_101691) hypothetical protein [Arabidopsis thaliana] (Eval 6e-040, coding 100%)
TIGR Annotation:	weakly similar to GP|10177075|dbj|BAB10517. gene_id:MKP11.15~unknown protein {Arabidopsis thaliana}, partial (34%)

GenBank Homologs:

Accession	Functional annotation	Identity	Evalue/Alignment
NP_173269	hypothetical protein [Arabidopsis thaliana] gi|173 ...	33%	6e-040 [Details]
AAF25987	F15H18.16 [Arabidopsis thaliana]	34%	2e-039 [Details]

Protein:

Accession: TC51862.p1	**Length:** 467 amino acids
Status: Full-length	**Mol Weight:** 60.01 kD
Expected pI: 5.68	**GRAVY Index:** -0.85

Pfam Domains:

Accession	Model Name	Description	Protein		Evalue/Alignment	More Sequences
			From	To		
PF00249	myb_DNA-binding	Myb-like DNA-binding domain	57	102	3e-014 [Details]	Search

Gene Expression Pattern:

Dataset Name	cDNA Libraries	Dataset Size	EST Counts	% Frequency
seedling, drought	Drought	9520	8	0.0840
leaf, developing	Developing leaf	9415	7	0.0743
root, nematode-infected	BNIR	3154	1	0.0317
root, mycorrhizal	MHAM; MtBC	15969	2	0.0125

EST Clone Information:

Most 5'-Proximal Clone: NF019D05IN1F1046	**GB Accession:** BF639668
5'-End ORF Coverage: +191 nt	**cDNA Library:** Insect herbivory

Available ESTs:

#	GB ID	Length	Contig		Clone	Library	Description
			Start	End			
1	BF639668	424	1	424	NF019D05IN1F1046	Insect herbivory	Medicago truncatula cDNA clone NF019D05IN 5'
2	BI263082	671	3	672	NF084C12PL1F1098	Phosphate starved leaf	Medicago truncatula cDNA clone NF084C12PL 5'

Figure 4. The query output from *MtGenes* provides four categories of gene information.

4.2 Protein Domain Search

The *MtGenes* warehouse can be queried using Pfam domain identifiers or key words from model descriptions. Pfam (http://pfam.wustl.edu/) is a database of profile Hidden Markov Models (HMMs) of known protein domains. The models are constructed from multiple sequence alignments, and curated manually (Bateman et al. 2002). The mapping between *M. truncatula* protein sequences and Pfam models were pre-computed and stored in the *MtGenes* warehouse (Figure 2). When queried using a Pfam model identifier or name, the system retrieves all the *M. truncatula* sequences that contains the protein domain. For example, a query using "Myb" results in 100 sequences that contain the Myb-like DNA-binding domain. Gene expression and other gene information can be further obtained by clicking on these sequence entries. Table 1 lists some of the *M. truncatula* Myb-like genes that may be involved in rhizobial and/or mycorrhizal symbioses, pathogen elicitation, drought stress, or development, based on their EST expression patterns.

Table 1. Selected *M. truncatula* Myb-like genes and their EST expression patterns.

Group/Gene	Expression pattern (EST count, % frequency)
Symbiosis	
TC54977	Nodulated roots (2, 0.011)
TC50907	Root nodules (1, 0.010), nodulated roots (1, 0.006)
TC50202	Root nodules (1, 0.010), mycorrhizal roots (1, 0.006)
TC46143	Mycorrhizal roots (2, 0.013), nodulated roots (2, 0.011)
TC58919	Mycorrhizal roots (2, 0.013)
Elicitation	
TC50497	Elicited cells (2, 0.020)
TC49031	Fungus-infected leaves (1, 0.011), elicited cells (1, 0.010)
TC58106	Nematode-infected roots (2, 0.063)
Stress	
TC45265	Drought seedlings (4, 0.042)
TC51861	Drought seedlings (5, 0.053), elicited cells (1, 0.010)
TC57421	P-starved leaves (2, 0.020)
Development	
TC48634	Developing flowers (3, 0.045)
TC50530	Developing stems (2, 0.019)
TC54194	Developing leaves (4, 0.043), insect herbivory leaves (1, 0.010)

4.3 BLAST Search

The local BLAST server (http://bioinfo.noble.org/blast.htm) is integrated with the *MtGenes* warehouse. Users can paste their sequences and search against datasets including *M. truncatula* ESTs, MtGI and protein sequences.

The output from the BLAST search is linked to *MtGenes* to provide all the gene information as described above. The BLAST server also supports searches against *M. truncatula* genomic sequences, GenBank NR, PDB sequences, and datasets from other species. In these cases, the BLAST hits are linked to the dataset sources to provide more information.

4.4 Expression-Based Target Selection

The Comparative EST Data Analysis (*CEDA*) tool was developed on top of the *MtGenes* database to support target selection based on gene expression (Figure 5). *CEDA*, available at http://bioinfo.noble.org/ceda.htm, can be used to compare two user-defined EST datasets (target and control datasets) to identify specific, up-regulated or down-regulated genes in the target dataset. It can also be used to retrieve the unigene set from a cDNA library. The output from *CEDA* is a list of gene entries, which are linked to *MtGenes* for detailed information.

The case study of *CEDA* uses is to select targets for functional genomics of rhizobial symbiosis. The worldwide EST sequencing projects of *M. truncatula* have generated eight cDNA libraries from nodulated roots and/or nodules. These libraries can be selected in *CEDA* as the target dataset. The control dataset consists of all the remaining 27 libraries from a variety of tissues and experimental treatments. Genes that are specifically expressed or up-regulated during rhizobial symbiosis can then be computed by comparing the target dataset with the control dataset. Table 2 lists some of these targets, which include a few genes that are known to be involved in nodulation (*e.g.*, nodulins and leghemoglobins). The target list also includes unknown proteins, enzymes, calmodulins and transcription factors. These genes may be interesting targets for functional genomics of rhizobial symbiosis.

5. COMPARATIVE SEQUENCE DATA ANALYSIS

MtGenes also provides the infrastructure necessary for computational genomics research. Genome-wide analyses of sequences or profiles often result in massive datasets, which need be integrated with the other available information to understand the biological problem under study. In many of these cases, an integrated database system is critically important.

One of our current interests is to search for legume-specific genes (LSGs). In this study, *M. truncatula* genes are selected as LSG targets if homologues are found only in legume species.

CEDA: Comparative EST Data Analysis in *Medicago truncatula*

CEDA helps you identify interesting gene targets through comparative analysis of EST expression datasets (EST counts). ***CEDA*** can be used to retrieve the unigenes and their EST abundance in a user-defined dataset (dataset #1 below); identify specific, up-regulated or down-regulated genes in your target dataset (dataset #1) when compared with a control dataset (dataset #2); and search for targets that are up-regulated in both datasets (datasets #1 and #2) when compared with the other cDNA libraries. The output from ***CEDA*** is a list of genes which are linked to *MtGenes* for more information. Please contact lwang@noble.org for any questions or comments about this system.

Select analysis type: Specific genes in dataset #1 (vs. dataset #2)

If for up or down-regulated genes, enter the threshold of minimum fold difference: 5

Select your target datasets below:

EST dataset #1:

- [ ] Developing root, no symbiosis (3054)
- [ ] MtRHE, root hair-enriched (899)
- [ ] KV0, not nodulated root (2752)
- [x] KV1, root - 1 day nod (2840)
- [x] KV2, root - 2 days nod (3330)
- [x] KV3, root - 3 days nod (4315)
- [x] MtBB, root - 4 days nod (7807)
- [x] GVN, root nodules (6468)
- [x] GVSN, senescent nodules (2788)
- [x] R108, root nodules (438)
- [x] Nodulated root, mixed (3299)
- [ ] MtBA, N starved root (7939)
- [ ] MHAM, mycorrhizal root (7368)
- [ ] rootphos(-), P starved root (1967)
- [ ] MHRP-, P starved root (2658)
- [ ] MtBC, mycorrhizal root (8601)
- [ ] MGHG, beta glucan elicited root (2687)
- [ ] DSIR, fungus-elicited root (2463)
- [ ] BNIR, nematode-infected root (3154)
- [ ] HOGA, oligogalac-elicited root (2861)
- [ ] Elicited cell culture (9859)
- [ ] Developing stem (10783)
- [ ] Developing leaf (9415)
- [ ] P starved leaf (10188)
- [ ] Insect herbivory leaf (10309)
- [ ] DSIL, *C. trifolii*-infected leaf (6003)
- [ ] Phoma-infected leaf (3281)
- [ ] DSLC, cotyledon and leaf (2143)
- [ ] Developing flower (6724)
- [ ] GPOD, developing pod (1915)
- [ ] GESD, developing seed (2672)
- [ ] GLSD, developing seed (2944)
- [ ] Germinating seed (1524)
- [ ] Drought seedling (9520)
- [ ] Irradiated seedling (6748)

EST dataset #2:

- [x] **All the other EST datasets except the above target datasets**

- [x] Developing root, no symbiosis (3054)
- [x] MtRHE, root hair-enriched (899)
- [x] KV0, not nodulated root (2752)
- [ ] KV1, root - 1 day nod (2840)
- [ ] KV2, root - 2 days nod (3330)
- [ ] KV3, root - 3 days nod (4315)
- [ ] MtBB, root - 4 days nod (7807)
- [ ] GVN, root nodules (6468)
- [ ] GVSN, senescent nodules (2788)
- [ ] R108, root nodules (438)
- [ ] Nodulated root, mixed (3299)
- [x] MtBA, N starved root (7939)
- [x] MHAM, mycorrhizal root (7368)
- [x] rootphos(-), P starved root (1967)
- [x] MHRP-, P starved root (2658)
- [x] MtBC, mycorrhizal root (8601)
- [x] MGHG, beta glucan elicited root (2687)
- [x] DSIR, fungus-elicited root (2463)
- [x] BNIR, nematode-infected root (3154)
- [x] HOGA, oligogalac-elicited root (2861)
- [x] Elicited cell culture (9859)
- [x] Developing stem (10783)
- [x] Developing leaf (9415)
- [x] P starved leaf (10188)
- [x] Insect herbivory leaf (10309)
- [x] DSIL, *C. trifolii*-infected leaf (6003)
- [x] Phoma-infected leaf (3281)
- [x] DSLC, cotyledon and leaf (2143)
- [x] Developing flower (6724)
- [x] GPOD, developing pod (1915)
- [x] GESD, developing seed (2672)
- [x] GLSD, developing seed (2944)
- [x] Germinating seed (1524)
- [x] Drought seedling (9520)
- [x] Irradiated seedling (6748)

Submit Query   Reset

Figure 5. *CEDA* is an expression-based target selection tool for functional genomics.

Table 2. Selected *M. truncatula* genes that are specifically expressed or up-regulated during rhizobial symbiosis.

Gene	Nodulation dataset (31,285 ESTs)		Control dataset (140,431 ESTs)		Annotation
	EST count	Frequency (%)	EST count	Frequency (%)	
TC51078	265	0.847	0	0	Leghemoglobin 1
TC51077	112	0.358	0	0	Unkown protein
TC43407	70	0.224	4	0.003	Early nodule-specific protein Enod8.1
TC43145	56	0.179	0	0	Leghemoglobin 2
TC51076	55	0.176	0	0	Similar to leghemoglobin
TC43512	51	0.163	0	0	Similar to leghemoglobin 29
TC43267	50	0.160	5	0.004	Nodulin 25
TC51133	48	0.153	0	0	MtN22
TC43017	45	0.144	0	0	Similar to carbonate dehydratase
TC51210	45	0.144	0	0	Unknown protein
TC51604	45	0.144	1	0.001	Similar to basic blue protein
TC51605	44	0.141	1	0.001	Similar to lectin-related polypeptide
TC51134	42	0.134	0	0	Similar to MtN22
TC51596	41	0.131	6	0.004	Similar to early nodulin ENOD40
TC51671	39	0.125	0	0	Similar to early nodulin ENOD20
TC51653	39	0.125	0	0	Unknown protein
TC51728	36	0.115	0	0	Unknown protein
TC43514	33	0.106	0	0	Unknown protein
TC51075	29	0.093	0	0	Unknown protein
TC51957	25	0.080	0	0	MtN29
TC52051	22	0.070	0	0	MtN1
TC44136	20	0.064	0	0	Similar to early nodulin ENOD18
TC51675	16	0.051	1	0.001	Similar to cysteine proteinase
TC51594	16	0.051	0	0	Similar to calmodulin
TC53459	8	0.026	0	0	CCAAT-binding transcription factor

We constructed a computational pipeline for identification and analyses of the LSG targets, and the essential components are shown in Figure 6. The local BLAST search of *M. truncatula* sequences against GenBank's NR and dbEST databases resulted in a very large output file. Since each of the *M. truncatula* sequences may have BLAST hits from many species, including legumes and non-legumes, it is not a feasible way to manually analyze the BLAST outputs for LSG identification. Instead, we used the *MtGenes* database system to integrate the sequence, BLAST output and GenBank's taxonomy datasets. More than 500 LSG targets were then selected in this way. Since these targets have at least one homologue in legume species other than *M. truncatula*, they should not be sequencing artifacts or species-specific genes. Interestingly, many of these LSG targets may be expressed only in root (based on EST counts), especially during rhizobial symbiosis. The results imply that some LSGs may have evolved specifically for root nodulation.

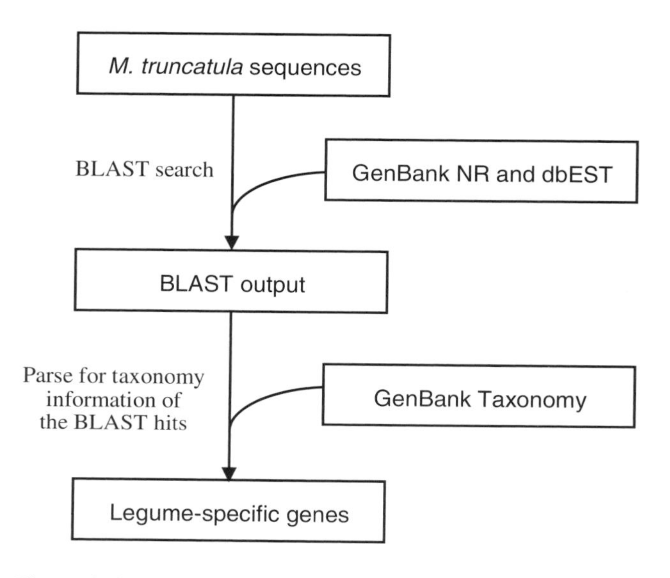

Figure 6. Searching for legume-specific genes using *MtGenes*.

We are also interested in understanding the function and evolution of legume transcription factors using a comparative analysis approach. Transcription factors function as switches of gene expression and thus may play important roles in plant development and response to environmental cues. Results from a previous study (Riechmann et al. 2000) suggest that the *A. thaliana* genome codes for more than 1,500 transcription factors. Interestingly, about 45% of these factors may be specific to plants. Analysis of the R2R3 Myb genes in maize indicates that this regulatory gene family may be amplified during evolution of land plants (Rabinowicz et al. 1999). We are using the protein domain information in *MtGenes* to identify *M. truncatula* transcription factors. These genes are being compared with the factors from *A. thaliana* and other plant species using phylogenetic analyses. Preliminary results suggest that some regulatory genes may have been amplified specifically in legumes. It will be interesting to demonstrate that these genes are involved in legume-specific biological processes. By further integrating the genomic sequence data and knowledge of *cis*-acting promoter elements into the *MtGenes* warehouse, we may also be able to understand the regulatory networks in *M. truncatula*.

6. CONCLUSIONS

We developed a local warehouse to integrate the sequence and related information from various public sources. The integrated database system permits efficient selection of targets for *Medicago* genomics, and supports large-scale comparative sequence data analyses. With the rapid accumulation of biological data in recent years, we believe that such an integrated database system is important for genomics research.

ACKNOWLEDGMENTS

We thank Drs. Richard A. Dixon, Gregory D. May and Lloyd W. Sumner for helpful discussions. Financial support for this project was provided by the Samuel Roberts Noble Foundation.

REFERENCES

Bateman A, Birney E, Cerruti L, Durbin R, Etwiller L, Eddy SR, Griffiths-Jones S, Howe KL, Marshall M, Sonnhammer EL (2002) The Pfam protein families database. Nucleic Acids Res. 30: 276-280.

Boisson-Dernier A, Chabaud M, Garcia F, Becard G, Rosenberg C, Barker DG (2001) Agrobacterium rhizogenes-transformed roots of *Medicago truncatula* for the study of nitrogen-fixing and endomycorrhizal symbiotic associations. Mol. Plant Microbe Interact. 14: 695-700.

Cook DR (1999) *Medicago truncatula* – a model in the making. Curr. Opin. Plant Biol. 2: 301-304.

Dixon RA, Steele CL (1999) Flavonoids and isoflavonoids – a gold mine for metabolic engineering. Trends Plant Sci. 4: 394-400.

Dixon RA, Sumner LW (2003) Legume natural products: understanding and manipulating complex pathways for human and animal health. Plant Physiol. 131: 878-885.

Ewing RM, Kahla AB, Poirot O, Lopez F, Audic S, Claverie JM (1999) Large-scale statistical analysis of rice ESTs reveal correlated patterns of gene expression. Genome Res. 9: 950-959.

Fedorova M, Van De Mortel J, Matsumoto P, Cho J, Town C, VandenBosch K, Gantt J, Vance C (2002) Genome-wide identification of nodule-specific transcripts in the model legume *Medicago truncatula*. Plant Physiol. 130: 519-537.

Frugoli J, Harris J (2001) *Medicago truncatula* on the move! Plant Cell 13: 458-463.

Harrison MJ (1999) Molecular and cellular aspects of arbuscular mycorrhizal symbiosis. Annu. Rev. Plant Physiol. Plant Mol. Biol. 50: 361-389.

Journet E, van Tuinen D, Gouzy J, Crespeau H, Carreau V, Farmer M, Niebel A, Schiex T, Jaillon O, Chatagnier O, Godiard L, Micheli F, Kahn D, Gianinazzi-Pearson V, Gamas P (2002) Exploring root symbiotic programs in the model legume *Medicago truncatula* using EST analysis. Nucleic Acids Res. 30: 5579-5592.

Leser U, Lehrach H, Roest-Crollius H (1998) Issues in developing integrated genomic databases and application to the human X chromosome. Bioinformatics 14: 583-590.

Paton NW, Khan SA, Hayes A, Moussouni F, Brass A, Eilbeck K, Goble CA, Hubbard SJ, Oliver SG (2000) Conceptual modeling of genomic information. Bioinformatics 16: 548-557.

Quackenbush J, Liang F, Holt I, Pertea G, Upton J (2000) The TIGR Gene Indices: reconstruction and representation of expressed gene sequences. Nucleic Acids Res. 28: 141-145.

Rabinowicz PD, Braun EL, Wolfe AD, Bowen B, Grotewold E (1999) Maize R2R3 Myb genes: Sequence analysis reveals amplification in the higher plants. Genetics 153: 427-444.

Riechmann JL, Heard J, Martin G, Reuber L, Jiang C, Keddie J, Adam L, Pineda O, Ratcliffe OJ, Samaha RR, Creelman R, Pilgrim M, Broun P, Zhang JZ, Ghandehari D, Sherman BK, Yu G (2000) *Arabidopsis* transcription factors: genome-wide comparative analysis among eukaryotes. Science 290: 2105-2110.

Shepherd AJ, Martin NJ, Johnson RG, Kellam P, Orengo CA (2002) PFDB: a generic protein family database integrating the CATH domain structure database with sequence based protein family resources. Bioinformatics 18: 1666-1672.

Shoop E, Silverstein KA, Johnson JE, Retzel EF (2001) MetaFam: a unified classification of protein families. II. Schema and query capabilities. Bioinformatics 17: 262-271.

Somers DA, Samac DA, Olhoft PM (2003) Recent advances in Legume transformation. Plant Physiol. 131: 892-899.

Stougaard J (2000) Regulators and regulation of legume root nodule development. Plant Physiol. 124: 531-540.

Stougaard J (2001) Genetics and genomics of root symbiosis. Curr. Opin. Plant Biol. 4: 328-335.

Population and Quantitative Genetic Aspects of Molecular Breeding

John W. Dudley
Department of Crop Sciences, University of Illinois at Urbana-Champaign, Urbana, IL 61801, USA. (Email: jdudley@uiuc.edu).

Key words: quantitative genetics, population genetics, plant breeding, molecular genetics, marker assisted selection

Abstract:

Molecular breeding is defined as the application of molecular genetic tools to plant breeding. In this paper the application of population and quantitative genetics principles to plant breeding and the implications of those principles for the use of molecular genetic tools are discussed. The ability to transform plants and thus introduce genes from any species into a crop plant or to develop entirely new genes has greatly broadened the germplasm base for plant breeders. At the same time, this ability has created a set of problems, the solutions of which require application of population genetic theory. The availability of molecular markers has brought a rebirth of interest in quantitative genetics. It is now possible to identify chromosome segments which control quantitative traits and follow those traits in breeding. In addition, use of quantitative genetic principles provides a way of utilizing gene expression data as a plant breeding tool.

A. Hopkins, Z.Y. Wang, R. Mian, M. Sledge and R.E. Barker (eds.), Molecular Breeding of Forage and Turf, 289-302.

1. INTRODUCTION

From the title of this paper one might infer that there is something unique about Molecular Breeding. While modern molecular techniques permit unique or more efficient approaches which were not possible in the past, the basic aspects of plant breeding remain the same. What has changed is our ability to tag and follow pieces of chromatin which were not previously identifiable, determine the sequence and function of genes, and design and insert into the genome a gene to do a specific function. This broadens the genetic variability available to the plant breeder and allows gene capture from species as diverse as bacteria and animals as well as development of genes previously unknown in any species.

Plant breeders are by nature integrators of information. In the past they have taken information from Cytogenetics, Statistics, Plant Pathology, Entomology, Plant Physiology, Population Genetics, Quantitative Genetics, Agricultural Engineering and other disciplines, identified the information and techniques from those disciplines which were useful and incorporated that information and those techniques into their breeding programs. Each discipline provided a tool for the plant breeder. In the 1950's, quantitative genetics was the hot new tool available to the plant breeder. Today, the areas of molecular genetics and genomics are the hot new tools.

In this paper, the emphasis will be on 1) the contributions of population and quantitative genetics to plant breeding and 2) the impact of genomics and molecular genetics on population and quantitative genetics and through them plant breeding. The approach will be to first discuss the plant breeding process and the contributions of population and quantitative genetics to that process. This discussion is followed by a discussion of the molecular tools now available followed by sections on the integration of those tools into plant breeding through population and quantitative genetics. Finally, there is a discussion of the role of functional genomics.

2. THE PLANT BREEDING PROCESS

The steps involved in plant breeding are relatively simple. The breeder identifies or creates a segregating population; selects elite types from the population; creates a stable variety from those elite types and then repeats the process. Where do population and quantitative genetics fit into this process?

2.1 Population Genetics

Population genetics is concerned with the behavior of simple systems: the frequency of a few genes conferring largely discrete phenotypes and the mechanisms that change these frequencies. From population genetic principles breeders have developed an understanding of the effects of selection on single genes and by extension, the effects of selection on multiple genes. In fact, the foundation of quantitative genetic theory is an extension of population genetic principles. Population genetics also deals with the effects of mutation, migration, inbreeding, and genetic drift on populations. Thus, population genetic concepts help provide the basis for such plant breeding decisions as the number of parents to include in synthetic cultivars of forage grasses and legumes and how many generations to inbreed to obtain homozygosity. The effects of mutation on genetic variability in populations provides an explanation for the continued response to selection for oil and protein in maize after 100 generations of selection (Walsh, 2004).

The application of population genetic concepts to predict the effects of cultural practices on development of resistance to new transgenes such as Bt will be discussed later.

2.2 Quantitative Genetics

Quantitative genetics is concerned with the behavior of complex systems: the frequencies of many genes, the interaction among those genes, the interaction of the whole genotype with the environment, the genetic basis for a given phenotype and the response of a population to selection on the phenotype. Quantitative genetics has a role in nearly every step in the plant breeding process. Consider the oldest rule in plant breeding: cross good x good and select something slightly better from the segregating progeny. The quantitative genetic basis for this rule was shown by Bailey (1977) when he demonstrated that the probability of obtaining a new line better than either parent was maximized when each parent contained similar numbers of loci containing favorable alleles for which the alternate parent contained unfavorable alleles. Schnell (1983), as discussed by Lamkey et al. (1995), expanded this concept when he presented the concept of usefulness of a cross. He defined usefulness as the highest performing predicted line from a cross. Thus

$$U=Y+\Delta_G$$

where U is usefulness, Y is the mean of a segregating population, and Δ_G is predicted gain from selection within the population. Quantitative genetics provided the basis for estimating and calculating Δ_G. In fact, the most important equation in quantitative genetics for use by plant breeders is the prediction equation

$$\Delta_G = ci\sigma^2_G/\sigma_P$$

where c is a pollen control factor (c = 1 if selection is prior to pollination and 0.5 if selection follows pollination), i is selection intensity, σ^2_G is the appropriate estimate of genetic variance and σ_P is the appropriate phenotypic standard deviation (Falconer 1989). This equation allows rapid comparison of efficiency of breeding procedures.

When plant breeders make selections, changes are likely to occur, not only in the trait for which selection is being practiced, but in other traits as well (correlated response). The extent of correlated response is a function of the heritabilities of the primary and correlated traits, as well as the genetic correlation between traits. The correlated response equation (Falconer 1989) takes the form:

$$CR_Y = ih_X h_y r_A \sigma_{PY}$$

where CR_Y is the correlated response in trait Y when selection is based on trait X, i is the standardized selection differential for X, h_X and h_y are the square roots of heritability of traits X and Y respectively, r_A is the additive genetic correlation between X and Y and σ_{PY} is the appropriate phenotypic standard deviation for Y. Multiplying CR_Y by c generalizes the equation to a form corresponding to that given for the prediction equation.

Given that most plant breeding programs are interested in response for net worth, often a function of several traits, methods of combining data from several traits into a selection index were developed. Smith (1936) was the first to present the concept of index selection. Smith presented an index of the form:

$$I = b_1X_1 + b_2X_2 + \ldots b_mX_m$$

where I is an index of merit of an individual and b_1 ... b_m are weights assigned to phenotypic trait measurements represented as X_1 ...X_m. The b values are the product of the inverse of the phenotypic variance-covariance matrix, the genotypic variance-covariance matrix, and a vector of economic weights. A number of variations of this index, most changing the manner of computing the b values, have been developed (Bernardo 2002).

The tools quantitative genetics provides for plant breeding allow for a statistical description of genetic variability, separation of genetic variability from environmental variability, and prediction of gain from various types of selection. These tools provide predictive power and have served the plant breeder well. For an excellent description of these techniques and their application to plant breeding see Bernardo (2002). These same quantitative genetics tools are an essential component of effective use of molecular technologies in breeding as will be seen later.

3. MOLECULAR TOOLS

The molecular genetics tools available to plant breeders may be roughly divided into two categories; 1) those that assist the breeder in identifying and manipulating chromosome segments most likely to contain quantitative trait loci (QTL) and 2) those that are useful in identifying, creating, and manipulating individual genes which may be of value in providing new sources of pest resistance, stress resistance, or quality characteristics.

The primary tools in the first category are molecular markers. The history of QTL markers evolves from the work of Sax (1923) who demonstrated linkage of seed coat color (a qualitative trait) to seed size (a quantitative trait) in beans. Molecular markers started with allozymes and have progressed to RFLPs (restriction fragment length polymorphisms), RAPDs (random amplified polymorphic DNA), AFLPs (amplified fragment length polymorphisms), SSRs (simple sequence repeats), SNPs (single nucleotide polymorphisms) and others (Liu 1998). Each marker type has advantages and disadvantages for use in plant breeding. The most useful marker characteristics for plant breeding include codominance, polymorphism in the species of interest, abundance, random distribution throughout the genome, relatively low operating cost, repeatability, and rapid turn around time from extraction of DNA to production of data. The relative advantage of different types of markers varies with the species. In some species certain marker types with major advantages are not very polymorphic. Thus, other less ideal marker types must be used.

Molecular markers allow plant breeders to identify and follow a particular segment of a chromosome during selection. By identifying linkage between a marker and a QTL, particular alleles of the QTL can be followed. In addition, markers have been used to determine the genetic relationship between individuals, populations, and lines. This information is useful in protecting germplasm, identifying new sources of genetic variability, and in predicting performance of crosses of untested lines when performance of tested lines is known and the relationship of the untested line to a tested line is known. The procedure used to predict performance is known as Best Linear Unbiased Prediction (BLUP). For an excellent description of BLUP and its application in plant breeding see Bernardo (2002).

Transgenic technology is a subject of great debate because use of transgenes has implications beyond plant improvement and production. The process involves identifying a gene with known value not currently known to be present in the species, cloning that gene, modifying it to allow expression in the species of interest, inserting it into the species of interest, stabilizing its expression, and then transferring it into adapted cultivars. Because each new transgenic event must be approved by the federal government, at a cost of millions of dollars, each transgenic event, once approved is then transferred by backcrossing to adapted cultivars rather than generating a new event for each cultivar. This process has major implications for breeders. The major advantage is that genes from species as diverse as bacteria and fish are now available to plant breeders. Gene constructs unknown in nature, for example the genes responsible for golden rice (Ye et al. 2000), can be made and inserted into a species. As illustrated in the golden rice example, more than one gene can be combined in a construct and inserted.

The latest tools from molecular genetics are whole genome sequencing of genes and gene expression detection using gene chip technology. Because of the vast amount of information generated from the sequencing efforts, bioinformatics (management of large amounts of data) has become a subject of extensive effort.

Given availability of these tools and what is known about population and quantitative genetics, what are the implications to plant breeding of combining population and quantitative genetics with the molecular genetic tools?

4. MOLECULAR BREEDING AND POPULATION GENETICS

Each transgene behaves as a single gene and thus is subject to the principles of population genetics. Corn (*Zea mays* L.) containing genes from *Bacillus thuringinsis* (Bt corn) will produce pollen containing the Bt gene. Control of this transgenic pollen is a major concern. Because the European Union has not approved the use of corn or corn products containing Bt genes in corn imported into Europe, great care must be taken to control the possible contamination of non Bt corn with pollen from Bt corn. Thus buyers may test their grain for presence of Bt and accept or reject truck loads of corn based on those tests. Pollination neighborhood studies should be used in determining the extent of such sampling. As another example, there are concerns that cultivars containing the Roundup Ready® gene may cross with weedy relatives thus providing resistance to a useful herbicide in a weed species. Risk assessment based on gene flow and fitness studies may be used to determine the potential for damage from such an event.

A specific example of use of population genetic principles in conjunction with molecular genetics is the development of regulations governing the planting of Bt and non Bt corn. One of the many kinds of Bt genes confers resistance to the European corn borer to an extent not seen by any single gene previously known. Government regulations provide that for each acre of Bt corn planted a particular acreage must be planted to non-Bt corn to provide a refuge for European corn borers. The bases for this regulation are population genetic models based on the hypothesis that resistance in the corn borers is recessive (Onstad and Guse 1999; International Life Sciences Institute, 1998). This hypothesis along with knowledge of the mating habits of the insect was used to develop models which predicted that the frequency of the recessive resistant gene would remain low due to the persistence of a large number of susceptible individuals. The weakness of this procedure is that the mechanism of resistance in the corn borers is only hypothesized and not known.

One of the most widely used transgenes is the Roundup Ready® (RR) gene in soybeans. Current estimates are that as much as 90% of the soybean acreage in the United States will be planted to Roundup Ready® beans in 2003. Unlike the Bt gene, this transgene has approval for use in the major export markets. Also unlike corn the soybean is self-pollinated instead of cross pollinated thus changing the population dynamics. A major concern in soybeans is the potential narrowing of the germplasm base because all the soybeans carrying the Roundup Ready® gene have been derived from one transgenic event which occurred in cultivar A5403. Subsequent cultivars have been derived from backcrossing and from direct breeding between crosses of cultivars derived from this original event. A legitimate question is how much has the germplasm base of the soybean been narrowed by use of this one event? Sneller (2003) using coefficient of parentage information found that use of Roundup Ready cultivars has had little impact on the genetic diversity available to farmers.

Because backcrossing is being widely used to incorporate transgenes into elite germplasm, an extensive body of theory has been developed to optimize the number of generations required to recover the recurrent parent, to balance the need for rapid recovery of the recurrent parent, and to minimize the cost of use of markers (Hospital 2002).

In summary, molecular approaches have created a new set of opportunities for breeders. Along with those opportunities have come a new set of problems the solution to which requires the use of population genetic principles. Fortunately many of these principles have been available for many years and are available for use in solving the problems.

5. MOLECULAR BREEDING AND QUANTITATIVE GENETICS

The development of molecular markers has caused a rebirth of interest and research in quantitative genetics. Examples of the possible integration of molecular genetics and quantitative genetics are discussed under the headings of selection of parents, marker based and assisted selection, and marker based yield prediction.

5.1 Selection of Parents

As noted earlier, the value of a breeding population can be defined as $U=Y+\Delta_G$. Using this equation as a starting point, molecular marker information has been used to help predict both Y and Δ_G . A detailed discussion of this topic is given by Dudley (2002). Panter and Allen (1995) suggested using best linear unbiased prediction (BLUP) methods to predict the midparent value, a good predictor of the mean of lines from a cross, of soybean crosses. BLUP methods take into consideration the performance of lines related to the line for which performance is being predicted (Bernardo 2002). Both Panter and Allen (1995) and Toledo (1992) found the coefficient of parentage between a pair of lines was related to genetic variance in the progeny. Based on these results, they suggested an effective method of choosing parents would be to identify pairs of lines with high midparent values estimated from BLUP and to select among such pairs those that were the most genetically diverse based on the genetic relationship matrix. The availability of molecular markers allows degree of relationship between lines to be established from molecular marker data (Lee 1995; Romero-Severson et al. 2001).

In corn breeding, lines are crossed based on heterotic patterns. The heterotic patterns have been established empirically to maximize performance of hybrids from crosses of lines belonging to different groups. Experimental data demonstrate the effectiveness of using molecular markers for assigning new germplasm to heterotic groups (Mumm and Dudley 1994).

5.2 Marker Based and Assisted Selection

Marker based selection (MBS) is defined as selection based entirely on marker information. Marker assisted selection (MAS) is selection based on a combination of marker information and phenotypic information. Dudley, 1993, showed gain from MBS relative to phenotypic selection could be expressed as:

$$G_m / G_p = [c_m i_m y_p / c_p i_p y_m] [R^{0.5} / h_p].$$

where G_m =gain from MBS, G_p = gain from phenotypic selection, c_m = a pollen control factor for MBS, i_m = selection intensity for MBS, y_p = number of years per cycle for phenotypic selection, c_p = pollen control factor for phenotypic selection, i_p = selection intensity for marker assisted selection, y_m = number of years per cycle for MBS, R =the proportion of the additive variance accounted for by the marker model being used, and h_p is the square root of the heritability for the trait being improved. From this equation, MBS will have an advantage when R > heritability, when number of years per cycle is less for MBS, and when i can be increased by MBS. Practically speaking, this means there is a real advantage to use of MBS for traits of low heritability and for traits of high heritability in environments where the trait cannot be measured such as off-season nurseries for traits such as grain yield in corn.

Both MBS and MAS require identification of QTL associated with marker genotypes. Such identification requires molecular marker data on a population in linkage disequilibrium along with precise phenotypic measurements. Paradoxically, evaluation of the phenotype in a QTL study involving low heritability traits will generally require more replication and measurement precision than traditionally used before the use of molecular markers. The most effective use of MBS comes from selection based on marker-QTL associations under conditions in which measurement of phenotypic traits is very expensive or impossible.

Identification of QTL-marker associations carries with it the possibility of errors. Dudley (1993) discusses the importance of Type 1 and Type 2 errors in plant breeding applications of MAS and MBS. In general, if the objective of identifying QTL is to locate a gene to clone or transfer to another cultivar by backcrossing, or the gene is one that controls a trait for which there is a minimum acceptable threshold, then very low Type 1 errors are necessary (Johnson 2001). If, on the other hand, the genes are for performance traits for which each of a number of genes of small effect contributes, then a certain number of Type 1 errors may be acceptable in order to reduce Type 2 errors (Johnson 2001). The most important type of error in marker-QTL identification is a Type 3 error, i.e., declaring an association significant but selecting the wrong marker allele as being linked to a favorable QTL. Fortunately Type 3 errors are rare.

In corn breeding programs, early testing is often used to reduce the number of lines to be evaluated in later generations. Eathington et al. (1997) and Johnson and Mumm (1996) demonstrated the effectiveness of MAS as an aid to early testing.

Johnson (2001) demonstrated the success of MAS in sweet corn breeding. He proposed six major factors which lead to success: 1) a focus on breeding, not QTL mapping, 2) a good blend of quantitative genetic theory and Mendelian genetics, 3) custom built ' highly interactive ' software applications to aid in the decision making process, 4) effective communication between the lab and field staff, 5) plant breeder endorsement (buy in), and 6) a specialty crop in which objectively measured quality traits are equally or more important than yield. In one example, they were able using MBS to select from a breeding cross lines yielding 4% more than conventionally selected lines and to have them available a year ahead of the conventional lines.

Johnson (2004) compared use of 3 cycles of rapid MBS with one cycle of combined phenotypic and marker based selection. In 43 populations testcross data in an early generation were used to identify marker-QTL associations. These data were then combined with phenotypic data to make selections. At the same time, the populations were subjected to 3 cycles of MBS using off-season nurseries. The MBS selected populations averaged 10 bushels per acre more yield than the lines selected based only on the index of marker and phenotypic data.

These results demonstrate the effectiveness of marker based selection. What they do not answer is the cost effectiveness of the procedures. Those comparisons are not available in the literature and will depend not only on the cost of marker data relative to phenotypic data but on the availability of sufficient laboratory capacity to provide high quality marker data in large quantities on a timely basis.

5.3 Marker Based Yield Prediction

In addition to use in selection, markers are being used in corn breeding programs to predict yield of untested hybrids using BLUP to combine data from tested and untested lines. Bernardo (2002) describes results from one program in a commercial company in which predicted yields could be used to reduce the number of hybrids to be tested by approximately 50%. Johnson and Mumm (1996) starting with the F_2 generation of two crosses, one from each of two different heterotic groups, evaluated 100 lines from each heterotic group. Along with the phenotypic data, marker genotypes were obtained. Each line from one heterotic group was crossed to one line from the other group. The 100 crosses were then evaluated for grain yield and the data combined with marker data to develop a genetic model for each marker locus. Using these data, the performance of 9900 hybrids was predicted and the highest 50 predicted hybrids selected and compared in performance trials to the 100 hybrids used to predict gain and 100 random hybrids. The 50 selected hybrids out yielded the random and predictor hybrids by approximately 6.5 bushels per acre.

6. MOLECULAR BREEDING AND FUNCTIONAL GENOMICS

Whether the discussion of functional genomics belongs in a separate section or as a part of a discussion of quantitative genetics and molecular breeding is not clear. What is clear is that a large amount of effort is going into sequencing genes in many different organisms. Along with that sequencing effort, has come the development of procedures for determining the functions of those genes and for measuring the changes in gene expression when placed in different environments. The *Arabidopsis thaliana* and rice (*Oryza sativa L.)* genomes have been sequenced and large numbers of expressed sequence tags (ESTs) are being generated from crops such as maize and soybean (*Glycine max* L.).

Simplistically there is a sense that if the sequences and functions of all the genes were known, it would be possible to design a plant from the ground up and produce the ideal plant. However, the physiology and biochemistry of plants do not lend themselves well to this reductionist approach. Genes interact with one another and if there are 10,000 genes there are potentially 49,995,000 two way interactions among them. Just knowing the function of genes does not provide information on their effect on important agronomic traits such as grain yield in corn or forage yield in alfalfa. Moreover, plant breeders do not try to select for all the genes at one time. Such an effort would be doomed to failure. Rather they cross two good parents or they start selection within a good synthetic and try to make incremental improvements working with parental material in which many genes are already fixed for a high level of performance.

Bernardo (2001) correctly points out that knowing all the genes, in and of itself, is of little value. Johnson (2004) suggests the problem is statistical. That is, estimation of allelic effects and prediction of breeding value based on these effects are two different things. Because many of the allelic effects will be correlated, models in which observed trait values are simultaneously fitted to allelic variables at closely linked loci will have poor predictive properties. In addition, the complexity of biochemical and physiological interactions among gene products, and of gene products with the environment makes complete specification of the phenotype from complete knowledge of the genome nearly impossible (Rosenberg 1985; Clark 1998). In recognition of this complexity, a new approach to study of biology called "systems biology" is emerging (Begley 2003). In this approach, interacting systems of genes are studied and the effects of every gene on every other gene taken into account.

Despite the complexity of the problem, there is evidence that knowing the function of specific genes and the incorporation of that information into selection indices designed to take advantage of it may have merit. Johnson (2004) described an experiment in corn in which photosynthetic rate was measured on a set of 98 homozygous lines under conditions of fully irrigated and limited irrigation conditions. Leaf tissue was sampled for gene expression analysis and net photosynthetic rate measured simultaneously at a late vegetative stage. Genetic correlations between net photosynthetic rate and gene expression responses were calculated. Expression of NAD(H)-dependent gluatamate dehyhdrogenase response had a near perfect correlation with response of net photosynthetic rate to irrigation treatment. Predicted gain in net photosynthetic rate from index selection using NAD(H)-GDH as a secondary trait was 250%. Johnson cautions that this is likely an over estimate of gain because the lines had been selected for a range of response to irrigation. Further, the potential increase in yield under drought stress from such a gain in photosynthetic rate is unknown. Despite these cautions, such a high predicted gain suggests use in selection of gene expression data coupled with phenotypic data may pay dividends in the future.

7. CONCLUDING REMARKS

Molecular genetics and genomics tools should be extraordinarily useful to plant breeders. However, these tools will only be useful in the context of breeding systems that make sense for the species of interest. Thus, they need to be integrated into current plant breeding procedures. To do this will require that the plant breeder of the future have an appreciation of molecular genetics and genomics as well as a background in quantitative and population genetics along with the other disciplines long held to be important for plant breeders. The plant breeder will continue to be the integrator of information because he/she is the individual who knows the phenotype of the plant and its strength and weaknesses.

To successfully use these new tools, the plant breeder will need to work closely with the genomics and bioinformatics specialists as well as the plant pathologists and entomologists with whom they have worked so successfully in the past. This collaboration will also require that the genomics and bioinformatics specialists be able to converse with and understand the constraints of the plant breeder.

ACKNOWLEDGMENTS

The author gratefully acknowledges the valuable comments of Jeanne Romero-Severson who reviewed the manuscript. This paper is a contribution from the Illinois Agricultural Experiment Station.

REFERENCES

Bailey TB Jr (1977) Selection limits in self-fertilizing populations following the cross of homozygous lines. In: Proc. Int. Conf. Quant. Genet., Pollak E et al. (eds.), Iowa State Univ. Press, Ames, IA.

Begley S (2003) Biologists hail dawn of a new approach: don't shoot the radio. Wall Street Journal. Feb. 21 p. B1, New York.

Bernardo R (2001) What if we knew all the genes for a quantitative trait? Crop Sci. 41: 1-4.

Bernardo R (2002) Breeding for quantitative traits in plants. Stemma press, Woodbury, MN.

Clark AG (1998) Limits to prediction of phenotypes from knowledge of genotypes. In: Limits to knowledge in evolutionary biology. Univ. California, Riverside.

Dudley JW (1993) Molecular markers in plant improvement: Manipulation of genes affecting quantitative traits. Crop Sci. 33: 660-668.

Dudley JW (2002) Integrating molecular techniques into quantitative genetics and plant breeding. In: Quantitative genetics, genomics and plant breeding, Kang MS (ed.), CABI publishing, New York.

Eathington SR, Dudley JW, Rufener GK II (1997) Usefulness of marker-QTL associations in early generation selection. Crop Sci. 37: 1686-1693.

Falconer DS (1989) Introduction to quantitative genetics. John Wiley & Sons, NY

Hospital F (2002) Marker-assisted back-cross breeding: a case study in genotype-building theory. In: Quantitative genetics, genomics, and plant breeding, Kang MS (ed.), CABI publishing, New York.

International Life Sciences Institute: Health and Environmental Sciences Institute (1998) An evaluation of insect resistance management in Bt field corn: A science-based framework for risk assessment and risk management. ILSI Press, Washington, D.C. Available at http://www.ilsi.org/file/h5_IRM.pdf (verified 2/21/03).

Johnson GR (2004) Marker assisted selection. Plant Breed Rev. (in press).

Johnson GR, Mumm RH (1996) Marker assisted maize breeding. In: Proceedings of the fifty-first annual maize & sorghum research conference. Amer. Seed Trade Assn, Washington, D.C.

Johnson L (2001) Marker assisted sweet corn breeding: A model for specialty crops. In: Proceedings of the 56th annual corn & Sorghum research conference. Amer. Seed Trade Assn, Washington, D.C.

Lamkey KR, Schnicker BJ, Melchinger AE (1995) Epistasis in an elite maize hybrid and choice of generation for inbred line development. Crop Sci. 35: 1272-1281.

Lee M (1995) DNA markers and plant breeding programs. Adv. Agron. 55: 265-344.

Liu BH (1998) Statistical genomics, linkage mapping, and QTL analysis. CRC Press, Boca Raton, FL.

Mumm RH, Dudley JW (1994) A classification of 148 U.S. maize inbreds: I. Cluster analysis based on RFLPs. Crop Sci. 34: 842-851.

Onstad DW, Guse CA (1999) Economic analysis of transgenic maize and nontransgenic refuges for managing European corn borer (Lepidoptera: Pyralidae) J. Econ. Entom. 1256-1265.

Panter DM, Allen FL (1995) Using best linear unbiased predictions to enhance breeding for yield in soybean. I: choosing parents. Crop Sci. 35: 397-404.

Romero-Severson J, Smith JSC, Ziegle J, Hauser J, Joe L, Hookstra G (2001) Pedigree analysis and haplotype sharing within diverse groups of *Zea mays* L. inbreds. Theor. Appl. Genet. 103: 567-574.

Rosenberg A (1985) The structure of biological science. Cambridge Univ. Press. Cambridge, UK.

Sax K (1923) The association of size differences with seed coat pattern and pigmentation in Phaseolus vulgaris. Genetics 8: 552-560.

Schnell FW (1983) Probleme der Elternwahl-Ein Uberblick. In: Arbeitstagung der Arbeitsgemeinschaft der Saatzuchtleiter in Gumpenstein, Austria. Nov. 22-24 1983. pp.1-11. Verlag and Druck der Bundesanstalt fur alpenlandische Landwirtschaft. Gumpenstein, Austria.

Smith H F (1936) A discriminant function for plant selection. Ann. Eug. 7: 240-250.

Sneller CH (2003) Impact of transgenic genotypes and subdivision on diversity within elite North American germplasm. Crop Sci. 43: 409-414.

Toledo JFF (1992) Mid parent and coefficient of parentage as predictors for screening among single crosses for their inbreeding potential. Rev. Brasil. Genet. 15: 429-437.

Walsh B (2004) Population and quantitative genetic models of selection limits. Plant Breed. Rev. (in press).

Ye XD, Al-Babili S, Kloti A, Zhang J, Lucca P, Beyer P, Potrykus I (2000) Engineering the provitamin A(beta-carotene) biosynthetic pathway into (carotenoid-free) rice endosperm. Science 287: 303-305.

AFLP-marker Analyses of Genetic Structure in Nordic Meadow Fescue (*Festuca pratensis* Huds.) – Tracing the Origin of Norwegian Cultivars and Local Populations

Siri Fjellheim, Zanina Grieg and Odd Arne Rognli
Agricultural University of Norway, Department of Chemistry and Biotechnology, P.O.Box 4050, 1432 Ås, Norway. (Email: odd-arne.rognli@ikb.nlh.no).

Key words: *Festuca pratensis*, AFLP, genetic diversity, germplasm

Abstract: Fifteen Norwegian wild populations and 13 Nordic cultivars of meadow fescue (*Festuca pratensis* Huds.) was analysed using AFLP markers in order to assess genetic diversity within and between populations, to compare the distribution of genetic variability within wild populations and cultivars, and to elucidate relationships between the populations and cultivars. AMOVA-analysis of 95 AFLP-markers showed that most of the variation was found within population (71.3%), whereas 25.1% was found between populations and only 3.6% between the two groups (cultivars and wild populations). Separate AMOVA-analyses of the two groups revealed a higher level of variation within registered cultivars (79.6%) than within wild populations (69.2%). A cluster-analysis based on average pairwise differences showed that the populations were divided into two clusters; one containing the cultivars and 7 wild populations, and one containing the rest of the wild populations, which were again divided into two clusters. These results were supported by PCA-analysis. The results indicate that the Nordic cultivars all together have a narrow genetic basis, and that the wild populations in Norway can be divided into three groups following the most probable routes of introduction of the species into Norway. The first group probably originate from feral populations established from cultivated land, and the second and the third group probably originate from human activity, e.g. trade, to the coastal western and northern parts of the country, and to the central parts of southern Norway, respectively.

A. Hopkins, Z.Y. Wang, R. Mian, M. Sledge and R.E. Barker (eds.), Molecular Breeding of Forage and Turf, 303-308.

1. INTRODUCTION

In the Nordic countries, grassland husbandry has always played an important part of agriculture because of the harsh climatic conditions, especially in northern parts, where grasslands become more and more dominant with higher latitudes. In Norway, 60 – 100 % of the cultivated land is being used for lays and pastures (Solberg et al. 1994). One of the dominating grassland species is meadow fescue (*Festuca pratensis* Huds.). It is, in contrast to the rest of Europe, still one of the most important grass species in Norwegian and Nordic leys due to its superior combination of winter-hardiness and quality. The species is distributed both in new and old meadows, and as feral populations in all parts of this region, however, it is less frequent in northern parts of Norway, Sweden and Finland. Only a single population is known originating from Iceland. Meadow fescue is, however, not indigenous to the Nordic area, but was probably first introduced to the Nordic countries as a forage grass in sown meadows (Lid and Lid 1994).

Meadow fescue is a diploid (2n = 2x = 14) outbreeder with a gametophytic self-incompatibility system controlled by two genes designated S and Z (Lundquist 1962). As a consequence of this, large genetic heterogeneity is expected to persist within populations. This issue has to be considered and the analysis must allow for quantification of variation within and between populations. We have used AFLP (Amplified Fragment Length Polymorphism, Vos et al. 1995) to assess genetic diversity within populations. Here we present the analysis of genetic diversity within 13 Nordic cultivars and 15 Norwegian wild populations of meadow fescue based on AFLP-markers. Relationships between the populations are elucidated in order to trace the origin of present Norwegian meadow fescue.

2. MATERIALS AND METHODS

Twenty-eight accessions of wild populations and cultivars obtained from the Nordic Genebank were analysed (Table 1). Of these, 15 were wild populations collected in Norway, and 13 were registered cultivars, of which three originated from each of the countries Norway, Finland, Sweden, and Denmark, and one from Iceland. The wild populations were selected to represent geographic diversity, both in latitude, longitude, and altitude. The cultivars were selected so that from each country, except from Iceland, both old and new cultivars were represented. Initially, 20 individuals were included from each population, but some of the samples were excluded from the analysis due to failure in amplification. The total number of plants analysed using three primer combinations of the restriction enzymes *Pst*1 and *Mse*1 were 582. Analysis of Molecular Variance (AMOVA) and calculations of genetic diversity indices were performed with Arlequin 2.0 (Schneider et al. 2000). Principal Component Analysis (PCA) and Unweighted Pair-group Method With Arithmetic Means (UPGMA) were performed using NTSYS-PC (Rohlf 2000).

3. RESULTS

A total number of 95 polymorphic AFLP-markers were scored. All haplotypes were unique. Genetic diversity indices are presented in Table 1. Generally, the cultivars showed a higher degree of diversity than the wild populations.

Table 1. Genetic diversity indices in 15 Norwegian populations and 13 Nordic cultivars of *Festuca pratensis* based on 95 polymorphic AFLP-markers. The indices are mean number of pairwise differences between populations and average gene diversity over loci. In parentheses are the cultivars country of origin and year of release indicated.

Population	# of loci	# of polymorphic loci	Average difference	Average gene diversity
NGB7546	93	44	14,07	0,1513
NGB7542	93	48	14,73	0,1583
NBG7539	89	44	14,60	0,1641
NGB7538	93	35	11,28	0,1213
NGB6768	94	34	10,59	0,1127
NGB6762	94	32	10,56	0,1123
NGB6638	92	42	13,15	0,1429
NGB6636	89	44	12,57	0,1412
NGB6631	93	43	12,80	0,1376
NGB5418	95	38	10,86	0,1143
NGB5415	91	60	18,09	0,1988
NGB4283	95	42	13,24	0,1393
NGB2912	95	50	13,84	0,1457
NGB7632	92	41	11,29	0,1228
NGB7702	88	35	12,42	0,1411
Norild (NOR 2001)	95	33	10,31	0,1066
Løken (NOR 1927)	94	47	15,91	0,1692
Fure (NOR 1989)	93	49	13,84	0,1488
Petursey (ICL)	90	45	13,30	0,1447
Leto Dæhnfeldt III (DEN1961)	86	42	12,70	0,1477
Pajberg (DEN 1961)	91	40	13,17	0,1447
Balder (DEN 1982)	90	50	14,66	0,1629
Tammisto (FIN 1929)	93	49	14,57	0,1567
Paavo (FIN 1948)	90	41	14,19	0,1576
Kalevi (FIN 1979)	90	52	16,65	0,1851
Svalöfs Sena (SWE 1917)	92	47	15,31	0,1664
Bottnia II (SWE 1955)	91	56	17,50	0,1923
Boris (SWE 1971)	88	45	16,09	0,1828

The AMOVA analyses showed that most of the variation is distributed within populations (71.3%) compared to between populations (25.1%). Only 3.6% of the variation was found between the two groups (cultivars and wild populations). Separate analyses of wild populations and cultivars showed that the variation within cultivars (79.6%) was larger than the variation within wild populations (69.2%).

The UPGMA analysis separates the populations into two clusters, except from the Norwegian cultivar Norild who came out separately from all the other populations (Fig. 1). Cluster 1 includes the populations NGB7539, 7538, 4283, 6638, 7702, 6762, 6768, and 5418. This cluster is again divided into two cluster, one containing the first five populations (Group 1a) and the second containing the last three populations (Group 1b). Cluster 2 includes all the cultivars (except Norild) and the populations NGB7546, 7542, 6636, 6631, 5415, 2912, and 7632 (Group 2).

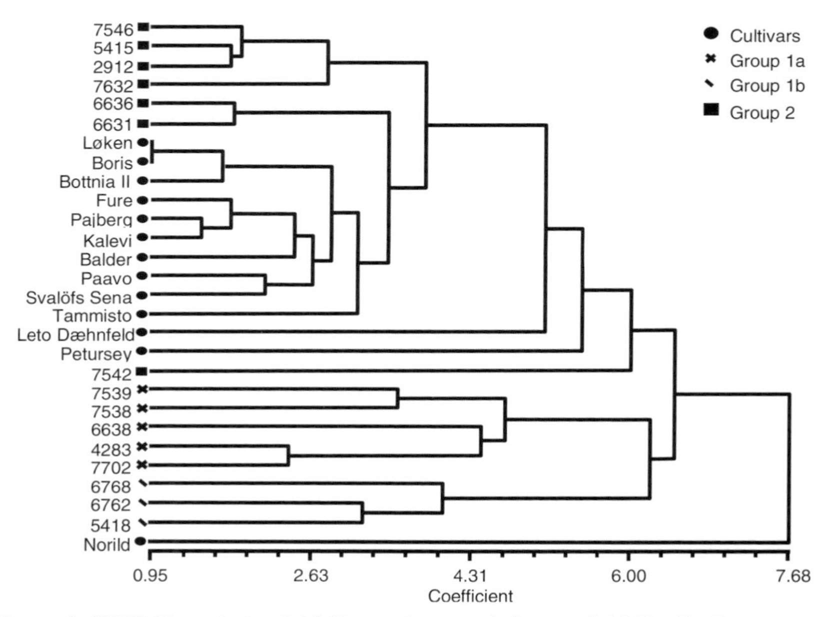

Figure 1. UPGMA analysis of 15 Norwegian populations and 13 Nordic *Festuca pratensis* cultivars based on 95 polymorphic AFLP-markers. Analysis is based on corrected average population pairwise F_{st}.

The PCA analysis of all 582 individuals reflects what was found in the UPGMA analysis, and is not shown. Group 2 from the UPGMA analysis is grouped close to the cultivars, and cluster 1 is again separated from this cluster and divided into two groups.

4. DISCUSSION

This analysis shows that the genetic heterogeneity is large both within and between Nordic meadow fescue cultivars and wild Norwegian populations, reflecting the outbreeding habit of this species. The PCA- and UPGMA-analyses reveal no clear groupings based on neither country of origin nor year of release of the cultivars. This reflect the fact that there has been substantial flow of seeds between the different Nordic countries, especially from Sweden to Finland, and from Denmark to Norway, both for breeding purposes and because both Finland and Norway have had problems with commercial seed production up to recent times (Hillestad 1990; Kivi 1965). In Sweden, the material that was used for breeding came in the beginning probably from Denmark, although Sweden also imported seeds from North-America (Sjödin 1986). Flow of breeding material between the Nordic countries would have reduced the variation between the cultivars. Compared to the wild populations, the AMOVA-analysis reveals a lower between-population variation in the cultivars compared to the wild populations.

There seems to be a general assumption that there is a risk of reduction in genetic diversity in cultivars when new methods of breeding are applied. This was not found in this study, where no significant differences were found between cultivars from different years. An exception from this was the Norwegian cultivar Norild. It shows the lowest genetic diversity of all the populations and cultivars analysed. This cultivar originates from a local population from Northern Norway and is a synthetic cultivar based on 11 clones selected among surviving plants following 3 years of field testing of half-sib families at a field testing station in Alta, Finnmark (Arild Larsen, pers. comm.). The breeding of this cultivar has involved strong selection for adaptation to forage production at higher latitudes, which might explain the restricted genetic base of this cultivar. Much of the breeding in the Nordic countries has been based on imported material. In this way, diversity in the newer cultivars could have been sustained. A higher level of diversity was found in the cultivars compared to the wild populations. This is contrary to the results obtained by Kölliker et al. (1998) who found that the level of variation in populations of meadow fescue from Switzerland was higher in natural populations than in cultivars. Our findings might be explained by at least two processes. Firstly, a narrow genetic base of meadow fescue in the Nordic countries could be due to the recent introduction of meadow fescue to Norway and possible associated bottlenecks. Secondly it could reflect changes over time in the breeding methods used in forage grass breeding.

The wild populations seem to separate into three groups in the PCA and UPGMA-analysis. In Flora for Norway from 1861 (Blytt 1861) meadow fescue is described as common in all parts of southern Norway up to the pine line, and in northern parts of Norway up to Bodø, but only in areas along the coast. The populations in group 2 correspond geographically to the areas in Norway where Blytt did not find meadow fescue. These populations also cluster with the cultivars, probably because the populations have been founded by spreading from cultivars sown in meadows in recent times. These populations also have a higher degree of genetic diversity than the other populations, which is expected since the cultivars have a higher degree of diversity than the wild populations. The rest of the wild populations (cluster 1) divide into two groups. One of these groups (1b) corresponds geographically to central parts of southern Norway and the second group (1a) to coastal areas of western and northern Norway. This might be a reflection of human activity, e.g. trading. In central parts of southern Norway, there have from the times of the Vikings been close connections to Denmark and Sweden through the Oslo fjord. This can be one of the ways that meadow fescue was introduced. The second group could have been introduced as a result of shipping activity along the western coast of Norway, both through the Vikings and later through the extensive Hanseatic trade.

ACKNOWLEDGMENTS

We would like to thank Vibeke Alm and Ingvild Marum for help with the DNA-extractions, and Nordic Genebank for financial support.

REFERENCES

Blytt MN, Blytt A (1861) Norges flora: eller Beskrivelser over de i Norge vildtvoxende Karplanter: Tilligemed angivelser af de geographiske Forholde, under hvilke de forekommer. Brøgger og Christie, Christiania.

Hillestad R (1990) Selskapet for Norges vels betydning for utnyttingen av foredlingsmateriale i kryssbefruktede vekster. In: Norsk planteforedling i nåtid og framtid: grunnleggende aspekter ved planteforedling, Norsk Landbruksforskning Supplement, Rognli OA (ed.), No. 9, pp. 103-109. Statens fagtjeneste for landbruket, Ås, Norway.

Kivi, EI (1965) Plant breeding in Finland. In: Acta Agriculturæ Scandinavica, Torsell R (ed.), Supplementum 12, pp. 52-69.

Kölliker R, Stadelmann FJ, Reidy B, Nösberger J (1998) Fertilization and defoliation frequency affect genetic diversity of *Festuca pratensis* Huds. in permanent grasslands. Mol. Ecol. 7: 1557-1567.

Lid J, Lid DT (1994) Norsk Flora, 6. ed. by Elven R. Det Norske Samlaget, Oslo, Norway.

Lundquist A (1962) The nature of the two-loci incompatibility system in grasses II. Number of alleles at the incompatibility loci in *Festuca pratensis* Huds. Hereditas 48: 169-181.

Rohlf FJ (2000) NTSYS-PC. Numerical Taxonomy and Multivariate Analysis System, version 2.1. Exeter Software, New York, USA.

Schneider S, Roessli D, Excoffier L (2000) Arlequin ver. 2.000: A software for population genetics data analysis. Genetics and Biometry Laboratory, University of Geneva.

Sjödin J (1986) Foderväxter. In: Svalöf 1886-1986 Växtforedling under 100 år, Olsson G, Hagberg A, Hummel-Gumaelius T (eds.), pp 157-165. Svalöf AB, Svalöv.

Solberg E, Rognli OA, Østrem L (1994). Potential for improving adaptation of *Lolium perenne* L. to continental climates in Norway. In: Breeding Fodder Crops for Marginal Conditions, Rognli OA et al. (eds.), pp 47-60. Kluwer Academic Publishers, Netherlands.

Vos P et al. (1995) AFLP – A new technique for DNA-fingerprinting. Nucl. Acid. Res. 23: 4407-4414.

Spatial Autocorrelation Analysis of Genetic Structure Within White Clover Populations

David L. Gustine
USDA-ARS, Pasture Systems and Watershed Management Research Unit, Curtin Road, Building 3702, University Park, PA 16802-3702, USA. (Email: d3g@psu.edu).

Key words: genetic diversity, clonal diversity, population dynamics, DNA markers, RAPD markers.

Abstract:

White clover (*Trifolium repens* L.) populations exhibit high genetic and clonal diversities, while existing for many decades in grazed swards at northern midlatitudes. Genetic structure might exist within rapidly changing populations and might be a factor in creating genetic diversity. Trifoliate leaf samples were taken monthly for two years from up to 37 specific stolon points in quadrats from May to September on three central Pennsylvania farm sites. Random amplified polymorphic DNA (RAPD) profiles for individuals within populations in quadrats were tested by analysis of molecular variance (AMOVA) and spatial autocorrelation. Genetic variance by quadrat population dates in the three pastures ranged from 15 to 74 % and 46 to 80% in 1997 and 1998, respectively. Significant ($P < 0.05$) overall spatial autocorrelation was found in 26 populations that had clones and in seven populations without clones. No significant autocorrelation was found in 27 and seven populations with and without clones, respectively. The estimated patch size did not change significantly over two growing seasons. Number of clones and patch size was less important in determining genetic structure than variable existence of spatial autocorrelation.

A. Hopkins, Z.Y. Wang, R. Mian, M. Sledge and R.E. Barker (eds.), Molecular Breeding of Forage and Turf, 309-314.

1. INTRODUCTION

White clover is an important functional component in temperate grazed ecosystems, fixes substantial nitrogen, and has high nutritional quality. White clover is an obligately outcrossing tetraploid species that flowers prolifically during the growing season. Although high seed counts have been found in the soil (Chapman and Anderson 1987; Tracy and Sanderson 2000), few seeds germinate under field conditions, and few seedlings establish as white clover plants (Barrett and Silander 1992; Brink et al. 1999; Fothergill et al. 1997).

White clover populations often exist for many decades in grazed swards at northern midlatitudes. Presumably this is due to rare seedling recruitment and prolific clonal growth (Barrett and Silander, 1992; Chapman, 1983; Fothergill et al., 1997; Gustine and Huff, 1999). Once a taproot has established, the life span is 1 to 2 yr (Pederson, 1995). Most plant growth and spreading during the growing season are through stoloniferous propagation (Chapman, 1983). As more stolon branches are produced, the plant, which is comprised of clonal members, expands to cover a greater surface area. Through decay of older stolons and environmental disturbance, a clone frequently fragments into smaller clones. Thus, physically separate but genetically identical clonal plants were found in quadrats as small as 1.67 by 1.67 m (Gustine and Sanderson 2001a, b). Any clone could potentially become a dominant genotype, producing patches covering large areas in grasslands (Cahn and Harper 1976; Harberd 1963). Under this scenario, the genetic variability within the white clover population would be reduced, thus increasing the likelihood of catastrophic plant loss in a large area from infestation by disease or insects.

Cahn and Harper (1976) did not find the expected low genetic variation nor did they find local domination by one or more clones. The maximum clonal patch width reported in various field studies is from several centimeters to several meters (Cahn and Harper 1976; Gustine and Sanderson 2001a; Harberd 1963). Gustine and Huff (1999) found high genetic variation within and among grazed white clover populations at 18 farms in three northeastern U.S. states using RAPD markers. Widén et al. (1994) surveyed genotypic diversity from data reported in 40 different studies on 45 clonal species and concluded that they were as variable as sexually reproducing plants. Even at a smaller scale, Gustine and Sanderson (2001a,b) found high genetic variances in 1.8-m^2 quadrats placed in paddocks on three Pennsylvania farms. Although white clover populations were genetically variable at this scale, RAPD profile analyses have shown that they nevertheless contain clonal plants (Gustine and Sanderson 2001a,b), suggesting genetic structure in the populations. Spatial structure resulting from clonal growth can also influence evolutionary processes in white clover by limiting gene flow through cross breeding of closely related individuals.

Does genetic structure exist within rapidly changing white clover clonal populations? Is genetic structure a factor in creating population genetic diversity? Answers to such questions could illuminate population growth mechanisms that make genetically diverse white clover persist over years of grazing. Because white clover spreads and propagates by vegetative and sexual reproductive means, clonality and isolation by distance are confounded in studies designed to separate genetic structure components created by either reproductive mode.

2. ANALYSIS OF MOLECULAR VARIANCE

In a series of studies, Gustine and Sanderson (2001a,b) and Gustine and Elwinger (2003) utilized RAPD profiles to follow physical and temporal positions of some genotypes and to characterize genetic variance (AMOVA, Excoffier et al. 1992) in white clover populations at the local scale. Trifoliate leaf samples were taken monthly for two years from up to 37 specific points in quadrats from May to September on three central Pennsylvania farm sites.

Some sampled clones were detected more than once in the same or different quadrats (Gustine and Sanderson 2001a). When sampled clones reappeared in a quadrat, member positions in the quadrat had changed and the member occurrences in the clone had changed. These results illustrate how of population genetic makeup can change rapidly due to combined temporal sampling effects and clonal spreading.

Gustine and Sanderson (2001b) found that within-population genetic variability of white clover populations containing clones ranged from 15% (highly clonal) to 80% (few clones). Thus, genotypic heterozygosity was maintained even though most populations had one or more clones present. No two of the three white clover populations in Pennsylvania pastures had similar genetic makeups when sampled on the same date. Similarly, Gustine and Huff (1999) demonstrated that genotypic composition changed in four 1996 Pennsylvania populations during a 6-wk period. Genetic composition in three Pennsylvania populations changed frequently throughout the 1997-growing season (Gustine and Sanderson, 2001a). Higher clonal member numbers and lower within-population variances in 1997 were consistent with reduced genetic diversity in highly clonal populations in 1996.

3. SPATIAL AUTOCORRELATION ANALYSES

Gustine and Elwinger (2003) used RAPD markers and spatial autocorrelation analysis to examine the genetic structure of white clover populations. The multivariate approach of Smouse and Peakall (1999), designed for use with data from PCR-based genetic markers, including RAPDs, is based on genetic distance methods and nonparametric permutational testing procedures. Applying this method, Gustine and Elwinger (2003) found that half the populations on two of the three farms studied in 1997 and 1998 displayed significant overall spatial overall autocorrelations. Interestingly, about half of the populations with significant genetic structure had clones and half did not.

Some white clover populations were analyzed both years in the same month (Gustine and Elwinger 2003). Some quadrat pairs did not have significant spatial autocorrelation in either year, while other quadrat pairs maintained significant spatial autocorrelation both years. Gustine and Elwinger (2003) found frequencies for significant spatial autocorrelation at both sites in 1998 was about half of that in 1997. Significantly ($P < 0.05$) more clones were found by Gustine and Sanderson (2001b) in1997 on these two sites plus a third site than in 1998.

In cases where overall significant autocorrelation was not detected in white clover populations with clones, different clonal patches may have been closely related genetically (e.g., siblings) and they may have overlapped sufficiently to break up structure. Therefore, they were not detected as geographically separate patches (Gustine and Elwinger 2003).

Gustine and Elwinger (2003) found the estimated patch size ranged from 46 to 80 cm (mean 66 cm) and did not change significantly over the two consecutive growing seasons. Patch size consistency over time may indicate that growth for clonal and closely related groups of individuals is limited, probably by environmental pressures typically found in pastures (Gustine and Elwinger 2003). Research is needed to elucidate the importance of factors such as herbivory by livestock or invertebrates, physical damage by livestock, and competition from more competitive pasture species.

4. GENETIC STRUCTURE

Restricted gene flow, whether in clonal or nonclonal populations, leads to genetic structure in plant populations as shown by spatial autocorrelation analysis (Bertorelle and Barbujani 1995; Hartl and Clark 1989; Smouse and

Peakall 1999; Sokal and Oden 1978), which makes spatial pattern analyses useful for examining gene flow in plant populations (Epperson and Allard 1989; Gustine and Elwinger 2003; Smouse and Peakall 1999).

For genetic structure to be detected, the genetic component must be spatially defined within the population and be geographically isolated from other genetic groups. Therefore, structure will not be evident if there are no distinctive clonal patches and random mating has occurred. Lack of genetic structure can occur when multiple clonal patches overlap and stolons intertwine or when there is only one genotype in the population.

Research results reviewed here are consistent with the notion that at any point in time, a significant unknown fraction of white clover genotypes in a field will not bear leaf samples at a sampling point. Later in the growing season or even the following year, old stolons or a stolon branch of the same genotype could bear new leaves at the same sampling point. Alternatively, a stolon of a different genotype could bear leaves at that sampling point. This provides a mechanism for dynamic changes in genotypes sampled at different harvest times. In this way, white clover populations can maintain temporally changing high genetic diversity due to its clonal growth habit. Additionally, rare seedling recruitment adds genetic diversity each year. As a result, any management schemes imposed by producers in the humid northeastern U.S. that promote white clover growth and maintains about 30% of this species in a grass–legume sward will ensure genetic diversity and persistence of white clover.

REFERENCES

Barrett JP, Silander JA, Jr (1992) Seedling recruitment limitation in white clover *(Trifolium repens*; Leguminosae). Am. J. Bot. 79:643–649.

Bertorelle G, Barbujani G (1995) Analysis of DNA diversity by spatial autocorrelation. Genetics 140:811–819.

Brink GE, Pederson GA, Alison MW, Ball DM, Bouton JH, Rawls RC, Steudemann JA and Venuto BC (1999) Growth of white clover ecotypes, cultivars, and germplasm in the Southeastern USA. Crop Sci. 39: 1809–1814.

Cahn MG, Harper JL (1976). The biology of the leaf mark polymorphism in *Trifolium repens* L. 1. Distribution of phenotypes at a local scale. Heredity 37:309–325.

Chapman DF (1983) Growth and demography of *Trifolium repens* stolons in grazed hill pastures. J. Appl. Ecol. 20:597–608.

Chapman DF, Anderson CB (1987) Natural re-seeding and *Trifolium repens* demography in grazed hill pastures. I. Flowerhead appearance and fate, and seed dynamics. J. Appl. Ecol. 24: 1025–1035.

Epperson BK, Allard RW (1989) Spatial autocorrelation analysis of the distribution of genotypes within populations of lodgepole pine. Genetics 121: 369–377.

Excoffier L, Smouse PE, Quattro JM (1992) Analysis of molecular variance inferred from metric distances among DNA haplotypes: Application to human mitochondrial DNA restriction data. Genetics 131: 479–491.

Fothergill M, Davies DA, Daniel JGD (1997) Morphological dynamics and seedling recruitment in young swards of three contrasting cultivars of white clover (*Trifolium repens*) under continuous stocking with sheep. J. Agric. Sci. 128: 163–172.

Gustine DL, Huff DR (1999) Genetic variation within and among white clover populations from managed permanent pastures of the northeastern U.S. Crop Sci. 39: 524–530.

Gustine DL, Sanderson MA (2001a) Quantifying spatial and temporal genotypic changes in white clover populations by RAPD technology. Crop Sci. 41:143–148.

Gustine DL, Sanderson MA (2001b) Molecular analysis of white clover population structure in grazed swards during two growing seasons. Crop Sci. 41: 1143–1149.

Gustine DL, Elwinger GF (2003) Spatiotemporal genetic structure within white clover populations in grazed swards. Crop Sci. 43: 337-344.

Harberd DJ (1963) Observations on natural clones of *Trifolium repens* L. New Phytol. 62:198–204.

Hartl DL, Clark AG (1989) Principles of population genetics. 2nd ed. Sinaur Associates, Sunderland, MA.

Pederson GA (1995) White clover and other perennial clovers. *In:* An introduction to grassland agriculture, Barnes, RF et al. (ed.), Forages: Vol. 1, p. 227–236, 5th ed. Iowa State Univ., Ames.

Smouse PE, Peakall R (1999) Spatial autocorrelation analysis of individual multiallele and multilocus genetic structure. Heredity 82: 561–573.

Sokal RR, Oden NL (1978) Spatial autocorrelation in biology. I. Methodology. Biol. J. Linnaean Soc. 10: 199–228.

Tracy BF, Sanderson MA (2000) Seedbank diversity in grazing lands of the Northeast United States. J. Range Manage. 53: 114–118.

Widén B, Cronberg N, Widén M (1994) Genotypic diversity, molecular markers and spatial distribution of genets in clonal plants, a literature survey. Folia. Geobot. Phytotaxon. 29: 245–263.

Dissection of Heterosis in Alfalfa Hybrids

H. Riday and E. C. Brummer
Department of Agronomy, Iowa State University, Ames, IA 50011 USA. (Email: brummer@iastate.edu).

Key words: alfalfa, heterosis, falcata, hybrids, geography, climate, DNA markers, yield

Abstract:

Progenies of *Medicago sativa* subsp. *sativa* by subsp. *falcata* crosses show hybrid vigor for biomass yield. Sativa-falcata hybrids represent a possible solution to current yield stagnation in alfalfa. In this study we characterized sativa-falcata hybrid biomass yield based on testcrosses of falcata germplasm from throughout its geographic range to elite sativa germplasm tester populations. Morphological, geographic, climate of origin, and molecular marker variables of the falcata parents were used to determine which were most predictive of hybrid biomass yield. European falcata and falcata with good autumn growth were consistent predictors of improved sativa-falcata hybrid performance. Molecular markers clearly separated the two subspecies; however, within falcata no clear genetic structure was found that correlated with geography or biomass yield heterosis. Based on this study, germplasm could be pre-selected before testcrossing based on geography and fall growth. This would alleviate some of the need for falcata-sativa testcrossing and evaluation, the most expensive and time-consuming part of hybrid cultivar development. In conjunction with this study, efforts are underway to dissect and map components of autumn growth and dormancy and determine their relationship with biomass yield. Mapping individual yield components could lead to QTL or candidate gene discovery, which would be useful in a marker assisted selection program.

A. Hopkins, Z.Y. Wang, R. Mian, M. Sledge and R.E. Barker (eds.), Molecular Breeding of Forage and Turf, 315-324.

1. INTRODUCTION

Alfalfa represents about 2.5% of the total agricultural hectarage and 6 billion dollars of annual production in the United States (USDA 2003). Primary traits of interest in alfalfa are yield, nutritive value, disease resistance, persistence, and winter hardiness. Current alfalfa breeding methods are almost exclusively based on recurrent phenotypic selection, involving intercrossing selected parents to produce synthetic varieties (Hill et al. 1988). Hybrid or semi-hybrid cultivars could be used to express hybrid vigor in farmer's fields (Brummer 1999). Implementing a hybrid breeding system requires the improvement of at least two independent and complementary populations, which in combination produce heterosis. *Medicago sativa* subsp. *falcata* (hereafter "falcata") has been identified as a subspecies that shows heterosis in crosses with elite *Medicago sativa* subsp. *sativa* (hereafter "sativa") breeding material (Riday and Brummer 2002ab; Riday et al. 2002, Riday et al. 2003). Currently few improved falcata breeding populations exist, slowing sativa-falcata semi-hybrid breeding system and cultivar development. Thus, a program to select improved falcata populations has merit. Unfortunately, little is known about which falcata germplasms show good heterosis with elite sativa breeding material and that could serve as the basis of a falcata population improvement program.

Falcata is yellow flowered and, compared to sativa, tends to be more winterhardy, to have more prostrate growth, to regrow slower, and to yield less in the late summer and early autumn (Lesins and Lesins 1979; Riday and Brummer 2002b). Geographically, falcata is distributed in the colder areas of Russia, Mongolia, Scandinavia and China, while sativa grows naturally in the Middle East, Southern Europe and Northern Africa (Hansen 1907; Lesins and Lesins 1979). Wild falcata and sativa germplasm overlap in some European regions and in Central Asia, where their natural hybrid, *M. sativa* subsp. *varia*, is found (Hansen 1907; Lesins and Lesins 1979). Both tetraploid ($2n = 4x = 32$) and diploid ($2n = 2x = 16$) populations of all subspecies are found (diploid sativa is denoted *M. sativa* subsp. *coerulea*), with the diploids presumed to be older than the tetraploids (Lesins and Lesins 1979). Currently there are 470 falcata accessions listed in the USDA National Plant Germplasm System's Germplasm Resource Information Network (GRIN, 2003). In addition to GRIN accessions, various germplasm centers throughout the world, as well as a few semi-improved North American populations, have been collected.

The best method to identify the most useful falcata germplasm for hybrid breeding programs is to make crosses with elite sativa germplasm, but making test crosses is time consuming and expensive. Thus, the identification of easily assayed traits, such as morphological or genetic markers, that are associated with hybrid progeny performance would streamline the selection process. Clear morphological differences exist between falcata and sativa germplasm (Crochemore et al. 1996; Jenczewski et al. 1999; Cazcarro 2000; Riday and Brummer 2002ab; Riday et al. 2002; Riday 2003). There are some indications that the morphological differentiation between falcata and elite sativa germplasm may partially explain the observed heterosis (Riday et al. 2002). Because most morphological traits are under environmental selection, certain environmental conditions may sculpt falcata morphology to give it a high probability of producing heterosis with elite sativa germplasm. We reasoned that because environmental factors undoubtedly mould the morphology and genetic constitution of falcata populations, identifying environments from which falcata genotypes produce superior sativa-falcata hybrids could help focus our falcata population improvement efforts. Given the large number of falcata accessions already present in GRIN, increased knowledge about the environments associated with hybrid performance could be used to prioritize accessions for evaluation and to target geographical regions for future collection activities.

In maize and sunflower, molecular markers have been successfully used to place genotypes into heterotic groups (Messmer et al. 1991; Cheres et al. 2000). In alfalfa, molecular markers generally show that sativa and falcata germplasm form distinct groups (Quiros 1983; Brummer et al. 1991; Kidwell et al. 1994), but the distinction between the subspecies is less clear when evaluating only wild (non-improved) germplasm (Crochemore et al. 1996; Ghérardi et al. 1998; Jenczewski et al. 1999; Cazcarro 2000). Based on a limited sampling of genotypes, we did not find an association of genetic distance with heterosis (Riday et al. 2003). Genetic clustering that established heterotic clusters would allow prescreening of germplasm sources to obtain falcata germplasm that showed good heterosis with elite sativa germplasm.

The primary objective of this study was to evaluate falcata genotypes from a broad range of wild and semi-improved populations for performance *per se* and in testcrosses with elite sativa tester genotypes. Our goal was to characterize the distribution of sativa-falcata hybrids for yield heterosis on a whole year and on a harvest basis to guide the development of improved falcata populations. Our secondary objective was to determine which types of parental falcata classification were most predictive of superior sativa-falcata hybrids in terms of heterosis and yield. Genotypes were classified based on eco-geographic origin, molecular genetic and physical markers, parental morphology, and parental selection status.

2. EXPERIMENT DESIGN

A total of 125 genotypes were used: 16 elite sativa genotypes from four populations; 3 wild sativa genotypes from 3 populations, and 106 falcata genotypes from 37 wild or semi-improved populations from throughout the native range of falcata (Fig. 1). The falcata genotypes we used were predominantly yellow flowered. However, some of these genotypes derived from accessions that also included plants with variegated flowers. Thus, the 37 wild populations could be subdivided into those in which all genotypes had yellow flowers and those that contained some individuals with variegated flowers. In addition, some of the 106 falcata genotypes had variegated flowers, so genotypes in the variegated populations could be either yellow or variegated. Falcata populations could be split into two classes based on improvement status (wild or improved). In some populations, both randomly chosen and visually selected genotypes were included in the crossing.

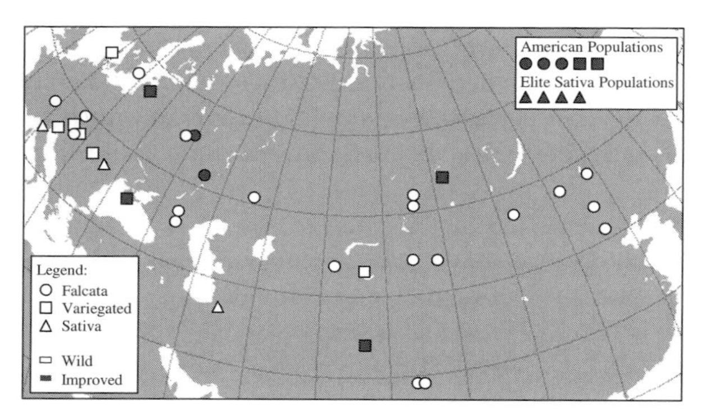

Figure 1. Geographic origin of 44 alfalfa populations sampled.

All 125 genotypes were testcrossed to the four elite sativa tester populations for a total of 500 testcross entries. Stem cuttings of the 125 parental genotypes were also made. Entries were established in sixteen plant, semi-sward plots in August 2000 at two Iowa locations. Harvests for yield were taken three times during 2001 and 2002 at both locations. Yield was determined on a per plant dry matter basis. Yield heterosis was estimated as the regression residual between the testcross progeny performance and the average of the parental clone and mean tester population performance.

Concurrent with yield harvests, maturity, plant width, and plant height were measured. Plant height was measured approximately on a weekly basis from plant emergence in the spring until the first damaging frost in the autumn. Vegetative density, growth angle, and regrowth were derived from other traits measured. In addition to field measurements, 139 polymorphic AFLP DNA fragments were scored on the 125 parental genotypes using a Li-Cor infrared visualization system (Vos et al. 1995; Keygene).

For wild populations, the following climate of origin variables were compiled from publicly available databases on a monthly mean basis: average, minimum, and maximum daily temperature, diurnal temperature range, cloud cover, vapor pressure, wet day frequency, precipitation, ground frost frequency, wind speed, and radiation (New et al. 1999; 1960-1990); snow cover (Change et al. 1993; 1978-1987); daylength (Forsythe et al. 1995; calculated); photosynthetically active radiation (Pinker and Laszlo 1997; 1984-1988); and maximum vegetative index and conversion factor (Tateishi and Kajiwara 1993; 1984-1988).

3. BIOMASS YIELD AND HETEROSIS

Comparison of hybrid performance of sativa (SS) and falcata (SF) genotypes crossed to sativa tester populations and of falcata clonal (FC) performance showed that SF and SS were equivalent for total, and first and third harvest yields, SF was inferior to SS at second harvest, and FC was lowest at all time points (Table 1). Of more interest than means were the testcross yield distributions. For all harvests SF variation was greater than SS variation. For total and individual harvest yields, some SF combinations were observed that were equivalent, if not superior, to the best SS. During second and third harvests, a number of SF were inferior to the lowest yielding SS.

Table 1. Biomass yield and heterosis, for sativa by sativa crosses, sativa by falcata crosses, and falcata clones, for year totals and 1st, 2nd, and 3rd harvests, averaged over 2001 and 2002 in two Iowa locations.

Entry Type	Yield				Heterosis			
	Year Total	Harvest			Year Total	Harvest		
		1st	2nd	3rd		1st	2nd	3rd
	g plant^{-1}							
Sativa x Sativa	219a	90b	75a	55a	-7.0b	-7.4b	0.7a	0.2a
Sativa x Falcata	219a	97a	70b	52a	1.2a	1.3a	-0.2b	-0.1a
Falcata Clones	166b	70c	56c	40b	---	---	---	---

Under an additive genetic model, no heterosis would be expressed and SF should fall midway between the SS and FC. Compared to SS, SF had superior total and first harvest yield heterosis, but at second harvest, SF heterosis was inferior to SS. These results confirm the sativa-falcata heterotic pattern we proposed previously (Brummer 1999; Riday and Brummer 2002a).

Subdivided groups of SF were compared to determine which were associated with superior hybrids. Comparisons of variegated and pure falcata populations in testcrosses (SV and SF, respectively) unexpectedly showed that SV had higher total yield than either SS or SF (Table 2). At first harvest, SV and SF had equivalent yield and both exceeded SS. The SV were intermediate to SS and SF during second harvest but were equivalent to SS and greater than SF during the third harvest. About half of selected genotypes from variegated populations had variegated flowers, while the other half had yellow flowers and, therefore, were indistinguishable from falcata. We compared variegated genotypes to yellow flowered genotypes for populations that had both types, but no differences were detected between flower color types for yield or heterosis during any harvest.

Table 2. Sativa by falcata testcross grouping comparisons for yield and heterosis, for year total and 1st, 2nd, and 3rd harvests, averaged over 2001 and 2002 in two Iowa locations.

Sativa by Falcata testcross Groupings	Yield				Heterosis			
	Year Total	Harvest			Year Total	Harvest		
		1st	2nd	3rd		1st	2nd	3rd
	g plant^{-1}							
Variegated	223a	97a	72a	54a	0.8a	0.5a	0.8a	0.2a
Falcata	217b	97a	69b	51b	1.4a	1.7a	-0.7b	-0.2b
Improved	224a	98a	71a	55a	1.4a	1.2a	-0.5a	1.6a
Wild	217b	96a	70a	51b	1.2a	1.1a	0.0a	-0.7b
Wild Variegated	223a	97a	73a	56a	1.1a	1.0a	0.9a	0.3a
Wild Falcata	215b	96a	69b	50b	1.1a	1.5a	-0.3a	-0.9a
European	229a	100a	74a	55a	9.2a	4.6a	3.3a	1.4a
Asian	204b	91b	65b	47b	-8.8b	-2.7b	-4.0b	-3.4b
Selected	233a	101a	75a	57a	10.8a	5.0a	3.1a	3.1a
Unselected	217b	96b	70b	51b	-2.5b	0.7b	-1.5b	-1.8b

Most variegated populations had been previously improved through selection, so their superior performance could have resulted from human selection. Improved populations of both varia and falcata had superior total and third harvest yield and heterosis and better autumn testcross performance compared to wild populations. The superiority of the improved germplasm mirrored variegated performance *per se*. To isolate variegated effects, wild variegated and falcata testcross performance was evaluated and again, variegated populations had higher testcross total and second and third harvest yield with no heterosis differences (Table 2). Although variegated germplasm offers no heterotic advantage, they offer superior testcross yield for all harvests except the first.

Visual field observation indicated European falcata and variegated germplasm created better testcrosses hybrids compared to Asian germplasm. Therefore wild falcata and variegated populations east and west of 60°E longitude were compared. European germplasm had superior yield and heterosis for total and individual harvests (Table 2). Comparisons of wild European variegated and falcata populations revealed equivalent total yield with higher falcata heterosis. At first harvest, wild European SF had superior heterosis and yield. During third harvest, variegated outyielded falcata testcrosses, but produced equivalent heterosis.

Finally random versus selected genotypes from within populations were compared. Selected genotype testcross performance was superior in all aspects of yield and heterosis (Table 2). The selected group had the highest yield and heterosis of any SF subgroup.

4. BASIS OF SATIVA-FALCATA HETEROSIS

Compilation of parental genetic, morphological, and climate of origin data allowed us to assess the data types potentially predictive value for hybrid performance. This information would enable breeders to weigh the costs and benefits of different types of data. Genotypes were clustered three times based on parental data type. For total hybrid yield, climate clusters explained 34% of the variation, followed by morphological (28%) and genetic (23%) (Table 3).

Table 3. R^2 and semi-partial R^2 (%) of parental cluster types (morphologic [M], genetic [G], and climate of origin [C]) predictive of year total and 1st, 2nd, and 3rd harvest yield, averaged over 2001 and 2002 in two Iowa locations

		Separately			Combined							
		M	G	C	T	M	G	C	M-G†	M-C	G-C	M-G-C
		%										
Biomass Yield												
Year Total		28***	23***	34***	44***	10*	0NS	5NS	0	6	12	12
Harvest	1	15*	8NS	29***	37**	8*	0NS	19*	0	3	2	5
	2	31***	16*	36***	49***	11**	1NS	9*	2	13	8	6
	3	24**	38***	26***	43***	5NS	15*	0NS	0	0	4	22
Biomass Heterosis												
Year Total		0NS	16*	26**	24*	0NS	0NS	8NS	0	0	16	0
Harvest	1	4NS	4NS	27**	29*	2NS	0NS	23*	0	0	3	1
	2	5NS	16*	31***	32**	1NS	1NS	11NS	0	4	15	0
	3	0NS	22**	11*	22NS	0NS	13NS	0NS	0	0	9	2

*, **, and *** significant at the $P = 0.05$, 0.01, and 0.001 level, respectively. NS not significant.
† Combinations represent collinearity between groupings.

Semi-partial R^2 identified the amount of variation for hybrid yield and heterosis that each of the three parental cluster types (morphological, genetic, and climate) explained uniquely and the amount that was collinear among them. The combined model explained 37% to 49% of yield across all measurement periods (Table 3). Morphological clusters explained ~10% of total and first and second harvest yield. Climate clusters explained 19% of first and 9% of second harvest yield. Genetic clusters explained 15% of third harvest yield. Collinearity of 12% and 22% among all three data types was observed for year total and third harvest hybrid yield (Table 3). Large collinearities for year total and second harvest yield were noted between genetic and climate clusters and between morphological and climate clusters. Almost no collinearity was observed between morphological and genetic clusters.

Yield heterosis R^2 values ranged from 22% to 32% for the combined model, across all measurement periods (Table 3). Parental morphological clustering contributed little toward explained heterosis variation. Climate clusters uniquely represented half of the total R^2 for year total and second harvest biomass heterosis; the remaining variation was collinear between genetic and climate clustering. First harvest climate clustering accounted for almost all explained variation (Table 3). Third harvest genetic clusters alone accounted for over half the R^2, the rest being collinear between genetic and climate clusters.

5. CONCLUSIONS

Based on the collinearity estimates, climate selection and genetic differentiation are key factors causing the sativa-falcata heterotic pattern. Although superior parental morphology is associated with higher yielding hybrids, it is not associated with increased heterosis. Based on the SF group comparisons, it is clear that European falcata, whether from variegated or pure yellow flowered populations, creates the best hybrids.

Molecular fragment data shows an Asian and European radiation. Skinner (2000) reported a similar east-west falcata division. The most within population divergent germplasm was diploid and tetraploid genotypes from the Caucasus Mountains in Russia. Sativa germplasm formed a distinct cluster from the falcata germplasm.

Single molecular fragments were correlated with parental and hybrid biomass yield and heterosis adjusting for each population. The strongest *P*-values were 0.01 – 0.001 for individual markers. A permutation test showed none of these to be significant. None of the low *P*-value fragments were polymorphic between the parents of a tetraploid mapping population currently being used for mapping biomass yield (Robins et al. 2003).

Climate data showed that warmer winter climates and decreased summer (June) photosynthetically active radiation were associated with increased hybrid yield and heterosis. The first association could likely be confounded with east-west genetic drift, especially, since the Asian populations sampled generally had colder winter climates than European populations. The second association suggests that photoperiod effects select for a falcata genetic architecture that in sativa combination leads to heterosis.

REFERENCES

Brummer EC (1999) Capturing heterosis in forage crop cultivar development. Crop Sci. 39: 943-954.

Brummer EC, Kochert G, Bouton JH (1991) RFLP variation in diploid and tetraploid alfalfa. Theor. Appl. Genet. 83: 89-96.

Cazcarro PM (2000) Thesis: Differentiating *M. sativa* subsp. *sativa* and subsp. *falcata* using molecular markers. Iowa State University, Ames, IA.

Chang A, Foster JL, Hall DK, Powell HW, Chien YL (1993) Monthly nimbus-7 SMMR derived global snow cover and snow depth data set (Oct 1978 - Aug 1987). Digital raster data on a half-degree geographic (lat/long) 720 by 360 grid. In: Global ecosystems database version 1.0: Disc B. Boulder, CO: NOAA National Geophysical Data Center.

Cheres MT, Miller JF, Crane JM, Knapp SJ (2000) Genetic distance as a predictor of heterosis and hybrid performance within and between heterotic groups in sunflower. Theor. Appl. Genet. 100: 889-894.

Crochemore ML, Huyghe C, MC Kerlan, Durand F, Julier B (1996) Partitioning and distribution of RAPD variation in a set of populations of the *Medicago sativa* complex. Agronomie 16: 421-432.

Forsythe WC, Rykiel Jr EJ, Stahl RS, Wu H, Schoolfield RM (1995) A model comparison for daylength as a function of latitude and day of year. Ecol. Modeling 80: 87-95.

Ghérardi M, Mangin B, Goffinet B, Bonnet D, Huguet T (1998) A method to measure genetic distance between allogamous populations of alfalfa (*Medicago sativa*) using RAPD molecular markers. Theor. Appl. Genet. 96: 406-412.

GRIN (2003) http://www.ars-grin.gov/cgi-bin/npgs/html/tax_acc.pl?104918 (verified on: 01/23/03).

Jenczewski E, Prosperi JM, Ronfrort J (1999) Evidence for gene flow between wild and cultivated *Medicago sativa* (Leguminosae) based on allozyme markers and quantitative traits. Am. J. Bot. 86: 677-687.

Kidwell KK, Austin DF, Osbron TC (1994) RFLP evaluation of nine *Medicago* accessions representing the original germplasm sources for North American alfalfa cultivars. Crop Sci. 34: 230-236.

Lesins K, Lesins I (1979) Genus *Medicago* (Leguminasae): a taxogenetic study. Kluwer Academic Publishers, Dordrecht, Netherlands.

Messmer MM, Melchinger AE, Lee M, Woodman WL, Lee EA, Lamkey KR (1991) Genetic diversity among progenitors and elite lines from the Iowa Stiff Stalk Synthetic (BSSSS) maize population: comparison of allozyme and RFLP data. Crop Sci. 83: 97-107.

New M, Hulme M, Jones PD (1999) Representing twentieth century space-time climate variability. part 1: development of a 1961-90 mean monthly terrestrial climatology. J. Climate 12: 829-856.

Pinker RT, Laszlo I (1997) Photosynthetically active radiation (PAR) and conversion factors (CF). Digital raster data on a 2.5 degree geographic (lat/long) 144x72 pixel grid. In: global ecosystems database Disc-B. Boulder, CO: NOAA National Geographical Data Center.

Quiros CF, Bauchan GR (1988) The genus *Medicago* and the origin of the *Medicago sativa* complex. In: Alfalfa and alfalfa improvement, Hanson AA, Barnes DK, Hill RR Jr (eds.), pp.93-124. ASA-CSSA-SSSA, Madison, WI.

Riday H, Brummer EC (2002a) Forage yield heterosis in alfalfa. Crop Sci. 42: 716-723.

Riday H, Brummer EC (2002b) Heterosis of Agronomic Traits in Alfalfa. Crop Sci. 42: 1081-1087.

Riday H, Brummer EC, Moore KJ (2002) Heterosis of Forage Quality in Alfalfa. Crop Sci. 42: 1088-1093.

Riday H, Brummer EC, Campbell TA, Luth D, Cazcarro PM (2003) Comparisons of genetic and morphological distance with heterosis between *Medicago sativa* subsp. *sativa* and subsp. *falcata*. Euphytica 131: 37-45.

Robins JG, Luth D, Santra M, Alarcón-Zúñiga B, Riday H, Brummer EC (2003) Construction of a genetic map of an intersubspecific cross between *Medicago sativa* subsp. *sativa* and *Medicago sativa* subsp. *falcata*. In: Abstracts of Plant & Animal Genome XI Conference, January 11-15, San Diego, CA, USA.

Skinner DZ (2000) Non random chloroplast DNA hypervariability in Medicago sativa. Theor. Appl. Genet. 101: 1242-1249.

Tateishi R, Kajiwara K (1993) Monthly maximum global vegetation index and land cover classifications from NOAA-9 (Jan 1986 - Dec 1989). Digital raster data on a 10-minute geographic (lat/long) 1080x2160 grid. In: Global ecosystems database version 1.0: Disc B, Boulder, CO: NOAA National Geophysical Data Center.

USDA National Agricultural Statistical Service (2003) State level data for field crops: hay http://www.nass.usda.gov:81/ipedb/main.htm (verified 02/02/03).

Vos P, Hogers R, Bleeker M, Reijans M, van de Lee T, Hornes M, Frijters A, Pot J, Peleman J, Kuiper M, Zabeau M (1995) AFLP: a new technique for DNA fingerprinting. Nuc. Acids Res. 23:21:4407-4414.

From Models to Crops: Integrated *Medicago* Genomics for Alfalfa Improvement

G. D. May
Plant Biology Division, The Samuel Roberts Noble Foundation, Ardmore, OK 73402 USA

Key words: *Medicago truncatula*, alfalfa, functional genomics

Abstract: The whole-system or global nature of genomics lends great potential to the identification of novel genes or gene classes that underlie the biology that is unique to specific plant families. Legumes are important crops for human nutrition world-wide and also serve as an important source of nutrition for animal and dairy production. Seed legumes suchs as soybean, peanut, chickpeas and lentils contain from 20 to 50 percent protein - two to three times that of cereal grains or meat. In addition, the often complex interaction of legumes with microorganisms have resulted in the evolution of a wide variety of plant natural products involved in symbiosis and defense interactions. Many of these compounds have anti-microbial activities and, additionally, positive effects on human and animal health. Exploitation of the diverse gene makeup of the legumes for the benefit of human-kind requires in-depth knowledge of legume genomes.

A. Hopkins, Z.Y. Wang, R. Mian, M. Sledge and R.E. Barker (eds.), Molecular Breeding of Forage and Turf, 325-332.

1. INTRODUCTION

Legumes are second only to grasses in economic importance worldwide. In comparison with other crops, the production of legumes reduces economic and environmental costs given their ability to fix nitrogen. Each independent origin of agriculture can also be traced back to systems based on the domestication of legumes and cereals such as soybean and rice in Asia and beans and maize in America. With more than 18,000 species, members of the pea family (*Leguminosae*) are second only to grasses in economic importance worldwide. Forage and pasture legumes are an important source of nutrition for animal and dairy production. Seed legumes such as peanut, soybeans, chickpeas and lentils contain approximately 20 to 50 percent protein – two to three times that of cereal grains and meat. Legumes therefore serve as an excellent source of protein and dietary fiber that is often deficient in the diets of individuals in developing nations.

Among crops, legumes are unique in their ability to fix atmospheric nitrogen through a novel symbiotic relationship with bacteria known as Rhizobia. The capacity of legumes to fix nitrogen in partnership with Rhizobia leads to another astonishing fact: legume nodules produce more ammonia fertilizer each year than total human industrial production worldwide. Legumes generate more than 17 million tons of agricultural nitrogen each year with an equivalent value of $8 billion dollars (US). Because of their central role in nitrogen cycling, legumes occupy a key place in most ecosystems.

Medicago truncatula is an omni-Mediterranean species and is closely related to the world's major forage legume, alfalfa. Unlike alfalfa, which is a tetraploid, obligate outcrossing species, *M. truncatula* has a simple diploid genome (two sets of eight chromosomes) and can be self-pollinated. Genes from *M. truncatula* share very high sequence identity to the corresponding genes from alfalfa and appear to be arranged in a similar order on the chromosomes to those of other legumes.

M. truncatula has been chosen as a model species for genomic studies in view of its small genome, fast generation time (from seed-to-seed) and the ability to be transformed (Cook 1999). *M. truncatula* has many strengths as a model legume, including:

- A relatively small genome (between 500 and 550 Mbp)
- Diploid and self-fertile
- Short seed-to-seed generation time and an abundant seed set
- Large collections of phenotypic mutants, especially in nodule formation and symbiotic nitrogen fixation

- A host to a highly characterized species of rhizobium, *Sinorhizobium meliloti*
- Vast collections of diverse, naturally occurring ecotypes — including easily accessible native populations
- A close relative to alfalfa, the most important forage crop worldwide

Genes from *M. truncatula* share high sequence identity to their counterparts from alfalfa (e.g. 98.7 and 99.1% at the amino acid levels for isoflavone reductase, and vestitone reductase, respectively), so it serves as an excellent genetically tractable model for alfalfa. Studies on syntenic relationships are establishing links between *M. truncatula*, alfalfa, and pea, as well as Arabidopsis.

As a legume, and unlike the most studied genetic model plant, Arabidopsis, *M. truncatula* establishes symbiotic relationships with nitrogen fixing Rhizobia. Roots of *M. truncatula* are also colonized by beneficial arbuscular mycorrhizal fungi (Harrison and Dixon 1993). In addition, the complex interactions of legumes with microorganisms have resulted in the evolution of a rich variety of natural product biosynthetic pathways impacting both mutualistic and disease/defense interactions. Of these, the isoflavonoid pathway, which is not present in Arabidopsis, leads to nodulation gene inducers and repressors, pterocarpan phytoalexins involved in host disease resistance, and isoflavones with anticancer and other health promoting effects for humans. This pathway has been well characterized in alfalfa, and in other legumes such a soybean and chickpea, at the metabolic, enzymatic and genetic levels (Paiva et al. 1994; Dixon et al. 1995; Dixon 1999). Exploitation of this diverse but complex chemistry for the benefit of humankind requires in-depth knowledge of the legume genome

2. AN INTEGRATED APPROACH TO *MEDICAGO* FUNCTIONAL GENOMICS

The *Medicago* functional genomics program at the Samuel Roberts Noble Foundation is a systematic approach in the study of the genetic and biochemical events associated with the growth, development, and environmental interactions of *M. truncatula*. It is anticipated that the majority of genetic mechanisms discovered in *M. truncatula* will be directly transferable to better understanding the genetics that underlie complex traits in *M. sativa*. Our methods include the development and integration of EST, transcript, protein, and metabolite datasets. We are dovetailing these cross-discipline data types to provide an integrated set of tools to address fundamental questions pertaining to legume biology. These questions include the analysis and understanding of: 1) the biosynthesis of natural products that affect forage quality and human health; 2) the function of legume-specific genes; 3) ecotype variation at the phytochemical level; 4) the cellular and molecular basis for the directional growth response of roots to gravity and the role of the cytoskeleton in this process; 5) legume root development and molecular mechanisms of polar auxin transport; 6) non-host pathogen resistance; 7) ABC transporters; 8) the RNA silencing pathway; 9) the interaction of *M. truncatula* with the arbuscular mycorrhizal fungus *Glomus versiforme* for analyses of the AM symbiosis; and 10) the function of members of the cytochrome P450 and glycosyltransferase multigene families. Our aim is to develop a program that will integrate gene expression, protein and metabolite profiling in conjunction with *M. truncatula* genetics to provide a global view of *Medicago* biology.

2.1 Expressed Sequence Tags

Since January 2000, more than 100,000 *M. truncatula* ESTs have been characterized at the Noble Foundation and a total of approximately 190,000 worldwide. Unidirectional cDNA libraries representing different stages of *M. truncatula* development and exposure to biotic and abiotic stresses have been generated. The international *Medicago* research community has characterized ESTs from more than 24 different cDNA libraries. The goal of the Foundation's EST project is to identify and characterize 20,000 to 40,000 unique *Medicago* cDNA isolates. Complete DNA sequencing of 3,000 abundant full-length cDNAs is being performed, in part, to assist our proteomics program.

2.2 Expression Profiling

Changes in gene expression underlie many biological phenomena. The use of DNA microarrays and serial analysis of gene expression will provide insights into tissue- and developmental-specific expression of genes and the response of gene expression to environmental stimuli. *M. truncatula* genome-wide microarrays are being generated using the *Medicago* Array-Ready Oligonucleotide Set (GS-1700-02) Version 1.0 (Operon). Approximately 16,000, amino-linked, 70-mer oligonucleotides are being printed onto aminosilane-coated "Superamine" slides (Telechem), using Telechem type SMP3 printing pins in Dr. David Galbraith's laboratory at the University of Arizona. Operon has agreed to update the *Medicago* oligonucleotide genome set as additional *M. truncatula* EST and genome sequence information becomes available. Preliminary results in our labs and those of other (Dr. Ian Ray, personal communications) suggest that *M. truncatula* oligonucleotide arrays hybridize well with targets synthesized using *M. sativa* mRNA as a template. These arrays should provide a valuable tool to study complex traits in alfalfa.

2.3 Protein and Metabolite Profiling

The protein complement of the genome, the proteome, serves as a biological counterpart to the *Medicago* EST and gene expression analyses. Given that many biological phenomena lack the requirement for *de novo* gene transcription, proteomics studies provide a mechanism to study proteins and their modifications under developmental changes and in response to environmental stimuli. An automated system has been established for the electrophoretic separation of complex protein mixtures and differential analysis to discover changes in proteome content. A state-of-the-art biological mass spectrometry laboratory has been established as part of the *Medicago* genomics activities. Instrumentation within the laboratory includes LC/MS, GC/MS, MALDITOF/MS and Q-TOF/MS. *M. truncatula* ecotypes and elicited cell cultures are being screened for changes in the levels of a wide range of primary and secondary metabolites

2.4 Bioinformatics

The Bioinformatics program at the Noble Foundation is building an integrated informatics platform to support all phases of *Medicago* genomics research, including target selection, tracking and data analysis. Two database systems have been developed for target information and tracking. MtGenes, a data warehouse for *Medicago* genomics, integrates internal and public databases to provide a unified view of gene function, protein information, EST expression and clone information. AIM is a microarray tracking and information system with preliminary visual analysis tools. The prototypic features of AIM include slide tracking, experiment tracking and integrated data analysis. Available data analysis tools include CEDA, ESTAnalyzer and MSFACTs. CEDA prompts comparative analysis of user-defined EST datasets for *Medicago* functional genomics. EST-Analyzer is an automated pipeline for the analysis of EST sequences. Features of EST-Analyzer include functional annotation, template based translation and detection of possible sequencing errors. MSFACTs is a software package for the automated import, alignment, reformatting, and export of large chromatographic data sets to allow for visualization and interrogation of metabolomic profiles.

2.5 Approaches to Forward and Reverse Genetics in *M. truncatula*

Reverse and forward genetic systems for *M. trucatula* are being developed at the Noble Foundation and elsewhere in the *Medicago* research community. Reverse genetics systems enable the isolation of mutations in genes of known sequence, while forward genetic systems facilitate efficient identification of genes underlying phenotypic traits of interest. Fast-neutron irradiation induces DNA damage and chromosomal deletions. Deletions that occur in known genes can be detected by a shift in the size of PCR amplification products of genes of interest. Of the approximately 10,000 fast-neutron irradiated M1 *M. truncatula* plants generated thus far, two percent display a visible mutant phenotype. It is anticipated that 100,000 M1 *M. truncatula* plants will be screened within the next three years. In collaboration with Dr. Pascal Ratet, CNRS, Gif sur Yvette, France, we are developing a large-scale, transposon-tagged mutant library of *M. truncatula* using the tobacco retrotransposon *Tnt1*. Approximately 20,000 tagged *M. truncatula* lines will be generated during the next five years. Transposon-plant genome junctions will be isolated and characterized through DNA sequence analyses. A database of these junction sequences will be created for a reverse genetics approach to determine gene function.

A transient gene knockout system for *M. truncatula* that utilizes a virus vector is also being developed. RNA expressed inappropriately within the cell leads to the induction of a host enzyme pathway that recognizes and destroys all RNA containing that sequence. RNA silencing is the general term used to describe this phenomenon. We are inserting sequences from genes of interest into virus vectors that infect *M. truncatula*. Plants are being inoculated with transcripts from chimeric viral vectors and observed for visible and biochemical phenotypes specific to the inserted genes. Induction of RNA silencing is being determined by analysis of accumulation of small RNAs with sequence identity to the target transcript and by transcript accumulation. A correlation between decreased transcript expression and altered phenotype will indicate a function for the specific gene in host development. These results will provide the basis for further research to fully understand the role of the specific gene in a particular biochemical or developmental pathway.

3. *MEDICAGO* GENOME SEQUENCING PROJECT

The Noble Foundation and the University of Oklahoma announced the initiation of an exploratory genome sequencing project in October of 2001. The *M. truncatula* genome project is a collaborative effort between Dr. Bruce Roe's laboratory at the University of Oklahoma, Drs. Douglas R. Cook and Dong Jin Kim at the University of California Davis and scientists at the Noble Foundation. As predicted earlier, a majority of the *M. truncatula* coding regions resides within only a small portion of the genome. A goal for the first round of the project's funding is to determine the DNA sequence of 1,000 *M. truncatula* bacterial artificial chromosomes (BACs) up to a level of seven-fold coverage. This number of BACs is sufficient to provide sequence information for a significant portion of the gene rich regions of the *M. truncatula* genome.

The *Medicago* genome sequencing project will produce a genetic blueprint that will provide scientists with the tools to better exploit legume genetic biodiversity and create varieties with enhanced, high impact characteristics. The knowledge that results from the *M. truncatula* genome sequence will benefit the entire field of biology by elucidating the molecular and cellular basis of plant symbiotic interactions, such as the formation of nitrogen-fixing nodules and phosphate-scavenging endomycorrhizae. It is also felt that legumes occupy an excellent place in the evolutionary tree of plants to act as a sequenced genome partner with *Arabidopsis.* Together, the genome sequences of *Arabidopsis* and *Medicago* will help to reveal the processes that have led to modern-day dicots and angiosperms. Though separated by a significant phylogenetic distance, legumes reside in the same order as *Arabidopsis* (Rosidae). Studies have already demonstrated genome regions with extensive micro-synteny with *Arabidopsis*, as well as other more prevalent regions where the genomes are highly diverged.

4. ADDITIONAL *MEDICAGO* GENOMICS RESOURCES

The rapidly expanding body of *Medicago* genomics data requires advanced bioinformatics tools and resources. These include:

University of Oklahoma Advanced Center for Genome Technology

A data resource providing detailed information about characterized *M. truncatula* BAC clones, updated information and ftp access to sequenced BAC clones, and information about the complete *M. truncatula* chloroplast genome sequence.

National Center for Genome Resources (NCGR)

A data analysis pipeline for *M. truncatula* ESTs providing relational database storage of sequence data and analysis results plus sequence quality control. NCGR is also home to the Legume Information System, which will provide initial processing and analysis for future genomic sequence data from the University of Oklahoma.

The Institute for Gene Research (TIGR)

The *M. truncatula* Gene Index (MtGI) is based on EST sequence data providing a non-redundant view of *M. truncatula* genes and their predicted expression patterns, cellular roles, functions, and evolutionary relationships.

University of Minnesota Center for Computational Genomics and Bioinformatics (CCGB)

A wealth of information about *M. truncatula*, including the home of MtDB, a data-mining resource that enables complex querying of *M. truncatula* EST sequences and their BLAST report output, plus up-to-date information on EST library sources, genetic markers and maps, and a variety of sequence analysis tools.

Center for Genome Research, University of Bielefeld, Germany

Data analysis pipeline for *M. truncatula* ESTs and expression profiles.

CNRS-INRA in Toulouse France

Database focused on the EST analysis of the *M. truncatula* root symbiotic interaction with rhizobium.

5. SUMMARY

We believe that *M. truncatula* is the most developed model legume, has the largest and most able research community and best serves as a model for developing new forage varieties. The long-term impact of our program will be the integration of transcript, protein, and metabolite data with plant forward and reverse genetic system and natural variants to advance all aspects of fundamental and applied legume research. This information will be used to develop agronomically important legume species that (1) are more resistant to fungal and viral diseases, and drought, (2) will provide higher crop yields with less need for chemical inputs, such as fertilizers and pesticides, and (3) will produce natural chemicals that promote human and animal health (nutraceuticals). Higher yields and lower production costs will enhance the economy of rural agriculture while reduced chemical usage will benefit the environment.

REFERENCES

Cook, DR, (1999) *Medicago truncatula* - a model in the making! Cur. Opin. in Plant Bio. 2:301-304.

Dixon, RA, Harrison, MJ, Paiva, NL (1995) The isoflavonoid phytoalexin pathway: from enzymes to genes to transcription factors. *Physiologia Plantarum.* 93:385-392.

Dixon, R.A. (1999) Isoflavonoids: biochemistry, molecular biology, and biological functions, in *Comprehensive Natural Products Chemistry* (Vol. 1) (Sankawa, U., ed.), pp. 773-823, Elsevier, Oxford.

Harrison, MJ, Dixon, RA (1993) Isoflavonoid accumulation and expression of defense gene transcripts during the establishment of vesicular arbuscular mycorrhizal associations in roots of *Medicago truncatula*. Mol. Plant-Microbe Int. 6:643-654.

Paiva, NL, Oommen, A, Harrison, MJ, Dixon, RA (1994) Regulation of isoflavonoid metabolism in alfalfa. Plant Cell, Tissue Organ Cult. 38:213-220.

Sequencing Gene Rich Regions of *Medicago truncatula*, a Model Legume

B. A. Roe and D. M. Kupfer
Advanced Center for Genome Technology, Department of Chemistry and Biochemistry, The University of Oklahoma, Norman, OK 73019 USA. (Email: broe@ou.edu).

Key words: genomic DNA sequence, *Medicago truncatula*, legume genes, euchromatic gene rich regions, heterochromatic pericentromeric repeat rich regions

Abstract: *Medicago truncatula*, barrel medic, is an important forage crop that also is considered a model legume for laboratory studies. It is genetically tractable with a relative small genome of ~470 million base pairs, has simple Mendelian genetics, a short seed-to-seed generation time, a relatively high transformation efficiency, an excellent collection of phenotypic mutants, and several large collections of diverse, naturally occurring ecotypes. The recent work of D. Cook and D.J. Kim, University of California at Davis has resulted in constructing an ~20-fold coverage BAC library and fingerprinting it to a depth of ~12-fold, and the Noble Foundation and The Institute for Genome Research have generated over 180,000 expressed sequence tags (ESTs) representing genes expressed in almost every *M. truncatula* tissue, developmental stage and growth condition. To complement these efforts, we recently began to sequence the *M. truncatula* genome. By collecting sample sequence data through an initial whole genome shotgun approach, we confirmed earlier cytogenetic data that indicates the eight chromosomes of *M. truncatula* are organized into distinct gene-rich euchromatic and separate pericentromeric repeat-rich regions. We now have sequenced almost 1000 gene-rich bacterial artificial chromosome (BAC) clones. The results of these studies indicate that the gene density in *M. truncatula* is of the order of one gene in every 6-7 kilobase pairs (kbp). The ~200 Mbp of euchromatic regions therefore encodes ~30,000 to 33,000 genes, of which ~66% are represented by ESTs. Following in the tradition of other genome projects, all our sequence data is freely available through the international databases.

A. Hopkins, Z.Y. Wang, R. Mian, M. Sledge and R.E. Barker (eds.), Molecular Breeding of Forage and Turf, 333-344.

1. INTRODUCTION

1.1 Importance of Legumes

Legumes, with more than 650 genera and 20,000 species, are one of the two most important crop families. Among cultivated plants, legumes have the unique ability to fix atmospheric nitrogen through a symbiotic relationship with species-specific Rhizobia bacteria. This property gives legumes an extremely high protein level (Vazquez 2002) that supplies nearly 33% of the human nutritional requirement for nitrogen. In many developing countries, legumes often serve as the single most important source of consumed protein. Legumes also synthesize a wide array of secondary compounds, including numerous isoflavonoids and triterpene saponins that possess anti-cancer and other health promoting properties. Legumes also are widely used in nearly all crop rotation schemes and are universally viewed as essential for secure and sustainable food production.

All major crop legumes are found in the monophyletic subfamily Papilionoideae. Within this subfamily, the tropical legumes include the economically important soybean (*Glycine max*), common bean (*Phaseolus* spp.), cowpea (*Vigna unguiculata*), and mungbean (*Vigna radiata*), while temperate legumes include species such as pea (*Pisum sativum*), alfalfa (*Medicago sativa*), lentil (*Lens culinaris*), and chickpea (*Vicia arietinum*). Papilionoid legumes first appeared around 65 million years ago based on fossil records (reviewed in Doyle 2001), the same time as other important crop families. Because they form a compact monophyletic evolutionary group, comparative genomics among Papilionoid species has huge potential to increase our understanding of this vitally important group of plants. A growing body of evidence demonstrating micro- and macrosynteny suggests that discoveries made in one Papilionoid species can often be extended to other members of the subfamily (Connor 1998, Foster-Hartnett 2002, Gualtieri 2002, Yan 2003).

1.2 *Medicago truncatula* as a Reference Legume

Among legume species, *M. truncatula* is widely considered the preeminent model for genomic research. Unlike most Papilionoid species, *M. truncatula* has a compact genome of approximately 470 million base pairs (Mbp) (www.rbgkew.org.uk/cval/homepage.html), simple Mendelian genetics, short seed-to-seed generation time, relatively high transformation efficiency, excellent collections of phenotypic mutants, and large collections of diverse, naturally occurring ecotypes (Cook 1999). Given these many desirable genetic features, several research programs are already committed to *M. truncatula*, and their work is leading to a wealth of excellent genomic resources and interesting biology.

The most important single genomic resource for any species is its nuclear genome sequence, and despite their central role in plant biology and world agriculture, no legume genome has yet been sequenced. Both *M. truncatula* (B. Roe, unpublished) and *Lotus japonicum* (Nakamura 2002) have growing bodies of genome sequence data, while soybean has an impressive set of other genomic tools, especially expressed sequence tags (ESTs) (Shoemaker 2002).

Given the importance of legumes and the value of *M. truncatula* as a model legume, we propose to sequence nearly all of its gene space by focusing on euchromatic regions. Some of the best cytogenetic work in any plant has been performed in *M. truncatula* (Kulikova 2001). These results, based on extensive fluorescence *in situ* hybridization (FISH) analysis of pachytene chromosomes, provide detailed insight into the organization of hetero- and euchromatic regions. Therefore, at present it is believed that the genome of *M. truncatula* is organized into clearly distinct regions of pericentromeric heterochromatin, rich in repeated sequences, and gene-rich euchromatic regions interspersed with smaller heterochromatin-like repeats. However, the exact genomic organization and structure of the *M. truncatula* genome awaits additional DNA sequence data.

In this present report, we will discuss the status of our *M. truncatula* genomic sequencing and give a preliminary analysis of this data.

1.3 Genomic Resources for *M. truncatula*

Numerous genomic resources have been developed for *M. truncatula* that provide a firm foundation for our genomic sequencing. These resources include:

- An emerging physical map (~15-fold BAC) that is based on a combination of BAC fingerprinting and EST tagged sites. The clones in this map come from three distinct BAC libraries, including one library with relatively large (>170 kbp) inserts. All of the BAC clones currently being sequenced by us are anchored to this physical map.
- The sequences of more than 170,000 ESTs plus associated microarray resources (Federova 2002) publicly available.
- Two efficient transformation systems in development, in which cut flowers are infiltrated with *Agrobacterium* and sepals are cultured to produce

up to 70-100 embryos, making possible medium-throughput gene tagging protocols (Chabaud 1996; Trieu 2000; Trinh 1998).

- A gene tilling initiative already shown to be effective in discovering mutants associated with target sequences (D. Cook, UC-Davis, personal communication).
- Extensive comparative genomic data demonstrating that the *M. truncatula* genome is highly conserved with alfalfa and pea (Gautieri 2002; Endre 2002), and moderately conserved with soybean at both the macro- and micro-syntenic level (Yan 2003).
- A strong bioinformatics community, including several actively collaborating centers for database development and data-mining initiatives (Bell 2001; Lamblin 2003).
- Additional biological resources, including a growing number of defined pathosystems and candidate resistance genes (Zhu 2002), recombinant inbred populations, high-density maps (Kulikova 2001), and collections of defined mutants and natural ecotypes (Penmetsa 2000).

1.4 Benefits of the *M. truncatula* Genome Sequence

The sequence and analysis of the gene space in *M. truncatula* will not only transform genomic research in *M. truncatula* but also will benefit legume researchers. These groups, from plant biologists to plant breeders, will have access to a reference genomic sequence representing most legume-specific properties. The *M. truncatula* sequence will facilitate positional cloning in legumes based on microsynteny and detailed studies of legume-specific gene family organization and evolution. In addition, this sequence will provide insights into legume developmental and biochemical pathways and allow for the discovery of pan-legume markers. Our ongoing *M. truncatula* genomic sequencing also will provide the genomic sequence information needed for evolutionary comparison with the other plant genomes either presently underway or being contemplated.

2. Genomic DNA Sequencing

2.1 Overview of Our Sequencing Approach

The Advanced Center for Genome Technology is a high throughput DNA sequencing and research facility located in the Department of Chemistry and Biochemistry on the Norman campus of the University of Oklahoma. Through our involvement in the human genome project (Dunham 1999; Lander 2001; Waterston 2002) we have established a semi-automated DNA sequencing and analysis pipeline which incorporates robotics into all levels of this process, from preparing shotgun libraries, growing and isolating DNA sequencing templates, as well as for pipetting the DNA sequencing reactions, cycle sequencing incubation, and post-reaction clean-up prior to sequence data collection on ABI 3700 fluorescence-based capillary sequencers (Deschamps 2003; Chissoe 1995; Pan 1994). Once the DNA sequence data is collected, it is transferred to Sun computer workstations for automated base-calling with Phred (Ewing 1998a; Ewing 1998b), assembly with Phrap (Green, unpublished) and viewing and analysis with Consed (Gordon 1998) and Exgap (Hua 2003). The closure and finishing process is based on custom synthetic primer synthesis, and either direct sequence extension off the shotgun clones or on PCR generated templates.

Finished sequences then are processed through a series of Perl scripts, that automate the processes of gene prediction using Genscan+ (Burge 1998), Genemark.hmm (Lukashin 1998), Glimmer (Salzberg (1999), Fgenesh (Solovyev 1994), splice sites prediction with GeneSplicer (Pertea), NetPlantGene (Vignal 1996) and tRNA gene analysis with tRNAScan-SE (Lowe 1997), as well as GenBank and EST database searches using the Blast program suite (Altschul 1990). The results of the analysis then are made publicly available through a Genome Browser (Stein 2002) interface on our Genome Center web site at URL: http://www.genome.ou.edu/medicago.html.

2.2 Overview of Our Sequencing Approach and Initial Analysis

Originally, 25,000 whole genome shotgun (WGS) sequence reads were generated in a pilot project to test the efficiency of this approach. Much to our surprise, this data was assembled by Phrap into several contigs as shown in Table 1.

When these contigs were analyzed, we discovered that ~10% of the whole genome shotgun sequence data represented *M. truncatula* chloroplast genomic sequence that was a contaminant of the originally isolated *M. truncatula* leaf genomic DNA. The chloroplast genomic sequence subsequently was completed, analyzed, and submitted to GenBank (Accession Number AC093544). A comparison of the *M. truncatula* chloroplast genome to that of *A. thaliana* is shown in Figure 1. Here it can be seen that the overall synteny of these genomes is highly conserved except for the presence of only one copy of the ~25 Kbp chloroplast repeat sequence in the *M. truncatula* chloroplast genome and two inverted copies of this repeat sequence in the *A. thaliana* genome (Sato 1999).

Table 1. M. truncatula whole genome shotgun sequence data assembly

Contig Size	Total Number	Total Length	% of Cumulative
0 - 1 kb	2088	1435654	62.4%
1 - 2 kb	427	524409	22.8%
2 - 3 kb	29	69217	3.0%
3 - 4 kb	14	48452	2.1%
4 - 5 kb	4	18249	0.8%
5 - 10 kb	19	134757	5.9%
10 - 20 kb	6	68716	3.0%
20 - 30 kb	0	0	0.0%
30 - 40 kb	0	0	0.0%
40 - 50 kb	0	0	0.0%
50 - 100 kb	0	0	0.0%
>100 kb	0	0	0.0%
Cumulative	2587	2299454	Consed_Err/10KB = 1705.64
Cumulative>1 kb	499	863800	Consed_Err/10KB = 920.93
Cumulative>2 kb	72	339391	Consed_Err/10KB = 200.64
Phrap Coverage: 3.2, Phrap Avg. Confirmed Length: 320.1			
Confirmed Reads: 5496, Entries: 23544			

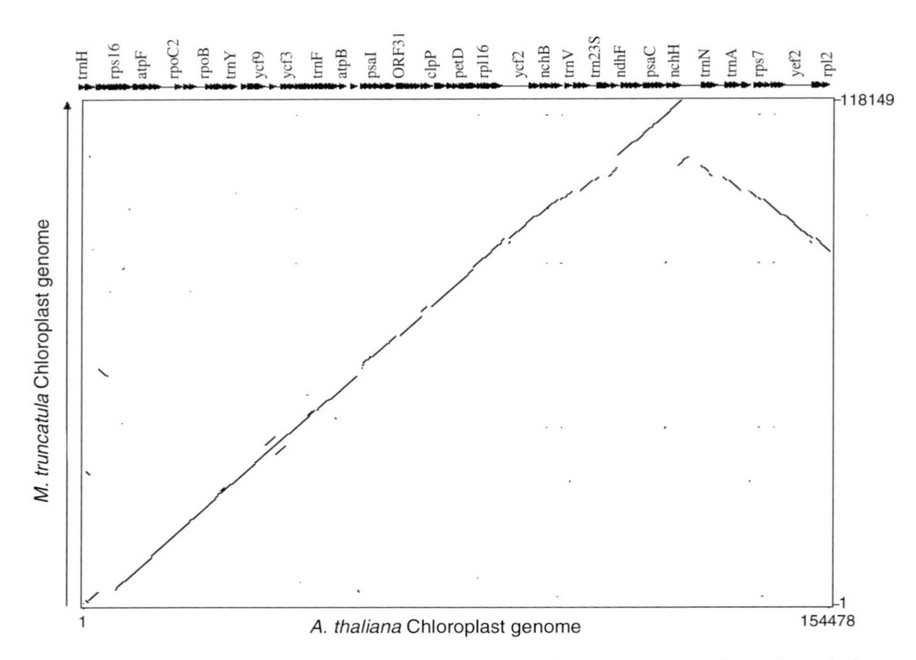

Figure 1. A dotplot comparison (Sonnhammer 1995) of the *M. truncatula* and *A. thaliana* chloroplast genomic sequences.

In addition, the initial WGS sequence data contained more than 1,000 sequence reads with similarity to *M. truncatula* ESTs and several novel *M. truncatula*-specific repeat families. These observations led us to conclude that a WGS sequencing approach should be abandoned and replaced by a BAC-based approach that would be more cost-effective in exploring the gene space of *M. truncatula*. As the physical map became more detailed, we begun to choose BACs for sequencing that were either well-spaced or extend existing sequence contigs.

As of mid-August 2003, approximately 700 BAC clones are in the pipeline at University of Oklahoma for sequencing. These BAC clones were fingerprinted and provided to us by Drs. D. J. Kim and D. Cook at the University of California, Davis. Of these 700 BACs, approximately 147 have been completely sequenced while the remaining are either in shotgun or closure-finishing phase. To date we have deposited almost 70,000 bp of BAC-based *M. truncatula* genomic sequence data into the publicly available GenBank database. The most recent statistics are available on our web site at URL www.genome.ou.edu/medicago_table.html.

Preliminary analysis of the Phase 2 and 3 clones based on Genscan analysis identified approximately 10,000 genes, suggesting a gene density within the genome region represented by these BACs of 1 gene per 6 kbp (adjusting for overlaps among sequenced BAC clones). Approximately 63% of the predicted genes show strong similarity to one or more *M. truncatula* ESTs. The G-C content is 34%, the average gene size, 3,980 bp, the average exon, 782 bp, and the average intron, 254 bp.

The goal of the current project has been to focus sequencing on 1000 BACs supplied by the UC Davis group from the gene rich euchromatic regions as defined by known biological markers and regions of biological interest. When the BAC-containing bacterial glycerol stocks are received from the UC Davis group they are individually grown in 200 ml liquid cultures and after the BAC DNA is isolated, random subclone libraries are created. Individual shotgun clones then are isolated and sequenced to approximately 7-fold coverage followed by closure and finishing. Progress on the BAC sequencing can be found at the center's website http://www.genome.ou.edu/DailySequencingProgress.html where daily updates show the number of bases found in projects in phase I (shotgun) phase II (contiguous sequences ordered and oriented) and phase III (a single contiguous sequence at high quality, <1 predicted error in 10,000 bases). Additional information is available on our *M. truncatula* specific site, http://www.genome.ou.edu/medicago.html, giving background on the project, links to other sites of interest and direct links to specific projects. In addition this page contains keyword and blast search sites allowing direct searching of all our current *M. truncatula* data which assembles into contigs of 2Kbp or greater. To aid in the annotation process, a genome browser for *M. truncatula* data has been developed and also is accessible at http://dna8.chem.ou.edu/cgi-bin/gbrowse?source=medicago_new. This is an interactive site that allows viewing of current annotation in a series of lines with a selectable menu. Options include ORF calling results from both FgeneSH and Genscan, repeat elements, both EST and cDNA matches and blast hits with Arabidopsis and the GenBank nr database. The ORFs are linked to pages with both DNA and protein sequence in fasta format and the blast homologs are linked directly to the GenBank accession.

As shown in Table 2 below, we have evaluated several conserved gene characteristics found during our examination of the *M. truncatula* genomic sequences. These features include several cis elements typically found in or near introns and exons. To accomplish these analyses, a series of scripts has been written, FELINES (Drabenstot, 2003 manuscript submitted), that creates a database of intron and exon sequences defined by alignment of ESTs with corresponding genomic regions. In addition, FELINES contains tools for examining conserved motifs in the resulting intron and exon sequence databases. We have focused on 5' and 3' splice sites and branch sites. Our initial analysis of ~50 Mbp of BAC-based genomic sequence data, as of the April 14, 2003 data freeze, contained regions with homology to 181,444 *M. truncatula* ESTs with a minimum sequence identity of 90%. As shown below in Table 2, over 12,000 exons and almost 9000 introns were identified by FELINES analysis.

Table 2. Gene characteristics in the sequenced *M. truncatula* genomic data

Feature	Total Number	Range (nt)	Mean Length (nt)
Exons	12,541	6-5789	268
Intronless Exons	1936	300-5789	680
Introns	8862	20-3921	429

The size ranges are similar to what has been determined by FgeneSH and Genscan. Interestingly, the single exons i.e. those without introns fell in the high end of the size range. Blast examination of these showed 12% with homology to retroelements (data not included).

As shown in Table 3, three subclasses of 5' splice sites were detected with the GU class being the most predominant. A population of AU introns and U12-type branch sites were detected (Table 3). U12 dependent introns sometimes have 5' AU splice elements in addition to a characteristic branch site motif and have been associated with post-transcriptional regulation (Patel 2002) in a number of eucaryotes including *A. thaliana* (Zhu 2003). From our examination, a consensus *M. truncatula* splice site motif emerged with GUAAG$(U)_{15}$ as the 5' splice site consensus sequence, $(U)_{16}$GCAG as the 3' splice site consensus sequence and UUUWUUYUR$\underline{A}$$(U)_5$ as the consensus branch site sequence.

Table 3. Conserved sequence elements in characterized *M. truncatula* introns

Intron Conserved Sequence Elements	Number	Percent
Introns with 5' GU	8792	99.12%
Introns with 5' GC	32	0.36%
Introns with 5' AU	28	0.31%
Introns with U12-type branch sites	12	0.13%

Comparison of *M. truncatula* genomic data with the completed genome of *A. thaliana (*Arabidopsis Genome Initiative 2000), and other sequenced genomes, will be extremely valuable for identification of conserved genome elements.

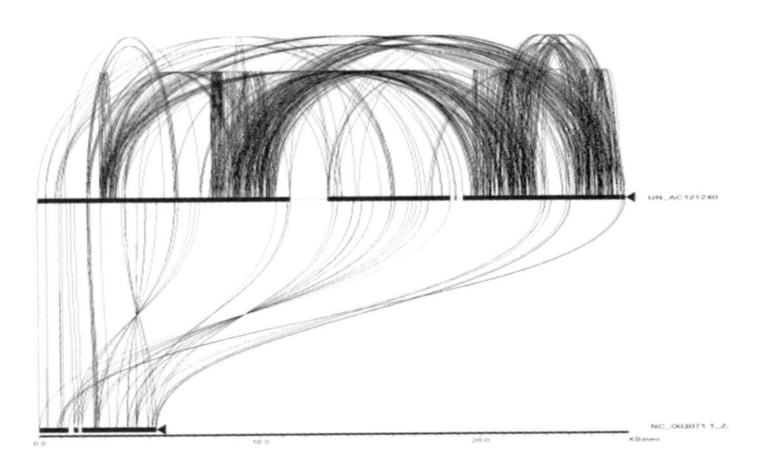

Figure 2. Printrepeats (Parsons 1995) comparison of syntenic regions of the *M. truncatula (upper panel)* and *A. thaliana (lower panel) genomes.*

Figure 2 shows such a comparison of a region of Chromosome 2 from *A. thaliana* with the syntenic region from *M. truncatula* (AC121240). Two genomic features immediately can be seen from this comparison. First, there is reasonable sequence conservation between the two genomes as indicated by the connecting lines connecting the two genomic regions. Second, the *M. truncatula* genome clearly has a ~5-fold expansion in this region, much of which is due to the presence of additional repeated sequences as indicated by the looping lines connecting several regions of the *M. truncatula* genomic region.

3. SUMMARY AND CONCLUSIONS

With funding from the Noble Foundation, we are well on our way to complete the working draft sequence of ~1000 mapped *M. truncatula* BAC colones and to finish a significant number of these BACs with additional funding from the DOE. Funding recently has been obtained from the NSF to complete the genomic sequence of 6 *M. truncatula* chromosomes as a collaborative effort involving our group in Oklahoma with groups at TIGR and the University of Minnesota. Additional funding to complete the 2 remaining chromosomes presently is anticipated from two European groups, making this truly an international effort to complete the gene rich regions of this model legume species in the near future.

ACKNOWLEDGMENTS

We thank our entire staff for their contributions to our *M. truncatula* project, with a special thanks to S. Lin, Y. Fu, and F. Z. Najar for leading the sequencing effort, and H. Lai, S. Kenton, J.D. White, and S. Drabenstot for their informatics help. We also are grateful for grant support for this project from the Noble Foundation, Ardmore, OK.

REFERENCES

Altschul SF, Gish W, Miller W, Myers EW, Lipman DJ (1990) Basic local alignment search tool. J. Mol. Biol. 215: 403-10.

Arabidopsis Genome Initiative. Analysis of the genome sequence of the flowering plant *Arabidopsis thaliana*. Nature (2000) 408: 796-815

Bell CJ, Dixon RA, Farmer AD, Flores R, Inman J, Gonzales RA, Harrison MJ, Paiva NL, Scott AD, Weller JW, May GD (2001) The *Medicago* Genome Initiative: a model legume database. Nucl. Acids Res. 29: 114-117.

Burge C, Karlin S (1997) Prediction of complete gene structures in human genomic DNA. J. Mol. Biol. 268: 78-94.

Chabaud M, Larsonneau C, Marmouget C, Huguet T (1996) Transformation of barrel medic (*Medicago truncatula* Gaertn.) by *Agrobacterium tumefaciens* and regeneration via somatic embryogenesis of transgenic plants with the MtENOD12 nodulin promoter fused to the *gus* reporter gene. Plant Cell. Rep. 15: 305-310.

Chissoe SL, Bodenteich A, Wang YF, Wang YP, Burian D, Clifton SW, Crabtree J, Freeman A, Iyer I, Jian L, Ma Y, McLaury HJ, Pan HQ, Sarhan OH, Toth S, Wang Z, Zhang G, Heisterkamp N, Groffen J, Roe BA (1995) Sequence and Analysis of the Human c-abl Gene, the bcr Gene, and Regions Involved in the Philadelphia Chromosomal Translocation. Genomics 27: 67-82.

Conner JA, Conner P, Nasrallah ME, Nasrallah JB (1998) Comparative mapping of the Brassica S locus region and its homeolog in *Arabidopsis*. Implications for the evolution of mating systems in the Brassicaceae. Plant Cell. 10: 801-812.

Cook DR (1999) *Medicago truncatula* – a model in the making! Curr. Opin. Plant Biol. 2: 301-304.

Deschamps S, Meyer J, Chatterjee G, Wang H, Lengyel P, Roe BA (2003) The mouse Ifi200 gene cluster: genomic sequence, analysis, and comparison with the human HIN-200 gene cluster. Genomics 82: 34-46.

Doyle JJ. (2001) Leguminosae. In Encyclopedia of Genetics. Brenner S, Miller JH (eds.) pp 1081-1085. Academic Press, San Diego.

Drabenstot SD, Kupfer DM, Buchanan K, Dyer D, Roe BA and Murphy DW (2003) A scheme for finding introns and exons and examining their structures (submitted).

Dunham I, Shimizu N, Roe BA, Chissoe S, et al. (1999) The DNA sequence of human chromosome 22. Nature 402: 489-495.

Endre G, Kereszt A, Kevei Z, Mihacea S, Kalo P, Kiss GB (2002) A receptor kinase gene regulating symbiotic nodule development. Nature 417: 962-966.

Ewing B, Hillier L, Wendl M, Green P (1998a) Basecalling of automated sequencer traces using phred. I. Accuracy assessment. Genome Res. 8: 175-185.

Ewing B, Green P. (1998b) Basecalling of automated sequencer traces using phred. II. Error probabilities. Genome Res. 8: 186-194.

Fedorova M, van de Mortel J, Matsumoto PA, Cho J, Town CD, VandenBosch KA, Gantt JS, Vance CP (2002) Genome-wide identification of nodule-specific transcripts in the model legume Medicago truncatula. Plant Physiol. 130: 519-537.

Foster-Hartnett D, Mudge J, Larsen D, Danesh D, Yan HH, Denny R, Penuela S, Young ND (2002) Comparative genomic analysis of sequences sampled from a small region on soybean (*Glycine max*) molecular linkage group G. Genome 45: 634-645.

Gordon, D, Abajian C, Green P (1998) Consed: A graphical tool for sequence finishing. Genome Res. 8: 195-202.

Gualtieri G, Kulikova, O, Limpens E, Kim DJ, Cook DR, Bisseling T, Geurts R (2002) Microsynteny between pea and *Medicago truncatula* in the SYM2 region. Plant Mol. Biol. 50: 225-235.

Gualtieri G, Kulikova, O, Limpens E, Kim DJ, Cook DR, Bisseling T, Geurts R (2002) Microsynteny between pea and Medicago truncatula in the SYM2 region. Plant Mol. Biol. 50: 225-235.

Hua A, Roe BA (2003) Exgap: a visualization of shotgun sequencing assembly results (manuscript in preparation).

Kulikova O, Gualtieri G, Geurts R, Kim DJ, Cook DR, Huguet T, de Jong JH, Fransz PF, Bisseling T. (2001) Integration of the FISH pachytene and genetic maps of *Medicago truncatula*. Plant J. 27: 49-58.

Kulikova O, Gualtieri G, Geurts R, Kim DJ, Cook DR, Huguet T, de Jong JH, Fransz PF, Bisseling T. (2001) Integration of the FISH pachytene and genetic maps of *Medicago truncatula*. Plant J. 27: 49-58.

Lamblin A. Crow J, Johnson J, Silverstein K, Kunau T, Kilian A, Benz D, Stromvik M, Endre G, VandenBosch K, Cook DR., Young ND, Retzel E (2003) MtDB: A database for personalized data mining of the model legume Medicago truncatula. Nucl. Acids Res. 31: 196-201.

Lander ES, Rogers J, Waterson RH, Hawkins T, Gibbs R, Sakaki Y, Weissenbach J, et al. (2001) Initial Sequencing and analysis of the human genome. Nature 409: 860-921.

Lowe TM, Eddy SR (1997) tRNAscan-SE: a program for improved detection of transfer RNA genes in genomic sequence. Nucl. Acids Res. 25: 955-64.

Lukashin AV, Borodovsky M (1998) GeneMark.hmm: new solutions for gene finding. Nucl. Acids Res. 26: 1107-15.

Nakamura Y, Kaneko T, Asamizu E, Kato T, Sato S, Tabata S (2002) Structural analysis of a *Lotus japonicus* genome. II. Sequence features and mapping of sixty-five TAC clones which cover the 6.5-Mb regions of the genome. DNA Res. 9: 63-70.

Pan HQ, Wang YP, Chissoe SL, Bodenteich A, Wang Z, Iyer K, Clifton SW, Crabtree JS, Roe BA (1994) The complete nucleotide sequences of the pSacBII P1 cloning vector and three cosmid cloning vectors: pTCF, svPHEP, and LAWRIST16. Genetic Analysis Techniques and Applications 11: 181-186.

Patel AA, McCarthy M, Steitz JA (2002) The splicing of U12-type introns can be a rate-limiting step in gene expression. EMBO J. 21: 3804-15.

Parsons, JD (1995) Miropeats: graphical DNA sequence comparisons, Comput. Applic. Biosci. 11: 615-619.

Penmetsa RV, Cook DR (2000) Production and characterization of diverse developmental mutants of *Medicago truncatula.* Plant Physiol. 123: 1387-1398.

Pertea M, Lin X, Salzberg SL (2001) GeneSplicer: a new computational method for splice site prediction. Nucl. Acids Res. 25: 1185-90.

Salzberg SL, Pertea M, Delcher AL, Gardner MJ, Tettelin H (1999) Interpolated Markov models for eukaryotic gene finding. Genomics. 59: 24-31.

Sato S, Nakamura Y, Kaneko T, Asamizu E, Tabata S (1999) Complete structure of the chloroplast genome of Arabidopsis thaliana. DNA Res. 6: 283-290.

Shoemaker R, Keim P, Vodkin L, Retzel E, Clifton SW, Waterston R, et al. (2002) A compilation of soybean ESTs: generation and analysis. Genome 45: 329-338.

Solovyev VV, Salamov AA, Lawrence CB (1994) Predicting internal exons by oligonucleotide composition and discriminant analysis of spliceable open reading frames. Nucl. Acids Res. 22: 5156-5163.

Sonnhammer EL, Durbin R (1995) A dot-matrix program with dynamic threshold control suited for genomic DNA and protein sequence analysis. Gene. 167: GC1-10.

Stein LD, Mungall C, Shu S, Caudy M, Mangone M, Day A, Nickerson E, Stajich JE, Harris TW, Arva A, Lewis S (2002) The generic genome browser: a building block for a model organism system database. Genome Res. 12: 1599-610.

Trieu AT, Burleigh SH, Kardailsky IV, Maldonado-Mendoza IE, Versaw WK, Blaylock LA, Shin H, Chiou T-J, Katagi H, Dewbre GR, Weigel D, Harrison MJ (2000) Transformation of Medicago truncatula via infiltration of seedlings or flowering plants with *Agrobacterium.* Plant J. 22: 531-541.

Trinh TH, Ratet P, Kondorosi E, Durand P, Kamaté K, Bauer P, Kondorosi A (1998) Rapid and efficient transformation of diploid *Medicago truncatula* and *Medicago sativa* ssp. falcata lines improved in somatic embryogenesis. Plant Cell. Rep. 17: 345-355.

Vazquez MM. Barea JM. Azcon R (2002) Influence of arbuscular mycorrhizae and a genetically modified strain of *Sinorhizobium* on growth, nitrate reductase activity and protein content in shoots and roots of *Medicago sativa* as affected by nitrogen concentrations. Soil Biol. Bioc. 34: 899-905.

Vignal L, d'Aubenton-Carafa Y, Lisacek F, Mephu Nguifo E, Rouze P, Quinqueton J, Thermes C (1996) Exon prediction in eucaryotic genomes. Biochimie. 78: 327-334.

Waterston RH, Lindblad-Toh K, Birney E, Rogers J, et al. (2002) Initial sequencing and comparative analysis of the mouse genome. Nature 420: 520-562.

Yan H, Mudge J, Kim DJ, Shoemaker RC, Cook DR, Young ND (2003) Estimates of conserved microsynteny among the genomes of *Glycine max, Medicago truncatula* and *Arabidopsis thaliana.* Theor. Appl. Genet. 106: 1256-1265.

Zhu HY Cannon SB Young ND Cook DR (2002) Phylogeny and genomic organization of the TIR and non-TIR NBS-LRR resistance gene family in *Medicago truncatula*. Mol. Plant-Microbe Interac. 15: 529-539.

Zhu W, Brendel V (2003) Identification, characterization and molecular phylogeny of U12-dependent introns in the Arabidopsis thaliana genome. Nucl. Acids Res. 31: 4561-72.

Agricultural Biotechnology and Environmental Risks: A Program Perspective

D. L. Sheely
U.S. Department of Agriculture, Cooperative State Research, Education, and Extension Service, Washington, DC 20250 USA. (Email: dsheely@csrees.usda.gov).

Key words: agricultural biotechnology, risk assessment, risk management, transgene escape, unintended effects, resistance evolution, viral recombination, regulation

Abstract:

The application of biotechnology to agricultural problems holds great promise and inspires lively debate. The promise of agricultural biotechnology includes crops that resist pests and diseases, thrive in spite of environmental stresses like drought or salinity, and provide increased nutritional value. The debate encompasses both biological and societal questions, many of which are not easily answered. This paper will discuss some of the more commonly cited environmental concerns associated with transgenic plants and animals. They are grouped into five categories: movement of transgenes, unintended effects, viral recombination, evolution of resistant pests, and whole organism characteristics. Since 1992, USDA has supported research to characterize and assess possible risks associated with the release of transgenic organisms into the environment. A goal of this research is to assist Federal regulatory agencies in making science-based decisions regarding biotechnology. A review of grants supported by USDA's Biotechnology Risk Assessment Research Grants Program shows how research to address environmental concerns has evolved over time.

A. Hopkins, Z.Y. Wang, R. Mian, M. Sledge and R.E. Barker (eds.), Molecular Breeding of Forage and Turf, 345-358.

1. INTRODUCTION

The application of biotechnology to agricultural problems holds great promise and inspires lively debate. The promise of agricultural biotechnology includes crops that resist pests and diseases, thrive in spite of environmental stresses like drought or salinity, and provide increased nutritional value. Since the deregulation of the FlavrSavr tomato in 1994, a growing number of genetically engineered crop plants have been approved for field release. These include crops with enhanced agronomic traits, such as herbicide tolerant corn, soybean, and cotton, insect-resistant corn, cotton, and potato, and virus resistant squash and papaya. A smaller number of crops with value-enhanced traits have also been approved. Among these are rapeseed with high-lauric-acid oil, and high-oleic-acid producing soybean, peanut, and sunflower. Despite some uncertainty about consumer acceptance of these foods, farmers have adopted them at a rapid rate. Herbicide-tolerant soybeans, which made up almost 70% of soybean acreage in 2001, have been commercially available only since 1996 (Figure 1). Likewise, the adoption of herbicide-tolerant cotton grew from 2% in 1996 to 31% in 2001. Insect-resistant crops--those that contain a gene from *Bacillus thuringiensis* (Bt)--also have shown significant growth in adoption during the years 1996 through 2001.

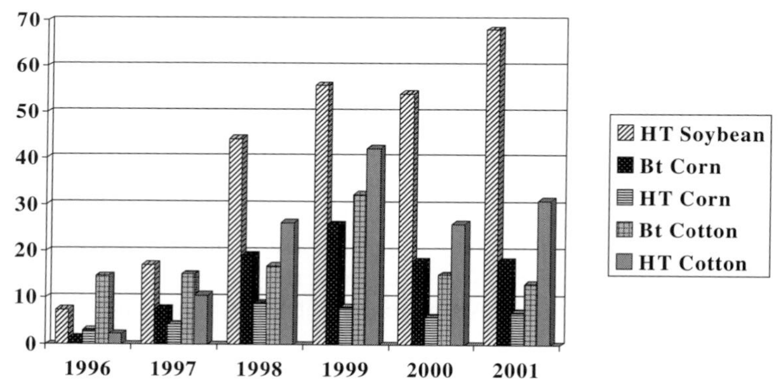

Figure 1. Percent of acres planted to herbicide tolerant and Bt crops. Source of Data: USDA Economic Research Service.

The debate on agricultural biotechnology encompasses both societal and biological issues, many of which are not easily resolved. A number of authors have discussed these issues (see Colwell 1985; Tiedje 1989; Paoletti and Pimentel 1996; Snow and Palma 1997; Stewart et al. 2000; and Wolfenbarger and Phifer 2000, for examples, as well as www.riskassess.org), and this paper will introduce a number of them. Societal issues include questions of consumer choice, legal questions about gene patenting and ownership, moral questions about transferring genes between species, and economic considerations such as international trade. Consumer acceptance of agricultural biotechnology varies around the world. Consumers in the United States have voiced little opposition to transgenic foods (Hoban 1998). This differs from consumers in Europe, who are more likely than Americans to view genetically modified crops as risky (Hoban 1997). High profile individuals have questioned the morality of using biotechnology for other than medical applications (The Prince of Wales 1996), and European countries have refused to buy transgenic crops since 1998. This practice was challenged recently by the United States in a lawsuit filed with the World Trade Organization (Washington Post 2003).

Biological questions focus around a small number of issues. An important question is whether transgenic food and feed are detrimental to human and animal health. This concern is based on fears of increased toxicity or allergenicity resulting from the introduction of one or more new genes. Risks to agricultural sustainability and possible risks to the environment are also acknowledged. For example, widespread adoption of herbicide tolerant crops may lead to increased herbicide use or to changes in production practices that harm non-target species.

Commonly cited environmental concerns include the transfer of genes to conspecifics or to wild relatives, increased fitness of transgenic organisms, unintended effects to non-target organisms and the environment, the evolution of new viral diseases, and the evolution of resistant pests. Transgene movement from a crop to wild or weedy relatives is a risk factor frequently discussed in both scientific and lay publications. Of primary concern is the transfer of a gene conferring a beneficial trait, such as herbicide resistance, and the possible creation of a superweed that could be either difficult or impossible to control. Transfer of such a gene to conspecifics grown for the organic market is also undesirable and potentially costly. The increased fitness of transgenic organisms is also cited as a possible risk to native populations. For example, salmon expressing a gene for increased growth hormone production will grow faster and larger than native fish. If these transgenic salmon were to escape confinement and enter natural environments, they could possibly outcompete native fish populations for food and mates. Over time, this could result in the decline or possible loss of native populations.

Unintended effects of a transgenic organism will include both direct and indirect non-target effects. Direct non-target effects may be experienced by susceptible beneficial or native organisms feeding on all or part of a Bt crop plant. For example, monarch butterfly larvae feeding on milkweed plants dusted with pollen from Bt corn may suffer direct effects on growth and survival. Indirect non-target effects may be experienced by species that feed on the pests or beneficial organisms controlled by Bt. Indirect effects may also be experienced by herbivorous insects or seed-eating birds that feed on weed species no longer found in fields producing herbicide-tolerant crops.

Evolutionary factors also contribute to environmental risks. Control of viral diseases through plant-expressed viral coat proteins provides an opportunity for evolution of new viruses. This may occur via recombination of a closely related viral pathogen with a plant-expressed coat protein gene. The possibility that new viruses may be more virulent or more infective than either parent virus is a possible risk. The evolution of resistant pests is also of concern, particularly with regard to the use of Bt to control pests. Bt may be used as a spray in organic production, or as a continuously expressed trait in transgenic crops. Losing Bt for control of pests may result in negative ecological consequences if other, less environmentally benign, pesticides are employed in the future.

The safe release and utilization of genetically modified plants, animals, and microbes is ensured by the U.S. Department of Agriculture Animal and Plant Health Inspection Service (APHIS), the Environmental Protection Agency (EPA), and the Food and Drug Administration (FDA). These three agencies coordinate to provide regulatory control over the products of biotechnology. APHIS regulates plant pests, plants, and veterinary biologics, and determines whether they are safe to grow, and safe for the environment. APHIS operates under authority of the Federal Plant Pest Act. EPA uses the authority of the Federal Insecticide, Fungicide, and Rodenticide Act to regulate plants and microbes producing pesticidal substances, and to regulate new uses of existing pesticides. In addition, EPA regulates novel microorganisms under the Toxic Substances Control Act. Its primary focus is on whether these substances are safe for the environment. EPA also oversees pesticidal substances on and in food and feed under the Federal Food, Drug, and Cosmetic Act (FFDCA). Similarly, FDA uses the authority of the FFDCA to regulate foods and feed derived from new plant varieties to determine whether they are safe to eat.

U.S. regulatory agencies are committed to science-based regulation of biotechnology. High quality research is critical to providing the data they need to carry out their responsibilities. This paper focuses on eleven years of environmental risk assessment research supported by the U.S. Department of Agriculture (USDA). The competitive grants program that supported the research is described below. This is followed by an analysis of the research supported by this program, showing how the focus of funded research has changed over time.

2. BIOTECHNOLOGY RISK ASSESSMENT RESEARCH GRANTS PROGRAM

2.1 Overview

Since 1992, USDA's Biotechnology Risk Assessment Research Grants Program (BRARGP) has been an important source of funding for research to address questions of regulatory interest. The program was authorized in Section 1668 of the 1990 Farm Bill to support environmental assessment research concerning the introduction of genetically-engineered plants, animals, and microbes into the environment. This competitive grants program is jointly administered by the Cooperative State Research, Education, and Extension Service (CSREES) and the Agricultural Research Service (ARS). The purpose of the BRARGP is to support research that will assist Federal regulatory agencies in making science-based regulatory decisions. To help achieve that goal, the program routinely invites members of the Federal regulatory agencies to serve on its peer review panels and to provide input into the annual Request for Applications (RFA). In the past, panelists from APHIS and EPA have provided valuable advice regarding the relevance and quality of submitted proposals. Scientists from academic research institutions also serve on peer panels to achieve an evaluation process that balances scientific rigor with regulatory relevance in selecting proposals for funding. Until 2003, the program's stated focus was on risk assessment research, and it did not support risk management studies. However, in the 2002 Farm Bill Congress expanded the scope of the program to include two new areas. These are: i) research to identify and develop management techniques to minimize physical and biological risks of transgenic organisms; and ii) research to develop methods to monitor the dispersal of transgenic organisms. Until Fiscal Year (FY) 2003, BRARGP was funded by a one percent set-aside from outlays of the Department of Agriculture for research in biotechnology. The 2002 Farm bill increased the set-aside to two percent. A complete listing of all studies supported by this program, organized by award year, may be found on the program's website (www.reeusda.gov/crgam/biotechrisk/biotech.htm).

Table 1 provides data on number, size, and duration of studies supported in Fiscal Years 1992 through 2002. In total, $17.6 million supported 98 studies. A small number of conferences were also supported during this eleven-year period, but are not included in this analysis. The funds available for research grants ranged from a low of $1.3 million in 1999 to a high of $2.1 million in 2001. The program achieved its largest funding levels in FY 2000 and 2001. BRARGP awarded between 7 and 10 new grants each year. Average award sizes ranged from a low of $139,424 in 1998 to a high of $234,039 in 2001, with an average over all years of $179,646. The duration of awards ranged from a low of approximately 29 months in both 1994 and 1996 to a high of 36 months in 1998, 2000, and 2001. Average award duration over all years is nearly 33 months. The average level of support provided to investigators per year is one indicator of the size and scope of studies selected for support. In this program, an average of $65,569 per year per award was provided.

Table 1. USDA Biotechnology Risk Assessment Research Grants Program award number, size, and duration for the years 1992 – 2002 (Values reported in this table do not include conference grants awarded in 1993, 2000, and 2001)

Year	Total No. of awards	Total award dollars (Millions)	Average award size ($)	Average award duration (Months)	Average $ per year
1992	8	$ 1.4	175,625	31.5	66,905
1993	11	1.6	149,396	31.6	56,667
1994	10	1.7	169,785	28.8	70,744
1995	10	1.6	159,083	32.4	58,920
1996	10	1.5	147,583	28.8	61,493
1997	10	1.5	153,679	34.8	52,993
1998	7	1.4	139,424	36.0	66,392
1999	7	1.3	192,307	34.3	67,308
2000	10	1.9	205,600	36.0	68,533
2001	10	2.1	234,039	36.0	78,013
2002	7	1.6	222,638	33.4	66,601
1992 – 2002	**98**	**$17.6**	**179,646**	**32.9**	**65,569**

2.2 Analysis of Funded Research

Over the years, the BRARGP has supported research in a number of environmental risk areas. For this paper, the 98 studies funded by the program between the years 1992 and 2002 are organized into the following five risk categories: 1) transgene movement; 2) unintended effects; 3) viral recombination; 4) evolution of resistant pests; and 5) whole organism characteristics. Each study is further classified according to the organism type that was the focus of the research. The 98 studies supported research on microbes (bacteria, fungi, and viruses), plants, fish, insects, and birds. Tables

2 and 3 show number of grants and total dollars awarded to support research in each risk category and for each focus organism.

Table 2. Number of studies funded by the Biotechnology Risk Assessment Research Grants Program during 1992 – 2002, shown by risk category and focus organism

	Microbes	Plants	Insects	Fish	Birds	Total
Transgene Movement	7	23	--	--	--	30
Unintended Effects	10	1	3	--	1	15
Viral Recombination	20	--	--	--	--	20
Resistant Pests	--	--	8	--	--	8
Whole Organism	14	2	2	7	--	25
Total	51	26	13	7	1	98

Table 3. Funding (in millions of dollars) provided for research by the Biotechnology Risk Assessment Research Grants Program during 1992 – 2002, shown by risk category and focus organism

	Microbes	Plants	Insects	Fish	Birds	Total
Transgene Movement	$1.29	$4.13	--	--	--	$5.42
Unintended Effects	1.75	0.18	$0.68	--	$0.27	2.88
Viral Recombination	3.50	--	--	--	--	3.50
Resistant Pests	--	--	1.82	--	--	1.82
Whole Organism	2.23	0.34	0.33	$1.09	--	3.99
Total	$8.77	$4.65	$2.83	$1.09	$0.27	$17.6

2.2.1 Results of Analysis by Risk Category

Transgene Movement. In total, 30 studies were supported across all years to evaluate the probability and consequences of genetic exchange between transgenic and native or wild organisms (Table 2). More than two-thirds of those studies focused on exchange between plant species, with the remaining studies addressing microbial exchange. This risk category garnered the largest amount of funding over the years, with $5.42 million, or nearly 31% of total program funds devoted to addressing this issue (Table 3). Interest in this area of research has remained consistently high over time, especially with regard to gene flow in crops. No new studies of microbial genetic exchange have been supported since 1998 (Figure 2). Microbially-focused grants have been nearly equally divided between studies of bacteria and viruses. Plant-focused studies centered on a number of different plants including alfalfa, canola, corn, creeping bentgrass, cucurbits, rice, sorghum, strawberry, and sunflower.

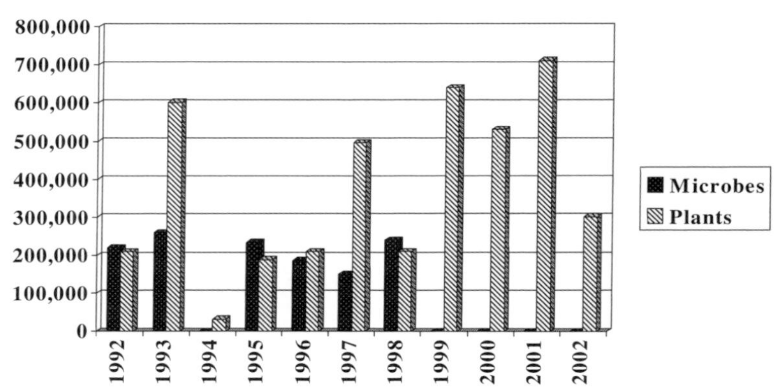

Figure 2. Dollars allocated to research on transgene movement shown by year and focus organism.

Unintended Effects. In total, 15 studies were supported across all years to assess direct and indirect non-target effects (Table 2). Ten studies assessed direct effects of a transgenic organism on a non-target, and five focused on indirect effects. Eight of the ten direct effects studies assessed the impact of a transgenic microbe on non-target microbial communities (4 studies), or the impact of a transgenic plant on microbial communities (4 studies). Two of the direct effects studies focused on the effects of transgenic plants on non-target soil dwelling insects. Of the five indirect effects studies, two focused on the effects of transgenic microbes on non-target microbial communities. The remainder assessed the impact of a transgenic plant on either a non-target microbe (1 study), plant (1 study), or bird (1 study) population or community. In total, more than 60% of dollars devoted to studies of unintended effects were focused on microorganisms (Table 3 and Figure 3). In addition, approximately 68% of all dollars spent for unintended effects research were awarded in the last three years of grants included in this study (Figure 3). Nearly all of the early studies supported in this category examined direct effects of a transgenic microbe or plant on a non-target microbe (Figures 3 and 4). This is in contrast to the studies supported in later years, which were evenly divided between those assessing direct and indirect effects (Figure 4). The more recent studies overwhelmingly focused on the effects of a transgenic plant on other organisms (microbes, insects, and birds). Overall, 16% of program funds were spent to address this risk category.

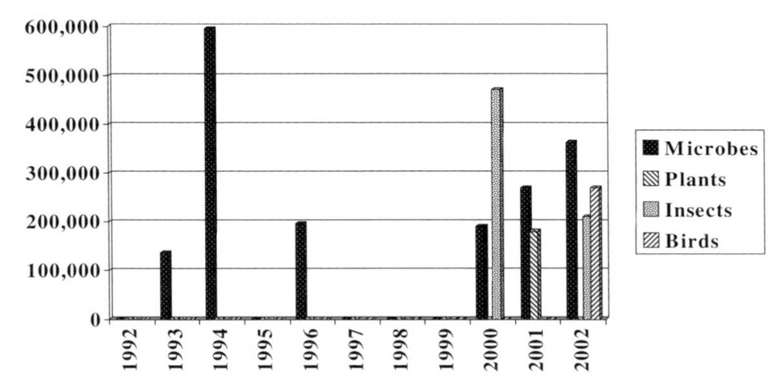

Figure 3. Dollars allocated to research on unintended effects shown by year and focus organism.

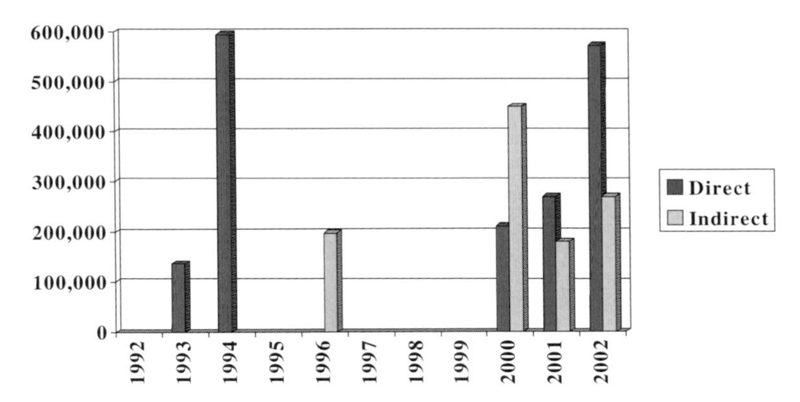

Figure 4. Dollars allocated to research on direct and indirect effects of transgenic organisms to non-target organisms.

Viral Recombination. Twenty studies were supported across all years to assess the probability and consequences of viral recombination (Table 2). Interest in this area of research was common during the early years of the program. It has begun to decrease in recent years (Figure 5). The primary subjects of these studies were plant viruses such as brome mosaic virus, papaya ringspot virus, tobacco mosaic virus and cucumber mosaic virus. Two studies on pseudorabies virus were also supported. With total funding of $3.5 million, this research category garnered approximately 20% of program funds over the past 11 years (Table 3).

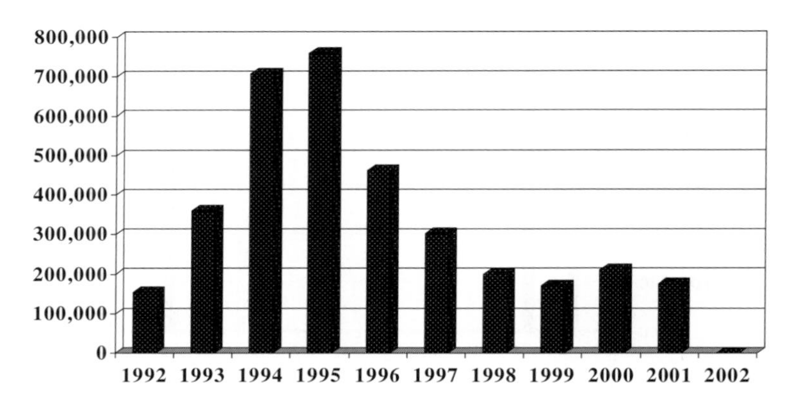

Figure 5. Dollars allocated to research on viral recombination shown by year. Viruses are the only focus organism in this risk category.

Evolution of Resistant Pests. Eight grants were awarded to assess the evolution of resistance by pests to Bt crops (Table 2). Although two early studies were funded in 1992 and 1994, six grants were awarded since 1999, indicating a relatively recent interest among researchers and regulatory agencies in this area of science (Figure 6). Studies focused on corn rootworm, tobacco budworm, bollworms, colorado potato beetle, and diamondback moth. With total funding of $1.82 million, this category of research received 10% of total program funds (Table 3).

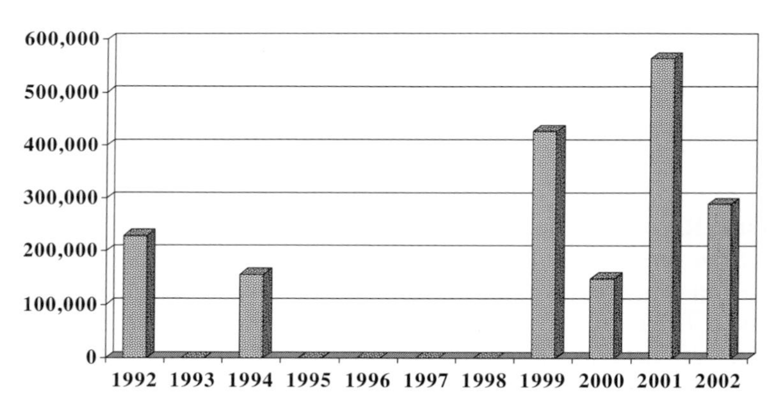

Figure 6. Dollars allocated to research on the evolution of resistant pests shown by year. Insects are the only focus organism in this risk category.

Whole Organism Characteristics. In total, 25 studies addressed whole organism and/or population characteristics (Table 2). This category includes all studies that do not fit the other four categories and includes research topics such as transgene expression, fitness, dispersal, survival, competitive advantage, behavioral fitness traits, and viral infectivity and transmission rates. Also included in this category are studies that test reproductive limitation strategies in fish and shellfish, assess methods for monitoring released fungal strains, and test models to estimate risk. This category contains $3.99 million in supported research, representing nearly 23% of total program funds (Table 3). More than half of all studies in this category focused on microorganisms, which were followed by studies on fish and shellfish, and then by plants and insects. Dollars allotted to these studies followed the same pattern. All supported studies that focused on fish are in this category. The microbial studies focused on bacteria and fungi in nearly equal numbers, with slightly fewer studies on viruses. Due to the diverse nature of the research placed in this category, no patterns can be seen over time in levels of funding or in funding by organism (Figure 7).

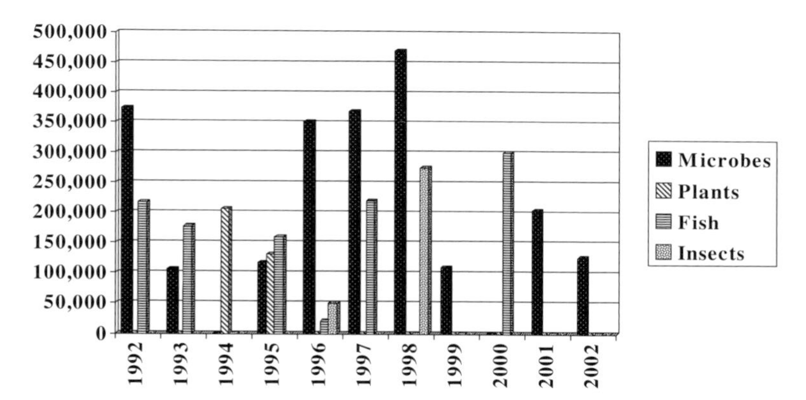

Figure 7. Dollars allocated to research on whole organism characteristics shown by year and focus organism.

2.2.2 Results of Analysis by Organism

It is useful to characterize these studies by grouping them according to the organism that received the focus of the research. Statistics on organism type discussed in this paper heretofore are based on the focus organism. Accordingly, Tables 2 and 3 show that half of all studies and program dollars were focused on microbial research. Plant-focused studies represent nearly a third of studies and dollars, with insects, fish, and birds following.

Another way to characterize these studies is to group them by risk organism. This organism is the cause of the possible risk and is, therefore, the subject of the risk assessment by the regulatory agency. Table 4 shows that more than 60% of studies and nearly two-thirds of research dollars were spent assessing risks associated with the release of transgenic plants. Twenty-five percent of dollars were spent to address possible risks associated with transgenic microbes, with much smaller values associated with fish, insects, and birds.

Table 4. Number of studies and total funding (in millions of dollars) provided, shown by risk organism.

	# of Studies	Total Funding ($ million)
Microbes	27	4.39
Plants	61	11.65
Fish	7	1.09
Insects	3	0.47
Birds	--	--
Total	98	17.6

3. DISCUSSION AND CONCLUSIONS

Analysis of studies by risk category and focus organism reveals patterns in the kinds of research supported by the BRARGP over time. Since 1992, studies of transgene movement and whole organism characteristics have received consistent support. They have also received the largest share of dollars awarded through the program over time. Studies to assess viral recombination also were of interest during the early years of the program, peaking in frequency in 1995 and tapering off in later years. These studies received the third largest share of funds. A small number of studies were supported in the early 1990's to examine unintended effects and the evolution of resistant pests, but the vast majority of studies to examine those possible risks were supported in recent years. In the early years of the program, studies addressing unintended effects focused almost exclusively on direct effects, with a rise in interest on indirect effects seen in later years.

These analyses also show patterns with regard to how support was apportioned to studies by focus organism, and how that has changed over time. Having received nearly half of all program dollars over the years, microbially-focused studies have been highly supported throughout much of the program's life. However, their frequency has begun to decline in recent years. In contrast, studies focused on insect species were rare in the early years of the program and have been increasing since 1998. This coincides with the rising interest in studies to assess the evolution of resistant pests and with increasing concerns about unintended effects on insects associated with transgenic plants. Research focused on plants and fish has received consistent funding over the years, with plant-focused studies exceeded only by microbes in frequency.

These results suggest a number of conclusions with regard to research needs and interests. As the number of deregulated genetically-modified plants grows, there continues to be a need for research on gene transfer, unintended effects, and the evolution of resistant pests, especially with regard to pesticide-producing and herbicide-tolerant plants. Indeed, one of the most commonly cited environmental concerns associated with transgenic crops is that of gene flow and the creation of superweeds. That concern is not likely to change in the near future, as agribusiness firms continue to put herbicide-tolerant genes into larger numbers of crop species, and begin to stack transgenes for herbicide, pest, and disease resistance. Likewise, the move toward engineering crop plants to produce industrial or pharmaceutical products is certain to sustain concerns about gene flow and unintended effects while increasing worries about food safety. Research to understand possible risks associated with the use of transgenic animals will grow in importance as development of these animals moves forward. A recent study by the National Research Council (NRC, 2002) examined a number of concerns associated with animal biotechnology, and recommended that environmental issues receive the highest priority for attention. Of particular concern is the environmental release of transgenic fish and shellfish. This is due to the difficulty of confining them and to the high probability of their escape into natural ecosystems.

The BRARGP was reauthorized in the 2000 Farm Bill with an expanded scope and a doubled set-aside for funding. Beginning with the current year's competition, the program will support research to develop risk management and monitoring techniques in addition to the risk assessment research it has traditionally supported. For this program to have maximum impact, it must fund research that addresses the questions regulatory agencies face today. What's more, it must attempt to identify the questions that will emerge as new and unique products of biotechnology are submitted for regulatory review tomorrow. A recent USDA workshop, "Future Directions and Research Priorities for the USDA Biotechnology Risk Assessment Research Grants Program" (see www.isb.vt.edu for more information) addressed this challenge by bringing together scientists from diverse disciplines and government agencies to discuss future research needs.

ACKNOWLEDGEMENTS

Many thanks to Gary Cunningham, Deb Hamernik, and Dan Jones for comments on an earlier version of this paper.

REFERENCES

Colwell RK, Norse EA, Pimentel D, Sharples FE, Simberloff D (1985) Genetic engineering in agriculture. Science 229: 111-112.

Hoban, TJ (1997) Consumer acceptance of biotechnology: an international perspective. Nature Biotechnology 15: 232-234.

Hoban, TJ (1998) Trends in consumer attitudes about agricultural biotechnology. AgBioForum Vol 1 No 1: 3-7.

National Research Council (2002) Animal biotechnology: science based concerns. The National Academies Press, Washington, DC.

Paoletti MG, Pimentel D (1996) Genetic engineering in agriculture and the environment. BioScience 46: 665-673.

Snow, AA, Palma PM (1997) Commercialization of transgenic plants: potential ecological risks. BioScience 47: 86-96.

Stewart CN, Richards HA, Halfhill MD (2000) Transgenic plants and biosafety: science, misconceptions and public perceptions. BioTechniques 29: 832-843.

Tiedje JM, Colwell RK, Grossman YL, Hodson RE, Lenski RE, Mack RN, Regal PJ (1989) The planned introduction of genetically engineered organisms: ecological considerations and recommendations. Ecology 70: 298-315.

The Prince of Wales (1996) The 50th anniversary of The Soil Association. www.princeofwales.gov.uk/speeches

Washington Post (2003) U.S. Attacks European Biotech Ban. May 14, 2003, Page E1. Washington, DC.

Wolfenbarger LL, Phifer PR (2000) The ecological risks and benefits of genetically engineered plants. Science 290: 2088-2093.

Field Evaluation of Transgenic White Clover with AMV Immunity and Development of Elite Transgenic Germplasm

M. Emmerling, P. Chu[1], K. Smith[2], R. Kalla, and G. Spangenberg
Plant Biotechnology Centre, Department of Primary Industries, La Trobe University, Victoria 3086, Australia. [1]CSIRO Plant Industry, Canberra, ACT 2601, Australia and [2]Pastoral and Veterinary Institute, Department of Primary Industries, Hamilton, Victoria 3300, Australia. (Email: German.Spangenberg@dpi.vic.gov.au).

Key words: transgenic white clover, field evaluation, alfalfa mosaic virus, virus resistance, transgene flow, transgenic germplasm

Abstract:

Viral diseases such as alfalfa mosaic virus (AMV) cause significant reductions in dry matter yield and persistency of white clover (*Trifolium repens* L.). Transgenic white clover plants expressing the AMV coat protein (AMV-CP) gene and showing immunity to AMV infection were evaluated in multi-site small-scale field releases. Two transformation events showing field immunity to aphid-mediated AMV infection This article outlines the development of transgenic elite white clover germplasm with AMV immunity, involving the world's first breeding nursery for transgenic white clover.were selected for elite transgenic germplasm development. Following top crosses with elite parental breeding lines, diallel crosses of heterozygous offspring plants and identification of AMV-CP homozygous T2 lines through quantitative PCR-based high-throughput screening, a breeding nursery with 1,300 transgenic white clover plants was established. Agronomically superior transgenic white clover plants were selected as parents for the production of world's first AMV immune transgenic white clover cultivars.

A. Hopkins, Z.Y. Wang, R. Mian, M. Sledge and R.E. Barker (eds.), Molecular Breeding of Forage and Turf, 359-366.

1. INTRODUCTION

Legumes such as white clover and alfalfa contribute to the agricultural industries by providing high-quality pastures for grazing and by fixing atmospheric nitrogen, important in the context of cropping. However, the performance of these important contributors to temperate pastures in Australia and worldwide is being significantly affected by representatives of the three largest families of plant viruses (Murphy et al. 1995).

Alfalfa mosaic virus (AMV), white clover mosaic virus (WCMV) and clover yellow vein virus (CYVV) are members of the *Bromoviridae*, potexvirus group and *Potyviridae*, respectively, and are estimated to cause combined losses to the Australian rural industries of more than $A800 million per year. Infections with these viruses result in reduced foliage yield, reduced nitrogen-fixing capacity and reduced vegetative persistence and can affect the production potential of white clover pastures by up to 30% (Campbell and Moyer 1984; Dudas et al. 1998; Garrett 1991; Gibson et al. 1981; Latch and Skipp 1987; Nikandrow and Chu 1991).

Even though potential sources of tolerance or resistance to AMV, CYVV or WCMV have been described in *Trifolium* species and *Medicago sativa* (Barnett and Gibson 1975; Crill et al. 1971; Gibson et al. 1989; Martin et al. 1997; McLaughlin and Fairbrother 1993), conventional breeding programs have not been very successful. This is mostly due to virus-strain limitations, lack of durability of natural resistance and barriers to interspecies sexual and/or somatic hybridisation. Chemical control of insect, fungal or nematode vectors is environmentally unacceptable and economically not viable in forage legumes.

Sanford and Johnson (1985) first suggested a genetic engineering approach of virus resistance by expressing whole, or parts of, viral genes in transgenic plants. A wide range of agronomically important crops protected against single or multiple virus infection, including tobacco, tomato, potato (Gielen et al. 1991; Jongedijk et al. 1992; Kaniewski et al. 1990; Powell et al. 1986) as well as squash and cucumber (Fuchs et al. 1998; Gonsalves et al. 1992; Tricoli et al. 1995) have since been developed. Here we report on the development and field evaluation of AMV-immune elite transgenic white clover germplasm.

2. AMV GENOME ORGANISATION AND GENE CONSTRUCT

AMV is a tripartite positive sense RNA virus (Figure 1A). RNA-1 and RNA-2 encode replicase A and B, respectively, necessary for viral replication. RNA-3 is bicistronic and encodes two gene products, but only the movement protein can be expressed. The second product, the AMV coat protein, is expressed from the non-replicating subgenomic RNA-4. The genomic RNAs are individually encapsidated by the coat protein resulting in bacilliform virions.

Resistance to viruses can be achieved using a number of different strategies, for example by expressing antiviral products *in planta*, by relying on RNA-based events like gene silencing, or by expressing parts or the whole of a viral genome. Expression of the viral coat protein results in high levels of the coat protein being present in the plant cells which in turn interferes with the virus life cycle leading to virus resistance.

The coat protein-mediated resistance is the most widely used approach to develop virus-immune plants (Hackland et al. 1994) and has been chosen for the development of AMV-immune white clover. The AMV RNA-4 cDNA expressing the coat protein was isolated from two Australian strains of the virus (Garrett and Chu 1997). The expression of the coat protein *in planta* is controlled by the strong constitutive double 35S CaMV promoter and the *Arabidopsis* rbcS terminator (Figure 1B).

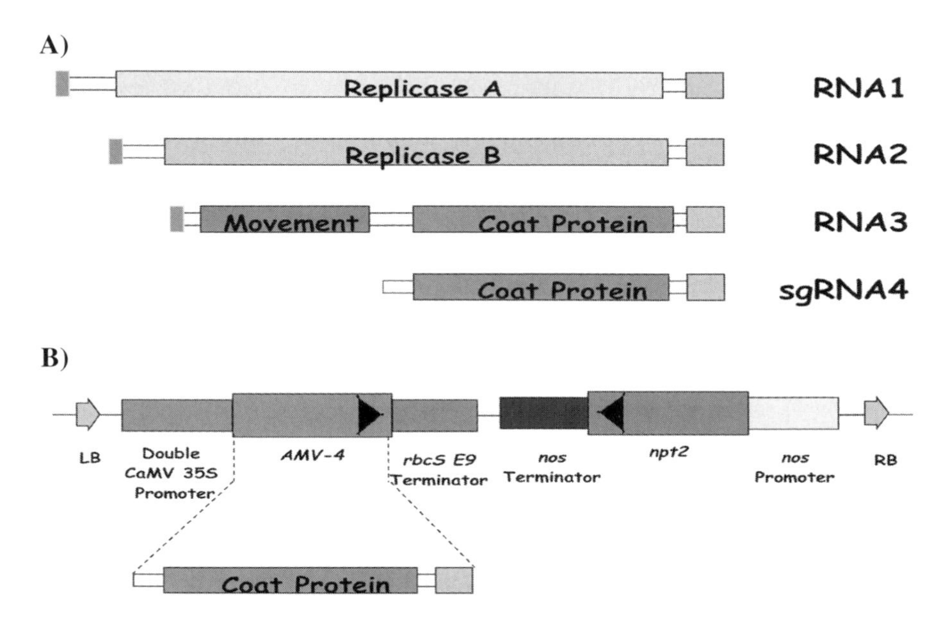

Figure 1. A) AMV genome structure composed of individual RNA molecules encoding essential viral gene products. B) Map of the T-DNA integrated into the genome of transgenic plants expressing the AMV4-encoded virus coat protein gene.

3. WHITE CLOVER TRANSFORMATION AND SELECTION

White clover was transformed with the binary vector shown in Figure 1B using *Agrobacterium* following a protocol developed for, and used with high rates of success in, the transformation of a wide range of legumes (Ding et al. 2003). After selection, putative transgenic plants were analysed for the presence and copy number of the transgene by Southern hybridisation as well as levels of expression of the transgene by Northern and Western hybridisation. The plants were also challenged with AMV under containment growth room and glasshouse conditions and showed immunity. The plants were clonally propagated and evaluated in a field trial in Hamilton, Victoria. Over a 2-year period, the plants proved to be immune to heavy natural aphid-mediated AMV challenges. White clover plants originating from 2 independent transformation events, H1 and H6, were chosen for the development of elite germplasm due to their high level of expression of the transgene (see Figure 2) and the high titre of coat protein (data not shown) as well as their AMV immunity phenotype and agronomical performance during the field trials.

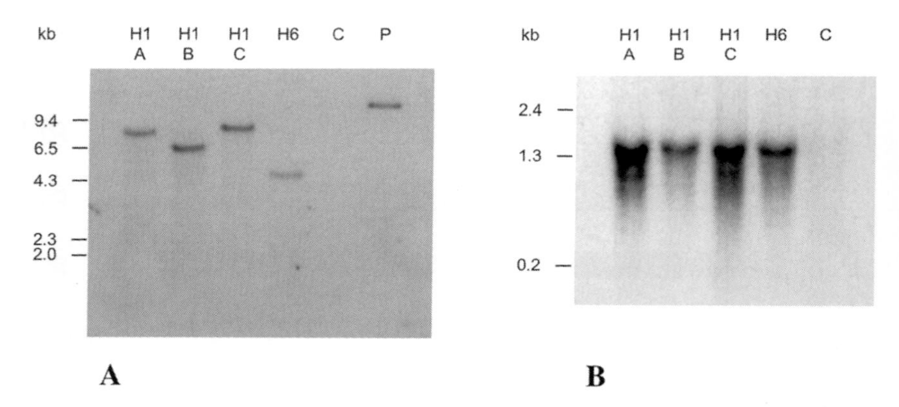

Figure 2. Molecular analysis of T_0 AMV-CP transgenic plants. A) Southern blot hybridisation of genomic DNA isolated from white clover plants obtained from 4 independent transformation events. C indicates wild type (negative) control, P indicates plasmid (binary vector) control. B) Northern blot hybridisation of RNA isolated from leaves of the same white clover plants. C indicates wild type (negative) control. Both blots were probed with the AMV CP cDNA (see Figure 1)

4. DEVELOPMENT AND FIELD EVALUATION OF AMV-RESISTANT WHITE CLOVER ELITE GERMPLASM

The two selected transgenic AMV-resistant white clover lines, H1 and H6, were crossed with the parents of the white clover cultivar “Mink” and subjected to an elite germplasm development strategy designed to bring the transgene to homozygosity while minimising inbred depression (Kalla et al., 2000). More than 8,000 T_2 offspring of these crosses were analysed by real time-PCR (RT-PCR), and a total of 1,300 plants homozygous for the AMV-CP transgene were identified, 888 derived from the H1 event, 412 from the H6 event.

A spaced plant field trial was subsequently established in Hamilton, Victoria, to evaluate the 1,300 transgenic white clover T_2 progeny (see Figure 3). The plants were assessed for virus infection with AMV four and five months after being established in the field. None of the transgenic plants showed any sign of virus infection whereas 28% of the non-transgenic wildtype control plants were infected with AMV.

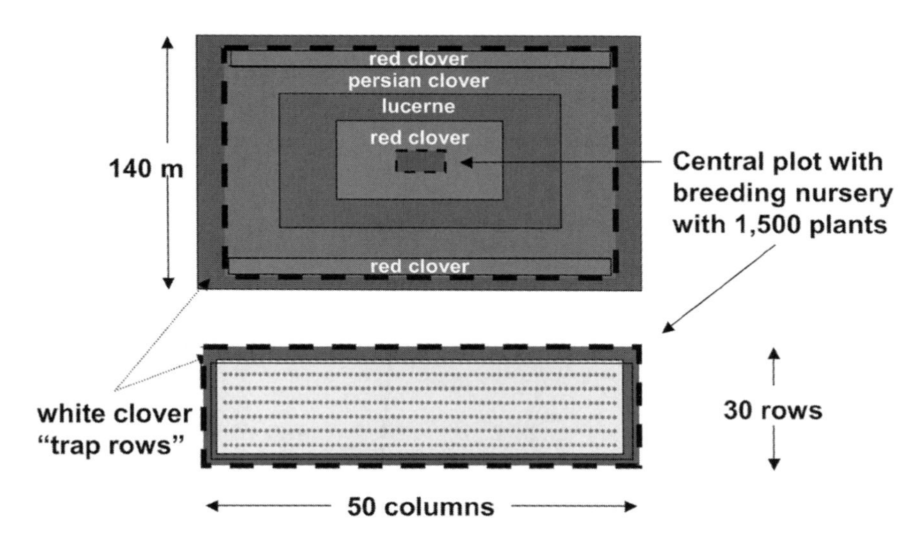

Figure 3. Layout of the spaced plant breeding nursery established in Hamilton, Victoria, to evaluate T_2 of spring of elite transgenic white clover lines homozygous for the AMV coat protein transgene (GMAC PR64X2). Checks of non-transgenic control plants (cv. "Mink", total of 200 plants) are uniformly distributed among the 1,300 transgenic T_2 white clover plants.

An initial selection of agronomically superior plants comprised 179 H1-derived and 104 H6-derived elite transgenic clover plants. The selection was based on plant height, stolon density, leaf length, internode length, flower number, summer growth and survival, and autumn and spring vigour. During the second growth season a further selection out of the initially selected plants led to the identification of 21 H1-derived and 16 H6-derived elite plants. These plants, resulting from the world's first breeding nursery for white clover, are the syn_0 parents for the production of agronomically superior transgenic AMV-immune white clover elite cultivars.

Polycrosses of these plants to produce the syn_2 generation are expected to be finalised in 2004 for final cultivar development to obtain a market-ready product by 2006.

5. TRANSGENE FLOW ANALYSIS

The intentional release, be it in a field trial or for commercial purposes, of an organism that has one or more new traits needs to be monitored closely to evaluate what impact, if any, the organism has on the environment. This applies to offspring of traditional breeding as well as transgenic plants. Monitoring the latter is relatively straightforward since the transgene, regardless of its origin, and the flanking sequences are fully characterised before generating the transgenic plant. Unique sequences contained within the construct can then be used to monitor the flow of the transgene by means of high-throughput methods like PCR.

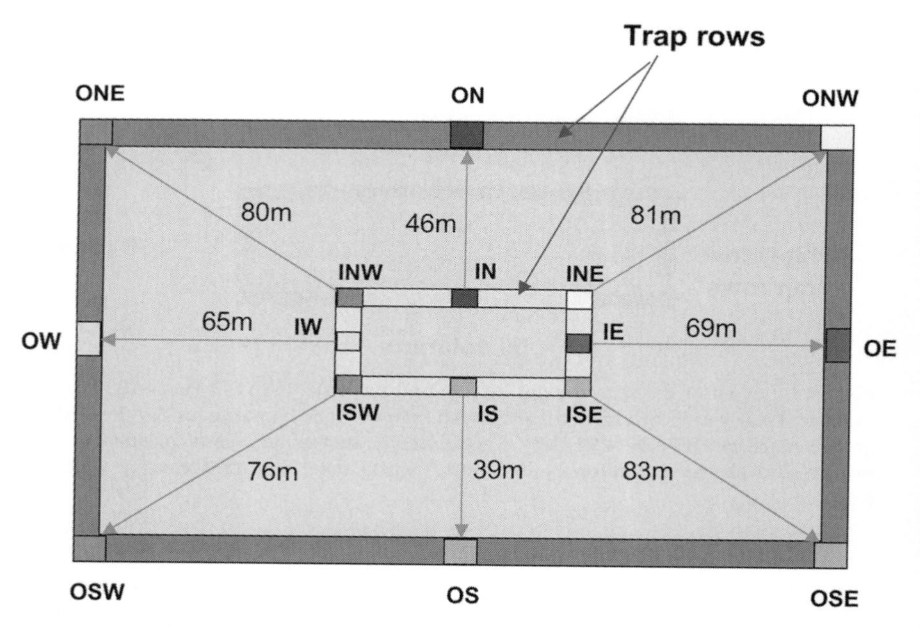

Figure 4. Schematic diagram of the sample areas for the analysis of gene flow. Seeds were sampled from 8 areas of the inner and outer trap rows, respectively.

This approach was used for the analysis of transgene flow from transgenic white clover to non-transgenic trap plants during the initial field trials with T_0

transgenic plants. Seeds were harvested from the inner and outer trap rows of non-transgenic white clover in order to provide an estimate of transgene flow from the trial plots to the surrounding trap rows. More than 86,000 seeds were germinated on medium containing 50 mg/l G418. Only 25 seeds germinated and were PCR-negative and Southern hybridisation negative for AMV coat protein transgene.

Woodfield et al. (1995) used the *Feathermark* phenotype to examine gene flow in white clover based on behaviour of pollinating bees. *Feathermark* is a single-locus, dominant mutation resulting in anthocyanin accumulation along the leaf midrib. 95% of all *Feathermark* progeny was found within 10 m of the source plant, and no consistent *Feathermark* progeny was recovered at distances greater than 30 m from the source plant.

These results confirmed the suitability of design used for the described field trials, particularly due to the inclusion of profusely flowering legumes that are highly attractive to bees in the buffer zones surrounding the central plots of transgenic plants (Kalla et al. 2000).

ACKNOWLEDGEMENTS

The original research described here was supported by the Dairy Research and Development Corporation, the Department of Primary Industries, Victoria, Australia, CSIRO Plant Industry, Canberra, Australia and Heritage Seeds Pty. Ltd.

REFERENCES

Barnett OW, Gibson PB (1975) Identification and prevalence of white clover viruses and the resistance of *Trifolium* species to these viruses. Crop Sci. 15: 32-37.

Campbell CL, Moyer JW (1984) Yield responses of 6 white clover clones to virus infection under field condition. Plant Dis. 68: 1033-1035.

Crill P, Hanson EW, Hagedorn DJ (1971) Resistance and tolerance to alfalfa mosaic virus in alfalfa. Phytopathology 61: 371-379.

Ding Y-L, Aldao-Humble G, Ludlow E, Drayton M, Lin Y-H, Nagel J, Dupal M, Zhao G, Pallaghy C, Kalla R., Emmerling M, Spangenberg G (2003) Efficient plant regeneration and *Agrobacterium*-mediated transformation in *Medicago* and *Trifolium* species. Plant Sci. (in press).

Dudas B, Woodfield DR, Tong PM, Nicholls MF, Cousins GR, Burgess R, White DWR, Beck DL, Lough TJ, Forster RLS (1998) Estimating the agronomic impact of white clover mosaic virus on white clover performance in the North Island of New Zealand. New Zealand J. Agr. Res. 41: 171-178.

Fuchs M, Tricoli DM, Carney KJ, Schesser M, Mcferson JR, Gonsalves D (1998) Comparative virus resistance and fruit yield of transgenic squash with single and multiple coat protein genes. Plant Dis. 82: 1350-1356.

Garrett RG (1991) Impact of viruses on pasture legume productivity. In: Proceedings of Department of Agriculture Victoria white clover conference, pp. 50-57.

Garrett RG, Chu PWG (1997) White clover expressing the coat protein of alfalfa mosaic virus: Field trial issues. In: Commercialisation of transgenic crops: Risk, benefit and trade considerations, pp. 125-136, Canberra.

Gibson PB, Barnett OW, Skipper HD, McLaughlin MR (1981) Effects of 3 viruses on growth of white clover. Plant Dis. 65: 50-51.

Gibson PB, Barnett OW, Pederson MR, McLaughlin MR, Knight WE, Miller JD, Cope WA, Tolin SA (1989) Registration of southern regional virus resistant white clover germplasm. Crop Sci. 29: 241-242.

Gielen JJL, de Haan P, Kool AJ, Peters D, van Grinsven MQJM, Goldbach RW (1991) Engineered resistance to tomato spotted wilt virus, a negative strand RNA virus. Bio/Technology 9: 1363-1367.

Gonsalves D, Chee P, Providenti R, Seem R, Slightom JL (1992) Comparisons of coat protein-mediated and genetically-derived resistance in cucumbers to infection by cucumber mosaic virus under field conditions with natural challenge inoculations by vectors. Bio/Technology 10: 1562-1570.

Hackland AF, Rybicki EP, Thomson JA (1994) Coat protein-mediated resistance in transgenic plants. Arch. Virol. 139: 1-22.

Jongedijk E, de Schutter AJM, Stolte T, van den Elzen PJM, Cornelissen BJC (1992) Increased resistance to potato virus X and preservation of cultivar properties in transgenic potato under field conditions. Bio/Technology 10: 422-429.

Kaniewski W, Lawson C, Sammons B, Haley L, Hart J, Delanny X, Turner NE (1990) Field resistance of transgenic Russet Burbank potato to effects of infection by potato virus X and potato virus Y. Bio/Technology 8: 750-754.

Kalla R, Chu P, Spangenberg G (200) Molecular breeding of forage legumes for virus resistance. In: Molecular breeding of forage crops, Spangenberg G (ed.), pp.219-237. Kluwer Academic Publishers, Dordrecht.

Latch GCM, Skipp RA (1987) Diseases. In: White clover, Baker MJ, Williams WM (eds.), pp. 421-446. CAB international, UK.

Martin PH, Coulman BE, Peterson JF (1997) Genetics to resistance to alfalfa mosaic virus in red clover. Can. J. Plant Sci. 77: 601-605.

McLaughlin MR, Fairbrother TE (1993) Selecting subclover for resistance to clover yellow vein virus. Phytopathology 83: 1421.

Murphy FA, Fauquet CM, Bishop DHL, Ghabrial SA, Jarvis AW, Martelli GP, Mayo MA, Summers MD (1995) Virus Taxonomy: Sixth Report of the International Committee on Taxonomy of Viruses, pp. 586. Wien, Springer Verlag.

Nikandrow A, Chu PWG (1991) Pests and diseases. The NSW experience. In: Proceedings White Clover Conference, pp. 64-67. Department of Agriculture Victoria.

Powell PA, Nelson RS, De B, Hoffman N, Rogers SG, Fraley RT (1986) Delay of disease development in transgenic plants that express the tobacco mosaic virus coat protein gene. Science 232: 738-747.

Sanford JV, Johnson SA (1985) The concept of pathogen-derived resistance: Deriving resistance genes from the parasite's own genome. J. Theor. Biol. 113: 395-405.

Tricoli DM, Carney KJ, Russell PF, Mcmaster JR, Groff DW, Hadden KC, Himmel PT, Hubbard JP, Boeshore ML, Quemada HD (1995) Field evaluation of transgenic squash containing single or multiple virus coat protein gene constructs for resistance to cucumber mosaic virus. Bio/Technology 13(13): 1458-1465.

Woodfield, DR, Clifford, PTP, Baird, IJ, Cousins, GR (1995) Gene flow and estimated isolation requirements for transgenic white clover. In: Proceedings 3rd International Symposium on the Biosafety Results of Field Tests of Genetically Modified Plants and Micro-organisms, Jones DD (ed.). Oakland, University of California.

Field Evaluation and Risk Assessment of Transgenic Tall Fescue (*Festuca arundinacea*) Plants

Zeng-Yu Wang, Andrew Hopkins, Robert Lawrence, Jeremey Bell and Megann Scott
Forage Improvement Division, The Samuel Roberts Noble Foundation, 2510 Sam Noble Parkway, Ardmore, OK 73401, USA. (Email: zywang@noble.org).

Key words: tall fescue, transgenic plant, field evaluation, risk assessment

Abstract:

Tall fescue (*Festuca arundinacea* Schreb.) is an outcrossing hexaploid grass species widely grown for forage and turf purposes. Transgenic tall fescue plants were generated by biolistic transformation of embryogenic cell suspension cultures of the commonly used cultivar Kentucky-31. T1 and T2 progenies were obtained after reciprocal crosses between transgenic and untransformed control plants. Molecular analysis of the progenies revealed stable meiotic transmission of transgenes following Mendelian rules in transgenic tall fescue. Agronomic performance of the primary transgenics and primary regenerants under field conditions were generally inferior to seed-derived plants, with primary transgenics having fewer tillers and lower seed yield. However, no major differences between the progenies of transgenics and the progenies of seed-derived plants were found for the agronomic traits evaluated. The addition of a selectable marker gene in the plant genome seems to have little effect on the agronomic performance of the regenerated plants. No indication of weediness was observed for the transgenic tall fescue plants. An experiment on pollen dispersal has also been carried out using transgenic tall fescue in a central plot, surrounded by untransformed recipient plants in a wagon wheel design. The highest transgene frequencies, 0.88% at 50 m and 0.59% at 100 m, were observed north of the central plot, the prevailing wind direction. Issues regarding experimental design for gene flow studies and future directions on risk assessment of forage and turf grasses are discussed.

A. Hopkins, Z.Y. Wang, R. Mian, M. Sledge and R.E. Barker (eds.), Molecular Breeding of Forage and Turf, 367-379.

1. INTRODUCTION

Tall fescue (*Festuca arundinacea* Schreber) is a wind-pollinated, highly self-infertile polyploid perennial cool-season forage and turf grass. It is used in pastures, parks, lawns, golf courses, football fields, highway medians and roadsides (Barnes 1990). Tall fescue is indigenous to Europe where it is well adapted. It was introduced to North America in the early to mid 1800s and has become the predominant cool-season pasture grass in the USA (Buckner et al. 1979; Barnes 1990). Its wide spread use in the USA is due to its adaptation to a wide range of soil conditions, tolerance of continuous grazing, high yields of forage and seed, persistence, long grazing season, compatibility with varied management practices and low incidence of pest problems (Hanson 1979; Sleper and West 1996).

Tall fescue is an allohexaploid (2n = 6x = 42 chromosomes) synthesized from two progenitor species, *F. pratensis* (2n = 2x = 14) and *F. arundinacea* var. *glaucescens* Boiss. (2n = 4x = 28) (Sleper and Buckner 1995). *Festuca* is a diverse and widely adapted genus which contains over 80 species, of which approximately 20 exist in the US (Barnes 1990). There has been little evidence on how theses species are interrelated genetically (Berg et al. 1979b; Terrell 1979; Sleper and West 1996). Tall fescue and meadow fescue are closely related to both perennial (2n = 2x = 14) and Italian ryegrass (2n = 2x = 14) as intergeneric crosses have been reported (Terrell 1979; Sleper and West 1996). Gene transfer by crossing among these taxa may occur freely and chromosome structural differentiation among them is slight (Sleper and West 1996).

Due to its agronomic importance, tall fescue has been a major target species for genetic and agronomic studies (Jauhar 1993; Spangenberg et al. 1998). The potential of biotechnology in the development of improved forage grass cultivars has been recognized in recent years (Spangenberg et al. 1998; Wang et al. 2001a). There have been many reports on the generation of transgenic plants in tall fescue (Wang et al. 1992; Dalton et al. 1995; Spangenberg et al. 1995a; Kuai et al. 1999; Cho et al. 2000; Wang et al. 2001b; Bettany et al. 2003; Chen et al. 2003; Wang et al. 2003a) and other related forage species, e.g. perennial ryegrass (Spangenberg et al. 1995b; Dalton et al. 1998; Dalton et al. 1999; Altpeter et al. 2000), Italian ryegrass (Wang et al. 1997; Ye et al. 1997; Dalton et al. 1998; Dalton et al. 1999; Bettany et al. 2003) and red fescue (Spangenberg et al. 1994; Spangenberg et al. 1995a; Altpeter and Xu 2000; Cho et al. 2000).

Although tremendous progress has been made on transformation techniques and transgenic grasses carrying marker genes or potentially useful agronomic genes have been obtained, there has been little information available on field evaluation and risk assessment of transgenic forage grass plants. Agronomic performance of transgenic plants under field conditions, potential of transgenic pollen dispersal and crossability with related grass species has become extremely important issues for any future release of value-added transgenic cultivars.

2. FIELD EVALUATION OF TRANSGENIC TALL FESCUE PLANTS

Plant regeneration and genetic transformation systems have been developed for the most widely cultivated tall fescue cultivar, Kentucky-31 (Wang et al. 2003a). Sterilized seeds/caryopses of Kentucky-31 were used as explants for callus induction. Embryogenic calluses were individually transferred to liquid culture medium to establish single genotype-derived cell suspension cultures (Wang et al. 1994; Wang et al. 2002). Cell clusters from the suspension cultures were used as direct targets for biolistic transformation to generate transgenic plants. A chimeric hygromycin phosphotransferase (*hph*) gene, which renders transformed cells resistant to hygromycin, was used as the selectable marker gene (Bilang et al. 1991). A chimeric β-glucuronidase (*gusA*) gene was co-transformed with the *hph* gene (Wang et al. 2003a). Hygromycin resistant calluses were obtained after microprojectile bombardment of suspension cells and subsequent selection in the presence of hygromycin (Wang et al. 2003a). Transgenic tall fescue plants were regenerated from the hygromycin resistant calluses and later transferred to the greenhouse (Wang et al. 2003a; Wang et al. 2003b). Transgenic nature of the regenerated plants was confirmed by polymerase chain reaction (PCR) screening, Southern and northern hybridization analyses and GUS staining (Wang et al. 2003a). Transgenic tall fescue plants were transferred to the field to study transmission genetics of the transgenes and to evaluate their agronomic performance.

Transmission of foreign genes to progenies is critical for any potential use of transgenic material in producing novel germplasm or cultivars. However, there has been only limited information on meiotic transmission of transgenes in *Festuca*, although transgenic tall, red and meadow fescues were obtained several years ago. This is primarily due to the outcrossing nature and vernalization requirement of these species, which make it difficult to obtain progenies under greenhouse conditions. A report touching transgene inheritance in tall fescue provided confusing information, in which no transgenes were detected in the progenies (Kuai et al. 1999). In our study, fertile transgenic tall fescue plants were obtained after vernalization under field conditions (Wang et al. 2003a). T1 and T2 progenies were obtained after reciprocal crosses between transgenic and untransformed control plants.

PCR and Southern hybridization analyses revealed a 1:1 segregation ratio for both transgenes in the T1 and T2 generations. Southern hybridization patterns were identical for T0, T1, and T2 plants. Our study unequivocally demonstrated the stable meiotic transmission of transgenes following Mendelian rules in transgenic tall fescue.

In order to comparatively evaluate agronomical performance of transgenic and non-transgenic tall fescue plants, primary transgenics from two genotypes, their corresponding regenerants from the same genotypes, and control seed-derived plants were transferred to the field and evaluated for two years. Progenies of these three classes of plants were obtained and evaluated together with seed-derived plants in a second field experiment (Wang et al. 2003b). The experimental design was a randomized complete block with three replications. The agronomic characteristics evaluated were: heading date, anthesis date, height, growth habit, number of reproductive tillers, seed yield and biomass. Factor analysis showed plants from the same genotype were more uniform than plants from seeds. This is not surprising since seed-derived plants would be expected to differ at a number of loci. Genotype differences were observed in this study, with both transgenic and regenerated plants from genotype 1 performing better than plants from genotype 2. The addition of a selectable marker gene in the plant genome seems to have little effect on the agronomic performance of the regenerated plants, since performance of the transgenics was very similar to the corresponding regenerants from the same genotype. In addition, progenies of the transgenics performed similarly to progenies of the regenerants (Wang et al. 2003b).

Agronomic performance of the primary transgenics and regenerants were generally inferior to the seed-derived plants, with primary transgenics having fewer tillers and less seed yield. However, no major differences between the progenies of transgenics and the progenies of seed-derived plants were found for the agronomic traits evaluated (Wang et al. 2003b). In a separate study on pollen viability of transgenic and control plants, progenies of primary transgenics and regenerants showed similar pollen viability when compared with that of seed-derived plants, although primary transgenic and regenerants had various levels of pollen viability in individual plants. The lower seed yield in the primary transgenics might relate to low pollen viability. The results indicate that once seeds are obtained from the primary transgenic plants, normal pollen viability and agronomic performance of the progenies can be expected. No indication of weediness of the transgenic tall fescue plants was observed. The field study provided evidence that outcrossing grass plants generated through transgenic approaches can be incorporated into forage breeding programs. The field trials using transgenic plants were approved by USDA-APHIS (Animal and Plant Health Inspection Service).

3. RISK ASSESSMENT OF TRANSGENIC GRASS PLANTS

Since wind-pollinated grass species have a high potential to pass their genes to adjacent plants, information regarding gene flow has become extremely important for deregulation and release of transgenic cultivars. In the case of transgenic forage and turf grasses, human consumption is indirect. Thus evaluation of the biosafety of transgenic grasses will likely focus on their environmental or ecological impacts. Two questions need to be answered concerning risk assessment of transgenic grasses. First, how far can grass pollen disperse and still remain viable, and second, what is the probability of transgene escape by crossing with related grass species under natural conditions?

For foundation seed production of cross-pollinated grasses, the isolation standard required by USDA (CFR201.76) is 900 feet (274 meters) isolation distance. Field trials of transgenic grasses are basically following the standard for foundation seed production. There are many forage and turf grasses that are cross-pollinated under natural conditions, thus current isolation standards are perhaps too general and broad. It is not clear if the isolation distance is sufficient to minimize contamination for either transgenic or non-transgenic seed production. Thus gene flow studies will not only answer questions regarding isolation distance for growing transgenic grasses and future release of transgenic cultivars, but also provide valuable information for regulation of the current seed production system in outcrossing grasses.

The use of morphological markers has produced conflicting results in grasses. Pollen comtamination in perennial ryegrass by an outside pollen source has been reported to be as little as 1% at 7.3 m (Copeland and Harding 1970), or as much as 10% at 182 m (Griffiths 1951). In smooth bromegrass (*Bromus inermis*), Johnson et al. (Johnson et al. 1996) reported pollen contamination of less than 5% between 22 and 27 m. Pollen trap study in perennial ryegrass showed that ryegrass pollen could travel 80 m, although the amount of pollen collected was much less than the traps near the center ryegrass field (Giddings et al. 1997a; Giddings et al. 1997b). The main drawback of pollen trap is that it's almost impossible to know whether the collected pollen still remains viable and competitive enough to pollinate plants. An isozme marker (*Pgi-2*) was used to investigate pollen dispersal in meadow fescue (Nurminiemi et al. 1998; Rognli et al. 2000). Pollen was captured by recipient plants at a distance of 250 m, but the authors noticed that the isozyme allele was also present in the surrounding areas of the experimetal plot. Transgenic creeping bentgrass has been the only transgenic turf grass species used for pollen flow study under field conditions. Herbicide resistant progenies were recovered at 292 m maximum distance from the source transgenic plants (Wipff and Fricker 2001). Frequencies of interspecific hybridization between transgenic creeping bentgrass and four related species were measured, with interspecific trangenic hybrids recovered between creeping bentgrass and *Agrostis capillaris* and *A. castellana* at frequencies of 0.044 and 0.0015%, respectively (Belanger et al. 2003). It should be noted that creeping bentgrass is a slow growing grass species exclusively used for turf purposes. This means that, unlike other more aggressive grasses, herbicide resistance (e.g. Roundup Ready® bentgrass) is not a major biosafety concern for this species. On the other side, the results of gene flow studies in creeping bentgrass cannot be directly applied to other major forage and turf species, e.g. fescues and ryegrasses.

3.1 A Preliminary Study on Transgene Flow in Tall Fescue

Transgenic plants provide unique material for studying pollen dispersal and gene transfer into related species. The detection of transgenes by molecular techniques offers distinctive advantages over the use of morphological or biochemical markers by providing clear-cut information on presence or absence of transgenes without subjective judgment or being influenced by levels of gene expression.

A misconception about using transgenic plants is that transgene frequency can be simply detected by selecting seeds or seedlings with antibiotics or herbicides, because the original transgenic plant(s) was transformed with a marker gene and selected with an antibiotic or an herbicide. This misconception oversimplifies the inheritance and expression of transgenes, because some introduced genes are not expressed as they should be. Epigenetic gene silencing inactivates expression of transgenes in higher plants at frequencies of up to 30% of independent transformants (Grant 1999). In the last decade, gene silencing has been recognized to be a general phenomenon reported for many species transformed with a variety of chimeric genes (Kunz et al. 2001). This epigenetic silencing may persist over many cell divisions or plant generations (Paszkowski and Whitham 2001). Therefore, even though antibiotic or herbicide selection may be an effective method in some cases, the seeds or seedlings from the recipient plants have to tested first by molecular analysis to make certain that the transgene(s) segregates in an expected manner, and the phenotype selection is closely correlated with the presence of transgenes. In other cases where gene silencing happens, the simple antibiotic or herbicide selection method will produce misleading results. In addition, under-estimation or over-estimation of transgene frequency may occure if not enough or too much selection pressure is applied. The most reliable method is to detect the presence of transgenes by molecular methods, such as polymerase chain reaction (PCR) and Southern hybridization analysis.

We initiated a small-scale pollen flow study using transgenic tall fescue plants. The experiment was carried out by growing transgenic tall fescue in a central plot, and surrounded by exclosures containing recipient plants. An exclosure is defined as a 2.0 x 2.0 m pen constructed of welded wire and metal posts, and is used to exclude any grazing animals, such as deer or livestock, from the tall fescue. The exclosures were 50 m apart and aligned in eight directions, much like spokes of wagon wheel, up to a distance of 200 m from the central source plants. The T1 and T2 generations of transgenic tall fescue were used as central donor plants. Seeds were collected from the recipient plants and germinated seedlings were used for high throughput DNA isolation and PCR analysis. More than 11,000 seedlings were analyzed by PCR. Transgenes were detected in recipient plants at 50 and 100 m with frequencies in the range of 0.29 – 0.88%. The highest transgene frequencies, 0.88% at 50 m and 0.59% at 100 m, were observed north of the central plot, i.e. downwind from transgenic plants based on the prevailing wind direction. Southern blot hybridization analysis confirmed the transgenic nature of the PCR positive plants. A supplement experiment demonstrated that transgene flow can be controlled by placing transgenic plantings downwind and long distances from non-transgenic seed increases, thus allowing tall fescue breeding and transgene development programs to be conducted concurrently at the same research station.

3.2 Future Research on Risk Assessment of Forage and Turf Grasses

Tall fescue, perennial ryegrass and Italian ryegrass are the most important and most intensively studied forage and turf species worldwide. Biotechnological aspects of research in these species include: large-scale EST sequencing, microarray and gene discovery, transformation, genomics, molecular markers and breeding methods. Transgenic materials carrying potentially useful transgenes are being developed and evaluated in many laboratories (Spangenberg et al. 2001; Wang et al. 2001a). Thus, there is an urgent need for gene flow studies in transgenic tall fescue and ryegrasses. Such studies should provide a body of knowledge regarding release of transgenic grasses for regulatory authorities.

Experiments on dispersal of transgenic pollen should be carried out in relatively larger scales. The inclusion of more exclosures should allow the modeling of gene flows. Regarding experimental designs, many risk assessment studies employed the wagon wheel or similar designs, simply because most information could be obtained from this type of design. Although there are other designs reported, none of them could provide more information or were less problematic than the wagon wheel design. While realizing wagon wheel may be excellent as a model small-scale experimental design, the main criticism is that the design may not accurately reflect the real world conditions of commercial production, which involves extensive, landscape level plantings of transgenic as well as conventional varieties. This argument sounds reasonable in theory, however, we have to face the very simple fact that it is impossible to have landscape plantings without extensive small-scale biosafety studies. In some countries, even small-scale field test of transgenic outcrossing grasses have been very difficult to carry out due to regulatory hurdles.

Since many grass species are self-infertile, we may take advantage of their outcrossing nature by using single genotype-derived plants as recipients in the exclosures. Plants of single genotype origin (obtained by vegetative propagation of a single plant) will not pollinate each other. This will maximize the possibility for transgenic pollen to pollinate recipient plants, and minimize the possibility of pollination within and between the exclosures. Thus the use of single genotype plants should reveal the worst case scenario that happens under natural conditions. It can also be argued that because grass plants are genotypically diverse with different growth patterns, the recipient plants should be a mixture of cross compatible recipient genotypes. However, preferential cross pollination between and within the exclosures will inevitably occur in this case, so such a study may only reflect certain field conditions, but certainly not the worst case scenario.

Research on pollen flow of outcrossing grasses not only involves the already complicated biological systems, but also environmental conditions. Important physiological factors found to influence pollen dispersal were gravity, wind speed, wind direction, turbulence, air density and air viscosity. Biological parameters that influence the effect of these factors include pollen density, pollen radius, pollen longevity and sedimentation velocity (Luna et al. 2001). It is unlikely that a specific design or experiment can answer all the questions. We have no other choice but to work step by step, from small to reasonably large scale field trials. In order to obtain the most information, we have to balance the potential outcome and the feasibility of the field experiment. Joint and collaborative efforts at the national and international level are needed to carry out biosafety research and to accumulate enough data for deregulation of certain transgenic grasses.

Because of the possibilities of cross hybridizations between different grass species, it is also very important to investigate the potential influences of releasing one transgenic species on another related species. In most of the important crops in the world, gene flow between cultivars and between wild and weedy relatives has always taken place (Messeguer 2003). But in forage and turf grasses, besides a few reports on creeping bentgrass, there has been little research on transgene flow between species.

Numerous attempts have been made to generate interspecific and intergeneric hybrids in *Festuca* and *Lolium* by conventional crossing (Berg et al. 1979a). Hybrids between related fescue and ryegrass species have demonstrated useful combinations of traits from both parents (Thomas and Humphreys 1991). Chromosome doubling is often necessary in these hybrids to restore fertility in synthetic amphiploids (Thomas and Humphreys 1991). Interspecific hybrids were generated in the combinations tall fescue x meadow fescue, tall fescue x giant fescue (*F. gigantea*) and tall fescue x *F. mairei*, and the hybrids were cytologically characterized (Berg et al. 1979a; Jauhar 1993). Intergeric hybrids of perennial ryegrass x tall fescue, Italian ryegrass x tall fescue, perennial ryegrass x meadow fescue, Italian ryegrass x meadow fescue as well as the perennail ryegrass x meadow fescue x tall fescue trispecies hybrids were obtained (Berg et al. 1979a; Thomas and Humphreys 1991). Fertile amphiploids between ryegrass and tall fescue were obtained by doubling either the chromosome numbers of the hybrids or of the parents before crossing (Berg et al. 1979a; Jauhar 1993).

For assessment of interspecific and intergeneric gene flow in transgenic tall fescue, the following species may be considered: *F. pratensis*, *F. rubra*, *F. ovina*, *F. idahoensis*, *F. gigantea*, *F. mairei*, *F. versuta*, *L. perenne* and *L. multiflorum*. It is known that *F. pratensis*, *F. rubra*, *F. ovina* and *F. idahoensis* are valuable species used for forage, turf or conservation purposes (Sleper and Buckner 1995). *F. gigantean*, *F. mairei*, *L. perenne* and *L. multiflorum* could hybridize with tall fescue as described above. *F. versuta* is a native species of southern US. Single genotype-derived plants should be used for each species to avoid cross-pollination of the same species by plants in the neighboring replicate. Otherwise, plants of the same species could pollinate each other between replicates, leaving little chance for transgenic plants to pollinate them.

In the context of risk assessment, we would like to comment briefly about transgenic traits. As forage scientists, we highly appreciate the important role and contribution of forage grasses in sustainable agriculture. In the mean time, we also need to realize that these species can be considered as 'weeds' if they start growing in fields of major cash crops (e.g. soybean, cotton, corn) or become invasive in natural habitats. Among the topics concerning the future of weed science, herbicide resistance has been at the forefront. Glyphosate resistance is one of the first traits commercialized in transgenic crops. Mainly due to the convenience of weed management, glyphosate resistant or Roundup Ready® crops have been widely adopted in recent years. Herbicide resistance in weeds could make a mode of action useless. The first naturally occurring glyphosate-resistant biotype of *L. rigidum* was identified in Australian orchards due to the continuous application of glyphosate herbicide (Lorraine-Colwill et al. 2001; Baerson et al. 2002). There have been many studies devoted to this new biotype since its discovery in 1996, and glyphosate resistance in this species has been considered as a major emerging threat to current weed management practices in Australia (Lorraine-Colwill et al. 2001; Baerson et al. 2002). Therefore, even though glyphosate resistance can be a beneficial trait in forage and turf grasses, we need to be extremely careful in employing the trait, particularly with aggressive grass species, although it should be acceptable with slow growing, non-aggressive species (e.g. creeping bentgrass). If we don't consider the issue seriously, damage could be caused to other production systems if an aggressive glyphosate resistant grass grows into Roundup Ready® soybean or cotton fields. Similarly, unwanted herbicide resistant grasses could cause problems for grass seed producers and natural resource managers. Although we believe that the situation can be controlled, the main problem is not just how big the economic damage could be. It will almost certainly encourage more talk of the already misleading word 'superweeds', and will provide ammunition to anti-biotech groups. The potential political consequence can be much larger than the actual economic consequence. As Vasil (Vasil 2003) pointed out: "The biotechnology community - academia and industry alike - must share at least some of the blame for the hostility of the anti-biotechnology groups and the difficulties being faced in the commercialization of transgenic crops". We need to be considerate of other production systems when choosing traits for engineering forage and turf grasses.

Gene flow is not unique to transgenic plants, it is a natural process that has happened in the past and will continue to happen in the future. The introduction of modern biotechnology has brought new attention to the process and raised both economic and ecological issues for scientists and policymakers to consider. As new transgenic grasses are tested and grown, preventing unwanted gene flow will present technical and regulatory challenges as well as possible economic conflict. Much more attention should be given and much more research should be done in the area of risk assessment of forage and turf grasses. Only by accumulating enough scientific knowledge can we make deregulation of transgenic grasses possible. Collaborative efforts are needed in this complicated research area.

ACKNOWLEDGEMENTS

The authors thank Mack Armstrong, Brian Motes, Dennis Walker and Brandi Williams for their valuable help with the field work.

REFERENCES

Altpeter F, Xu JP (2000) Rapid production of transgenic turfgrass (*Festuca rubra* L.) plants. J. Plant Physiol. 157: 441-448.

Altpeter F, Xu JP, Ahmed S (2000) Generation of large numbers of independently transformed fertile perennial ryegrass (*Lolium perenne* L.) plants of forage- and turf-type cultivars. Mol. Breed. 6: 519-528.

Baerson SR, Rodriguez DJ, Biest NA, Tran M, You JS, Kreuger RW, Dill GM, Pratley JE, Gruys KJ (2002) Investigating the mechanism of glyphosate resistance in rigid ryegrass (*Lolium ridigum*). Weed Sci. 50: 721-730.

Barnes RF (1990) Importance and problems of tall fescue. In: Biotechnology in tall fescue improvement, Kasperbauer MJ (ed.) pp. 1-12. CRC, Boca Raton.

Belanger FC, Meagher TR, Day PR, Plumley K, Meyer WA (2003) Interspecific hybridization between *Agrostis stolonifera* and related *Agrostis* species under field conditions. Crop Sci. 43: 240-246.

Berg CC, Webster GT, Jauhar PP (1979a) Cytogenetics and genetics. In: Tall fescue, Bush LP (ed.) pp. 93-109. ASA-CSSA-SSSA, Madison.

Berg CC, Webster GT, Jauhar PP (1979b) Cytogenetics and genetics. In: Tall fescue, Buckner RC, Bush LP (eds.), pp. 93-109. ASA-CSSA-SSSA, Madison.

Bettany AJE, Dalton SJ, Timms E, Manderyck B, Dhanoa MS, Morris P (2003) *Agrobacterium tumefaciens*-mediated transformation of *Festuca arundinacea* (Schreb.) and *Lolium multiflorum* (Lam.). Plant Cell Rep. 21: 437-444.

Bilang R, Iida S, Peterhans A, Potrykus I, Paszkowski J (1991) The 3'-terminal region of the hygromycin-B-resistance gene is important for its activity in *Escherichia coli* and *Nicotiana tabacum*. Gene 100: 247-250.

Buckner RC, Powell JB, Frakes RV (1979) Historical development. In: Tall fescue, Buckner RC, Bush LP (eds.), pp. 1-8. ASA-CSSA-SSSA, Madison.

Chen L, Auh C, Dowling P, Bell J, Chen F, Hopkins A, Dixon RA, Wang ZY (2003) Improved forage digestibility of tall fescue (*Festuca arundinacea*) by transgenic down-regulation of cinnamyl alcohol dehydrogenase. Plant Biotechnol. J. (in press).

Cho MJ, Ha CD, Lemaux PG (2000) Production of transgenic tall fescue and red fescue plants by particle bombardment of mature seed-derived highly regenerative tissues. Plant Cell Rep. 19: 1084-1089.

Copeland LO, Harding EE (1970) Outcrossing in ryegrasses (*Lolium* spp.) as determined by fluorescence tests. Crop Sci. 10: 254-257.

Dalton SJ, Bettany AJE, Timms E, Morris P (1995) The effect of selection pressure on transformation frequency and copy number in transgenic plants of tall fescue (*Festuca arundinacea* Schreb.). Plant Sci. 108: 63-70.

Dalton SJ, Bettany AJE, Timms E, Morris P (1998) Transgenic plants of *Lolium multiflorum*, *Lolium perenne*, *Festuca arundinacea* and *Agrostis stolonifera* by silicon carbide fibre-mediated transformation of cell suspension cultures. Plant Sci. 132: 31-43.

Dalton SJ, Bettany AJE, Timms E, Morris P (1999) Co-transformed, diploid *Lolium perenne* (Perennial ryegrass), *Lolium multiflorum* (Italian ryegrass) and *Lolium temulentum* (Darnel) plants produced by microprojectile bombardment. Plant Cell Rep. 18: 721-726.

Giddings GD, Hamilton NRS, Hayward MD (1997a) The release of genetically modified grasses .1. Pollen dispersal to traps in *Lolium perenne*. Theor. Appl. Genet. 94: 1000-1006.

Giddings GD, Hamilton NRS, Hayward MD (1997b) The release of genetically modified grasses. Part 2: The influence of wind direction on pollen dispersal. Theor. Appl. Genet. 94: 1007-1014.

Grant SR (1999) Dissecting the mechanisms of posttranscriptional gene silencing: divide and conquer. Cell 96: 303-306.

Griffiths DJ (1951) The liability of seed crops of perennial ryegrass (*Lolium perenne*) to contamination by wind-borne pollen. J. Agric. Sci. 40: 19-38.

Hanson AA (1979) The future of tall fescue. In: Tall fescue, Buckner RC, Bush LP (eds.), pp. 341-344. ASA-CSSA-SSSA, Madison.

Jauhar PP (1993) Cytogenetics of the *Festuca-Lolium* complex: relevance to breeding. Springer-Verlag, Berlin, New York.

Johnson RC, Bradley VL, Knowles RP (1996) Genetic contamination by windborne pollen in germplasm-regeneration plots of smooth bromegrass. Plant Genet. Resour. Newsl. 106: 30-34.

Kuai B, Dalton SJ, Bettany AJE, Morris P (1999) Regeneration of fertile transgenic tall fescue plants with a stable highly expressed foreign gene. Plant Cell, Tissue Organ Cult. 58: 149-154.

Kunz C, Schob H, Leubner MG, Glazov E, Meins F, Jr. (2001) beta-1,3-Glucanase and chitinase transgenes in hybrids show distinctive and independent patterns of posttranscriptional gene silencing. Planta 212: 243-249.

Lorraine-Colwill DF, Powles SB, Hawkes TR, Preston C (2001) Inheritance of evolved glyphosate resistance in *Lolium rigidum* (Gaud.). Theor. Appl. Genet. 102: 545-550.

Luna VS, Figueroa MJ, Baltazar MB, Gomez LR, Townsend R, Schoper JB (2001) Maize pollen longevity and distance isolation requirements for effective pollen control. Crop Sci. 41: 1551-1557.

Messeguer J (2003) Gene flow assessment in transgenic plants. Plant Cell, Tissue Organ Cult. 73: 201-212.

Nurminiemi M, Tufto J, Nilsson NO, Rognli OA (1998) Spatial models of pollen dispersal in the forage grass meadow fescue. Evol. Ecol. 12: 487-502.

Paszkowski J, Whitham SA (2001) Gene silencing and DNA methylation processes. Curr. Opin. Plant Biol. 4: 123-129.

Rognli OA, Nilsson NO, Nurminiemi M (2000) Effects of distance and pollen competition on gene flow in the wind-pollinated grass *Festuca pratensis* Huds. Heredity 85:550-560.

Sleper DA, Buckner RC (1995) The fescues. In: Forages, Barnes RF, Miller DA, Nelson CJ, Heath ME (eds.), pp. 345-356. Iowa State University Press, Ames, Iowa.

Sleper DA, West CP (1996) Tall fescue. In: Cool-season forage grasses, Moser LE, Buxton DR, Casler MD (eds.), pp. 471-502. ASA-CSSA-SSSA, Madison.

Spangenberg G, Kalla R, Lidgett A, Sawbridge T, Ong EK, John U (2001) Breeding forage plants in the genome era. In: Molecular breeding of forage crops, Proceedings of the 2nd International Symposium, Spangenber G (ed.), pp.1-39. Kluwer Academic Publishers, Dordrecht, Netherlands.

Spangenberg G, Wang ZY, Nagel J, Potrykus I (1994) Protoplast culture and generation of transgenic plants in red fescue (*Festuca rubra* L.). Plant Sci. 97: 83-94.

Spangenberg G, Wang ZY, Potrykus I (1998) Biotechnology in forage and turf grass improvement. Springer, Berlin, New York.

Spangenberg G, Wang ZY, Wu XL, Nagel J, Iglesias VA, Potrykus I (1995a) Transgenic tall fescue (*Festuca arundinacea*) and red fescue (*F. rubra*) plants from microprojectile bombardment of embryogenic suspension cells. J. Plant Physiol. 145: 693-701.

Spangenberg G, Wang ZY, Wu XL, Nagel J, Potrykus I (1995b) Transgenic perennial ryegrass (*Lolium perenne*) plants from microprojectile bombardment of embryogenic suspension cells. Plant Sci. 108: 209-217.

Terrell EE (1979) Taxonomy, morphology and phylogeny. In: Tall fescue, Buckner RC, Bush LP (eds.), pp. 31-39. ASA-CSSA-SSSA, Madison.

Thomas H, Humphreys MO (1991) Progress and potential of interspecific hybrids of *Lolium* and *Festuca*. J. Agric. Sci. 117: 1-8.

Vasil IK (2003) The science and politics of plant biotechnology — a personal perspective. Nat. Biotechnol. 21: 849-851.

Wang GR, Binding H, Posselt UK (1997) Fertile transgenic plants from direct gene transfer to protoplasts of *Lolium perenne* L. and *Lolium multiflorum* Lam. J. Plant Physiol. 151: 83-90.

Wang ZY, Bell J, Ge YX, Lehmann D (2003a) Inheritance of transgenes in transgenic tall fescue (*Festuca arundinacea* Schreb.). In Vitro Cell. Dev. Biol. Plant 39: 277-282.

Wang ZY, Hopkins A, Mian R (2001a) Forage and turf grass biotechnology. Crit. Rev. Plant Sci. 20: 573-619.

Wang ZY, Legris G, Nagel J, Potrykus I, Spangenberg G (1994) Cryopreservation of embryogenic cell suspensions in *Festuca* and *Lolium* species. Plant Sci. 103: 93-106.

Wang ZY, Lehmann D, Bell J, Hopkins A (2002) Development of an efficient plant regeneration system for Russian wildrye (*Psathyrostachys juncea*). Plant Cell Rep. 20: 797-801.

Wang ZY, Scott M, Bell J, Hopkins A, Lehmann D (2003b) Field performance of transgenic tall fescue (*Festuca arundinacea* Schreb.) plants and their progenies. Theor. Appl. Genet. 107: 406-412.

Wang ZY, Takamizo T, Iglesias VA, Osusky M, Nagel J, Potrykus I, Spangenberg G (1992) Transgenic plants of tall fescue (*Festuca arundinacea* Schreb.) obtained by direct gene transfer to protoplasts. Biotechnology 10: 691-696.

Wang ZY, Ye XD, Nagel J, Potrykus I, Spangenberg G (2001b) Expression of a sulphur-rich sunflower albumin gene in transgenic tall fescue (*Festuca arundinacea* Schreb.) plants. Plant Cell Rep. 20: 213-219.

Wipff JK, Fricker C (2001) Gene flow from transgenic creeping bentgrass (A*grostis stolonifera* L.) in the Willamette valley, Oregon. Int. Turfgrass Society Res. J. 9: 224-241.

Ye X, Wang ZY, Wu X, Potrykus I, Spangenberg G (1997) Transgenic Italian ryegrass (*Lolium multiflorum*) plants from microprojectile bombardment of embryogenic suspension cells. Plant Cell Rep. 16: 379-384.

Protecting Plant Inventions

Rob Hanson and Steve Highlander
FULBRIGHT & JAWORSKI, LLP[1], *Austin, Texas, USA*

Abstract:

Intellectual property protection is essential if plant breeders are to protect their investment in the development of new varieties. This is underscored by the ease with which plant inventions can be appropriated. Three general forms of intellectual property protection are available for plant inventions: plant patents, utility patents and certificates of protection under the Plant Variety Protection Act. Which form or combination of forms of protection can be obtained depends upon the nature of plant variety created and the resulting ability or inability to meet certain statutory requirements. This article discusses the characteristics of the different intellectual property regimes and the practical implications each has to the plant breeder and agriculture in general.

[1] The views expressed are solely those of the authors. This article is offered for informational purposes only and does not constitute legal advice.

A. Hopkins, Z.Y. Wang, R. Mian, M. Sledge and R.E. Barker (eds.), Molecular Breeding of Forage and Turf, 381-395.

1. INTRODUCTION

The United States seed and ornamental plant industries serve a multibillion dollar market. These numbers continue to increase as advances in genetic engineering allow the creation of plants with value-added traits, including herbicide, insect and disease resistance. Not surprisingly, intellectual property protection of plant varieties is of considerable interest to the agri-industry.

The need to protect against unauthorized use of plants arises not only from the inherent value of the plants, but also because they are so easily misappropriated. Once a plant is sold, it can be reproduced essentially in perpetuity, each time producing an identical copy of the original plant. A prime example of this is the world's most widely grown apple, the Red Delicious. Over 15 million Red Delicious apple trees have been sold since the creation of the first Red Delicious tree in 1893, each plant a clonal descendant of the first Red Delicious tree.[2] Plant breeders must therefore prevent copying of plants. Similarly, other types of appropriation must be prevented, including the use by competitors of proprietary plant varieties in plant breeding programs for the production of new varieties.

The importance of intellectual property protection for plants is underscored by the fact that each plant exists not merely as a discrete invention, but represents a source of elite genetic material, or "germplasm". Such germplasm may take decades of expensive and exhaustive breeding efforts to develop and often constitutes the major portion of a seed or biotechnology company's competitive advantage.[3] If a competitor is allowed to incorporate the proprietary germplasm and breed this into its own plant varieties, the competitive advantage will quickly be lost. Such examples further serve as a disincentive against investing the substantial resources needed to develop new plant varieties among plant breeders in general.

Traditional intellectual property remedies, when applied to plants, are subject to circumvention and fraught with complications. For example, traditional forms of intellectual property do not preclude the introduction of elite germplasm into new genetic backgrounds. On the other end of the spectrum, certain rights are too easy to obtain, as illustrated by the fact that individuals completely uninvolved in the creation of the original Red Delicious apple have been able to obtain plant patents on Red Delicious apple trees that contain simple, naturally-occurring mutations, including new

---

[2] *See* http://www.starkbros.com

[3] *See* John M. Poehlman, "Breeding Field Crops," 171-72, 705 (3rd ed. 1986).

flower color or leaf shape.[4] Thus, companies and practitioners alike must therefore consider all types of intellectual property protection, and select those approaches that best suit the ultimate commercial goals.

There are three major types of intellectual property protection available for plant varieties: the Townsend-Purnell Plant Patent Act (PPA), the Plant Variety Protection Act (PVPA), and the Utility Patent Act (UPA). Protection under state trade secret laws may also be available, but this will generally be limited to hybrid varieties where the parent lines can be maintained in "secret."

Protection under the PPA, a "plant patent," is limited to plants that are asexually reproduced, while PVPA protection is available only for plants that are sexually reproduced. A "utility patent" under the UPA is available for both sexually and asexually reproduced plants, but has only fairly recently been sanctioned by courts. However, given other exceptions and exclusions limiting the protection afforded by the PPA and PVPA, the UPA typically offers the broadest level of protection.

2. PLANT PATENTS

Prior to the enactment of the Plant Patent Act (PPA), it was deemed that two obstacles prevented the patentability of plants. First, it was believed that plants, even if artificially bred, were "products of nature" for purposes of the patent law and thus were unpatentable. Second, plant inventions were deemed incapable of meeting the written description requirement, which requires a patent applicant to fully describe the invention. In particular, it was believed that because new plants may differ from old plants only in such traits as fruit taste or smell, which are not capable of an adequate description by words, true and consistent differentiation by written description was impossible.

Recognizing the problems and the importance of protecting plant varieties, Congress passed the PPA in 1930. Under the PPA, the perceived limits on patentability of plants were overcome with express provisions making plants patentable subject matter and relaxing the written description requirement in favor of a description that need only be "as complete as is reasonably possible."

---

[4] *See e.g.*, Plant Patent No. PP4,819 (a plant patent directed to a mutant sport of the Red Delicious apple tree).

2.1. Requirements for PPA Protection

Plant patent protection under the PPA can be obtained for "any distinct and new variety that has been asexually reproduced." A variety will generally be found to be distinct if it possesses at least one unique characteristic, when compared to the "prior art."[5] Such characteristics can include habit, immunity from disease, flower color, fruit color, and fruit quality.

A plant patent must fulfill the same "novelty" requirement as any other patent. However, assessing novelty for plants presents a unique situation. Generally speaking, a written or oral description of a plant does not place the public in "possession" of the plant. In *In re LeGrice*, the Court of Customs and Patent Appeals noted that photographs and a description of a rose plant did not defeat novelty in a later application for a plant patent on the rose. The court reasoned that the prior descriptions were not sufficient to reproduce the plant in question exactly.[6] On the other hand, a sale or public use of a plant variety more than one year before the filing date would, of course, still constitute a novelty bar.

Express exceptions to requirements that a patent applicant provide "written description" and "enablement" for a plant variety are made under the PPA. First, the written description requirement is modified to require only "as complete [a description] as is reasonably possible." This requirement can be met by providing a verbal description and photograph of the variety sufficient to distinguish it from other varieties.[7] The second exception provided by the PPA relates to the absence of an "enablement" requirement, eliminating the need for biological deposits (unless specifically requested by the USPTO).

A plant patent specification must still "particularly point out where and in what manner the variety has been asexually reproduced." The specification must also include a registered variety name, and this name must not be confusingly similar to a previous variety name. Finally, a single claim must be included describing the variety and its unique characteristics.

Like an application for a utility patent, the most common type of patent, a plant patent application will be subject to an examination by the U.S. Patent and Trademark Office (USPTO). The examination involves a back and forth

---

[5] *See Ex parte Moore,* 115 U.S.P.Q. (BNA) 145 (P.T.O. Bd. App. 1957).
[6] (133 U.S.P.Q. (BNA) 365, 368, 372 (C.C.P.A. 1962)).
[7] *See* Patent Law Fundamentals, Sec. 13.07, at 3 (Peter D. Rosenberg, Ed., 2000)

between the applicant and USPTO in which compliance with the requirements of the PPA is established. Once all requirements are met, the plant patent issues and becomes effective.

2.2. Rights and Limitations Under the PPA

A plant patent entitles the patent owner to exclude others from asexually reproducing the plant or selling or using the plant so reproduced. A 1998 amendment of the PPA extended protection to include the use, sale or importation of parts of the patented plant as well.[8] The statutory term of a plant patent is the same as a standard "utility" patent, 20 years from the date of filing.[9] The asexual reproduction that is prohibited involves the reproduction of an isolated plant part into a new "clone" of the starting plant.[10] It is not essential that the infringer be aware of the source of the patented plant that is asexually reproduced, or that the infringer has notice of the patent.[11] Each separate prohibited act is an infringement.[12]

A major shortcoming of plant patents arises from the court decision in *Imazio Nursery v. Dania.*[13] In *Imazio*, the legislative history of the PPA was read as limiting the scope of a plant patent to only "a single plant," and its "asexually reproduced progeny."[14] Based on this "single plant" theory, the court held that to show infringement of a plant patent, the patent owner must prove that the alleged infringing plant is an asexually reproduced progeny of the patented plant.[15] A mere demonstration of morphological identity between the patented and alleged infringing varieties was found insufficient to meet this standard, and actual evidence of "copying" was required. In fact, independent creation was held to be a defense to plant patent infringement.[16]

---

[8] Plant Protection Amendment Act of 1998, PL 105-289, 112 Stat. 2780 (Oct. 27, 1998), *See also,* 35 U.S.C. §163 (2000) ("In the case of a plant patent, the grant shall include the right to exclude others from asexually reproducing the plant, and from using, offering for sale, or selling the plant so reproduced, or any of its parts, throughout the United States, or from importing the plant so reproduced, or any parts thereof, into the United States.")

[9] *See* 35 U.S.C. §271 (2000).

[10] *See* Poehlman, *supra.*

[11] *See Yoder Bros., Inc. v. California-Florida Plant Corp.,* 537 F.2d. 1347, 1382 (5th Cir. 1976)

[12] *See id.*

[13] *See Imazio Nursery, Inc.*, 69 F.3d at 1569, *See also* generally Vincent G. Gioia, *Plant Patents – R.I.P.,* 79 J. Pat. & Trademark Off. Soc'y. 615.

[14] *See id.* at 1569.

[15] *See id.* at 1568.

[16] *See id.* at 1570.

This ruling contradicted the findings of the lower court, which had held that allowing such a defense would cause the patent owner "great difficulties" in enforcing plant patent rights.[17] Indeed, such difficulties are easily envisioned.

In theory, a plant patent owner could use genetic testing to prove the required copying under the *Imazio* standard. However, this may be easier said than done. First, a set of genetic markers that could be used to show derivation may not exist for the species at issue. Second, even if genetic markers were available, they may not prove sufficiently variable to distinguish actual copying from mere genetic similarity. Third, obtaining such evidence and presenting it at trial will require a technical expert trained in genetic testing, further adding to the cost of the litigation. Finally, if all of the foregoing can be achieved, it is still not a trivial exercise to obtain a jury verdict based solely on genetic evidence.

3. THE PLANT VARIETY PROTECTION ACT

A conspicuous gap that remained following enactment of the PPA was for the protection of sexually reproduced plants. This gap was especially problematic to seed companies, as nearly all major field crops are sexually propagated.[18] Recognizing this shortcoming, Congress passed the Plant Variety Protection Act (PVPA) in 1970.[19] The purpose of the enactment was to "encourage the development of novel varieties of sexually reproduced plants and to make them available to the public, providing protection available to those who breed, develop, or discover them, and thereby promoting progress in agriculture in the public interest."[20]

The PVPA grants patent-like protection for plant varieties by the issuance of certificates of protection. Following a 1984 amendment to the PVPA, this protection also became available for first generation (F1) hybrids and to tuber-propagated plant varieties, both of which were previously unprotectable. The PVPA is administered by the Plant Variety Protection Office (PVPO), established by the Department of Agriculture and under the auspices of the Secretary of Agriculture, and is independent of the U.S. Patent and Trademark Office.

---

[17] *See id.*

[18] *See* Poehlman, *supra,* at 239-41, 706.

[19] 7 U.S.C. §2321, et seq. (2000), H.R. Rep. No. 91-1605 (1970), reprinted in 1970 U.S.C.C.A.N. 5082, available in 1970 WL 5934.

[20] *See* H.R. Rep. No. 1246, 91st Cong. 2d. Sess., 84 Stat. 1542 (1970), reprinted in 1970 U.S.C.C.A.N. 1793).

3.1 Requirements to Obtain PVPA Protection

PVPA protection is available for any "new ... distinct ... uniform ... and stable" sexually reproduced plant variety.[21] The requirement that a variety be "new" is met if, as of the filing date for PVPA rights, the variety has not been "sold or otherwise disposed of for commercial purposes in the U.S. more than 1-year prior to the PVPA application." A variety is "distinct" if it is "clearly distinguishable" from known varieties as of the filing date. Generally speaking, a difference in a single characteristic will suffice to demonstrate distinctness. Uniformity is present if any variations are "describable, predictable, and commercially acceptable."[22] To be "stable" the variety must, when reproduced, remain unchanged as to "essential and distinctive characteristics of the variety with a reasonable degree of reliability" in relation to the varieties in the breeding category of the variety.

Applications for rights under the PVPA are made directly with the PVPO. Upon receipt of the application, the PVPO conducts a substantive examination of the application similar to that carried out under the PPA, including a review of "all available documents, publications, or other material relating to varieties of the species involved in the application."[23] An application must include the applicant's name, a completed application form, and at least 2,500 seeds of the variety.[24] If the plant is a hybrid, evidence of deposit of both parent lines must be submitted, as the hybrid plant itself would not meet the "stable" requirement of the PVPA.[25] Upon completion of the requirements of the PVPA, the Secretary of Agriculture will issue a certificate of protection for the subject variety.[26]

3.2 The Scope of PVPA Protection

A PVPA certificate allows the right holder to exclude others from "selling the variety, or offering it for sale, or reproducing it" for a term of 20 years from the date of issuance of a PVPA certificate.[27] In addition to covering the individual protected variety, a 1994 amendment to the PVPA extended

---

[21] *See* 7 U.S.C. §2402(a)(1-4)
[22] *See* 7 U.S.C. §2401(a)(3).
[23] *See* Sec. 97.100(b) Plant Variety Protection Act. *See also* 7 U.S.C. §2323 (2000).
[24] *See* §97.5, §97.6 Plant Variety Protection Act.
[25] *See id.*
[26] *See* 7 U.S.C. §2323 (2000).
[27] *See* 7 U.S.C. §2483 (2000).

protection to varieties that are "essentially derived" from the protected variety.[28]

The definition of an "essentially derived variety" is provided in the 1994 amendment to the Act. Included under the definition are varieties that are (i) "predominantly derived from" the protected variety, yet retain essential characteristics of the variety; (ii), are not "clearly distinguishable" from the protected variety; and (iii) other than differences resulting "from the act of derivation," conforms to the protected variety in the expression of the essential characteristics resulting from the initial variety.[29] An essentially derived variety may be a natural or induced mutant of the protected plant or may be made by backcrossing or genetic engineering of the protected variety.[30]

If infringement is found, the PVPA owner may bring suit for damages. In such an action, the PVPA certificate will enjoy a presumption of validity. Damages available should be "adequate to compensate for the infringement," and not less than a reasonable royalty. However, a court may treble damages, or alternatively, the damages can be decreased in the court's discretion where the infringer was found to have innocent intentions. Injunctions may also be obtained. Attorney fees may be awarded to the prevailing party in exceptional cases. Defenses to a claim under the PVPA include noninfringement, invalidity of the PVPA certificate, and use under an otherwise valid adverse PVPA certificate.

3.3 Exceptions to the PVPA

The rights granted by the PVPA are limited by two important exceptions. First, under the "crop exception," a farmer can save seed and use the saved seed in the production of a "crop for use on the farm of the person."[31] Prior to 1994, this exception was viewed as permitting a farmer to sell any seed the farmer saved, provided that the amount sold did not exceed the amount saved for personal use, *i.e.*, planting on the farm.[32] In 1994 however, the PVPA was amended to clarify that seed could not be sold to another farmer for

[28] *See* 7 U.S.C. §2401, 7 U.S.C. §2541. The "essentially derived" provision is not applicable to variety that "is itself an essentially derived variety, or any variety that is not clearly distinguishable from a protected variety."
[29] *Id.*
[30] 7 U.S.C. §2401(a)(3)(B).
[31] *See* 7 U.S.C. § 2543.
[32] *See* 7 U.S.C. §2543, *See also Asgrow Seed Company v. Winterboer,* 513 U.S. 179 (1995).

further production of seeds of the variety.[33] The exception has thus not had a large impact on the seed industry. The impact of this exception is also minimized by the fact that several of the most important seed crops, most notably maize, are sold to farmers as hybrid seeds, the offspring of which will be non-uniform and will not stably maintain the desirable characteristics of the parent.[34] Thus, the seed of hybrid plants that the farmer reproduces on his land will be useless.

A second and potentially more troublesome exception to the PVPA is the "research exception," which allows use of the protected seed in breeding research for the development of new varieties.[35] This exception allows the "use and reproduction" of the protected variety for plant breeding or "other *bona fide* research" without infringement of the PVPA certificate. The dilemma for the PVPA certificate holder is that much of the value in a given variety may reside in its potential for making new varieties. As described above, this type of misappropriation can damage the certificate holder not merely with respect to the claimed variety, but also by giving a competitor access to the unique germplasm that may have been developed over many years and at great expense. Another concern under this exception is that it permits appropriation of genetically-engineered traits. More specifically, the research exemption allows a competitor to obtain a genetically altered plant and breed the plant with the competitor's varieties, thereby introducing the genetically-engineered trait into a new variety.

4. UTILITY PATENT PROTECTION FOR PLANT VARIETIES

With the enactment of the PVPA in 1970, intellectual property protection was available for the first time for both sexually and asexually reproduced plants.[36] However, it was still believed at that time that plants were not patentable under the utility patent statute.[37] Chief among the reasons for this were the continued belief that plants were products of nature. A major advance in the understanding of this issue was achieved by the U.S. Supreme

---

[33] *See* PL 103-349, Sec. 9(3), Oct. 6, 1994, 108 Stat. 3136.
[34] *See* Poehlman, *supra* at 239-41, 706.
[35] *See* 7 U.S.C. §2544 ("The use and reproduction of a protected variety for plant breeding or other bona fide research shall not constitute an infringement of the protection provided under this Act").
[36] *See* Robert J. Jondle, Overview and Status of Plant Proprietary Rights, in CROP SCIENCE SOC'Y OF AM., INC. , ASA SPEC. PUB. NO. 52, INTELLECTUAL PROPERTY RIGHTS ASSOCIATED WITH PLANTS, 5.5 (Billy E. Caldwell et al., eds., 1989).
[37] *Diamond v. Chakrabarty,* 447 U.S. 303, 309, 206 U.S.P.Q. at 197 (1980).

Court in *Diamond v. Chakrabarty.*[38] In that case, a patent applicant had claimed a bacterium genetically engineered to break down crude oil. The U.S. Patent and Trademark Office rejected the claim, alleging that the bacterium was a "product of nature" and therefore not patentable subject matter. The U.S. Supreme Court reversed, finding that the scope of the utility patent laws extended to "anything under the sun that is made by man."[39] The genetically engineered bacterium was found to have "markedly different characteristics from any found in nature," and its property of breaking down crude oil was not a characteristic naturally found in the bacterium.[40]

The first case officially upholding the patentability of plants under the utility statute was *Ex parte Hibberd.*[41] In *Hibberd*, a Patent Office appeal board reversed a rejection of claims to plants, seeds, and tissue cultures of a maize plant having increased levels of the amino acid tryptophan.[42] The examiner alleged that UPA protection of plants was pre-empted by the plant-specific PVPA and PPA. The Board disagreed, however, indicating that the UPA was not in conflict with the PVPA and UPA, and that Chakrabarty permitted patenting of anything made by man.

Though *Hibberd* was indeed a significant step, the issue of plant utility patents was somewhat unsettled until *J.E.M. Ag Supply, Inc. v. Pioneer Hi-Bred Int'l, Inc.*, 534 U.S. 124 (2001).[43] In *J.E.M.*, defendant J.E.M. Ag Supply, Inc. argued that protection of sexually and asexually reproduced plants was intended by Congress to be exclusively governed, and thus pre-empted by, the PVPA and PPA. The Supreme Court disagreed, determining that the PVPA and PPA are not in conflict with the UPA, and thus it was Congress' intent that effect be given to both acts and the UPA.[44] The Supreme Court thus opened the door for widespread use of utility patent protection for protecting potentially any economically important plant variety.

---

[38] *Chakrabarty,* 447 U.S. 303 at 309.
[39] *Id.*
[40] *Id.*
[41] *See Ex parte Hibberd*, 227 U.S.P.Q. 443 (Bd. Pat. App. & Int. 1985)).
[42] *Id.* at 443.
[43] *J.E.M. Ag Supply, Inc. v. Pioneer Hi-Bred Int'l, Inc.*, 534 U.S. 124 (2001)
[44] *J.E.M.* 534 U.S. at 154-155.

4.1 Requirements for Plant protection Under the UPA

To obtain utility patent protection, a plant variety must have utility[45] and be novel[46] and non-obvious.[47] Additionally, the application for utility patent must contain a written description of the invention and must enable "any person skilled in the art to which it pertains" to make and use the invention.[48] Each of these requirements can create issues that are unique to plant inventions and must be considered carefully by practitioners.

4.1.1 Plant Utility

For most plant inventions, the utility requirement of 35 U.S.C. § 101 no longer presents a serious obstacle to patentability provided the plant was "made by the hand of man."[49] However, protection cannot be obtained under the UPA for plants that exist in nature (*e.g.*, are "products of nature"). As described above, whether a plant is or is not a product of nature may depend on how the plant is made. Courts have held that plants made by plant breeding are not products of nature. However, plants developed by mere screening or selection of mutants may occur naturally. Therefore, whether a plant variety is patentable subject matter will likely turn on whether the extent of human intervention required to create the plant amounts to being "made by the hand of man." If a mutation is induced by treating a plant with a mutagen, a colorable argument can at least be made that the plant is not a product of nature. Alternatively, mere selection of a plant having a naturally-occurring mutation may be a true "product of nature."

Guidance regarding the patentability of plants selected as mutants, in the context of plant patents, was provided in *Yoder Brothers v. California-Florida Plant Corp.*[50] There, the court addressed the issue of whether a plant selected as a naturally-occurring mutant, or "sport," was patentable subject matter. In deciding for the patent owner, the court held that the mere fact that a mutation occurred naturally was irrelevant, given that the particular mutant variety would not persist in nature without the intervention of man.[51] The court emphasized that, absent intervention, the unique genetic background of the patented plant would be lost forever in the next sexual generation, given

[45] 35 U.S.C. §101.
[46] 35 U.S.C. §102.
[47] 35 U.S.C. §103.
[48] 35 U.S.C. §112, first paragraph.
[49] *Ex parte Allen,* 2 U.S.P.Q. 2d 1425, 1427 (Bd. Pat. App. & Interf. 1987).
[50] *See id.*
[51] *See id.*

that plants in the wild usually do not breed true-to-type. This reasoning invites the argument that even a variety selected as a naturally-occurring mutation, and asexually reproduced, is "made by the hand of man," and therefore patentable subject matter.

4.1.2 Novelty and Non-obviousness

As with other types of inventions, plant varieties must be novel and non-obvious under the UPA. These standards present unique issues when applied to plants. For example, although a sale, offer for sale, or public use of an invention more than 1 year before filing of a patent application remains a novelty bar like as with any other invention, a publication of a description of a plant may not defeat novelty. Plants cannot generally be reproduced simply by using a description of the plant. It is typically required that a copy of the plant itself be accessible to reproduce the plant.

In *In re LeGrice*, a court held that photographs and a description of a rose plant were not a novelty bar.[52] A different result was obtained, however, in *Ex parte Thompson.*[53] In that case, a utility patent for a cotton variety was barred based on a determination that the claimed plant was reproducible by those of skill in the art. The USPTO Board of Appeals noted that, in addition to a publication describing the variety more than one year before the filing date, there was evidence of concurrent commercial availability of the variety in Australia. Because of this, the publication was found to put the public in possession of the invention more than a year before the patent application was filed, thereby defeating novelty. Novelty will thus typically only be a problem when a variety has been placed in the public domain.

Even when a plant is novel, patentability may be barred because the variety is "obvious."[54] This is in contrast to the PVPA, where no equivalent of the UPA obviousness standard exists. A plant will be obvious if the differences between the variety and prior art varieties are so insignificant that they would be obvious to those of "ordinary skill in the art."[55] It is not enough that the prior art merely renders it "obvious to try" to make the claimed variety. The prior art also must provide an "enabling" disclosure relative to the claimed invention.[56] This will be a difficult standard patent examiners to meet, given the unpredictability of plant breeding. Even when

---

[52] *See In re LeGrice* (133 U.S.P.Q. (BNA) 365, 368, 372 (C.C.P.A. 1962))
[53] *Ex parte Thompson,* 24 USPQ2d 1618 (Bd. Pat. App. & Inter. 1992).
[54] *See* 35 U.S.C. §103
[55] *See Graham v. John Deere,* 383 U.S. 1, 17-18 (1965)
[56] *See In re Donohue*, 766 F.2d 531, 226 USPQ 619 (Fed. Cir. 1985)

the same two non-inbred parent plants are crossed repeatedly, it is almost impossible to obtain the same progeny plant twice. Therefore, absent a prior art plant that is both (a) in the hands of the public and (b) nearly genetically identical to the claimed plant variety, obviousness should not present a bar to patentability.

4.1.3 Written Description and Enablement

Compliance with the requirement that an applicant provide a written description of a plant and enable one of skill in the art to make and use the plant are potentially the most difficult hurdles to obtaining protection under the UPA. This is because plants fall within the category of "microorganisms" that have been recognized by the courts and USPTO as fundamentally different from other types of inventions. This was noted by the court in *In re Argoudelis*, stating that "a unique aspect of using microorganisms as starting materials is that a sufficient description of how to obtain the microorganism from nature cannot be given."[57] This problem has been largely overcome for most inventions by allowing applicants to make biological materials available to the public by a biological deposit with a public depository. Use of deposits of seed in the context of plant inventions has been specifically sanctioned for compliance with the enablement requirement. To satisfy the requirement, a seed deposit must be made with a depository meeting the terms of the Budapest Treaty, such as the American Type Culture Collection (ATCC) in Mansassas, VA. Unlike deposits made in accordance with the PVPA, such a deposit will be made available and unrestricted to the public following issuance of a patent.

A special problem is presented for plants that are not seed-propagated. Plants themselves cannot be deposited, as depositories are unwilling to continually grow and asexually reproduce plants during the minimum 30 year term of deposit.[58] An alternative is found in a rule providing that no deposit is necessary if the biological material in question is "known and readily available" to the public or can be made or isolated without undue experimentation."[59] In most cases, a plant could be made "known and readily available" by placing the variety on sale before filing a patent application on the variety, although care would have to be taken to not exceed the one year novelty bar date before filing a patent application.

---

[57] *In re Argoudelis*, 434 F.2d 1390, 1392 (CCPA 1970).
[58] *See* 37 C.F.R. 1.806.
[59] 37 C.F.R. §1.802.

There is currently no case defining when a biological material is "known and readily available." However, the USPTO's Manual of Patent Examining Procedure provides some guidance by listing a number of factors considered indicia that a particular biological material is known and readily available to the public.[60] Factors listed include the extent of commercial availability, references to the biological material in printed publications, declarations of accessibility by those working in the field and evidence of predictable isolation techniques. Where commercial availability is used as evidence of availability that a biological material is known and readily available, it is required that clear and convincing evidence of public access be given. Such evidence may include evidence that the price at which the materials were sold was reasonable and that the commercial supplier of the plant was truly independent of the variety owner.

A problem with placing a variety on sale before filing is that foreign rights in the invention would be destroyed under the absolute novelty standard that is used by most nations. One possibility to avoid this would be to file foreign before filing in the U.S. The danger in this strategy is that the applicant must be careful to obtain a petition for a foreign filing license before filing in a foreign country, if the invention was made in the U.S., or risk possible forfeiture of U.S. rights.[61] If none of the these options can be met, protection will not be available under the UPA.

4.2 The Scope of Protection Under the UPA

The flexibility of protection under the UPA is unparalleled by either PVPA or PPA. A utility patent permits the patent owner to prevent others from making, using, selling and offering to sell the invention.[62] The fact that multiple claims can be drafted, unlike the PPA and PVPA, allows drafting of claims specifically targeted to cover all possible infringements. Types of claims that can be advanced include plants, seeds, plant parts, tissue cultures and methods of breeding or growing the protected variety.

Method claims can be particularly useful, as they can cover the possibility that a competitor will use the protected variety in plant breeding programs. This is important because neither the PVPA nor the PPA reach this type of activity. Further, the plants themselves created by such breeding will likely not fall within the literal or even equivalent scope of composition claims. A risk also exists that breeding might be done overseas. However, by including

[60] *See* MPEP §2404.01.
[61] See 35 U.S.C. §184
[62] [145] 35 U.S.C. § 271(a).

claims to methods of breeding the plant, protection against importation of the resulting plants may be available.[63]

5. CONCLUSION

As a result of historical misconceptions regarding the patentability of plant varieties, use of utility patents has been limited primarily to major crop plants. Given recent legal clarifications, and the exceptions and exclusions that limit the effectiveness of the PPA and PVPA, it is likely that increasing use will be made of utility patents under the UPA. Neither the PVPA nor PPA prevent the use of varieties in plant breeding protocols. Further, under the PPA, a plant must be shown to have been clonally derived from the protected variety to infringe.

Admittedly, protection under the UPA is typically more difficult to obtain than under the PPA and PVPA. The main hurdle to obtaining UPA protection is the enablement requirement, *e.g.*, the requirement that the application itself place the invention in the hands of the public. A description of the technique used to make a plant will typically not enable "one of ordinary skill in the art" to make and use the invention, as required by the statute. For seed propagated varieties, this can be satisfied by making a deposit of seed with an approved depository. For asexually reproduced plants it will likely be required to make the plant "known and readily available." When this is not possible, protection may not be available under the UPA.

The strengths of the UPA are manifest, including: (i) the opportunity for broad protection; (ii) the ability to cover plants, seeds, progeny, and methods of using and creating plants; and (iii) applicability to both sexually- and asexually-produced plants. UPA protection may well ultimately dominate the plant intellectual property arena, although there will continue to be situations where the PPA and PVPA will be the preferred or only form of intellectual property protection available.

---

[63] 35 U.S.C. § 271 (g)

Developments in Plant Breeding

1. H. Schmidt and M. Kellerhals (eds.): *Progress in Temperate Fruit Breeding.* 1994
ISBN 0-7923-2947-3
2. O.A. Rognli, E. Solberg and I. Schjelderup (eds.): *Breeding Fodder Crops for Marginal Conditions.* Proceedings of the 18th Eucarpia Fodder Crops Section Meeting, Loen, Norway (August 1993). 1994 ISBN 0-7923-2948-1
3. A.C. Cassells and P.W. Jones (eds.): *The Methodology of Plant Genetic Manipulation: Criteria for Decision Making.* Proceedings of the Eucarpia Plant Genetic Manipulation Section Meeting held at Cork, Ireland (September 11–14, 1994). 1995 ISBN 0-7923-3687-9
4. P.M.A. Tigerstedt (ed.): *Adaptation in Plant Breeding.* 1997
ISBN 0-7923-4062-0
5. H. Guedes-Pinto, N. Darvey and V.P. Carnide (eds.): *Triticale: Today and Tomorrow.* 1996 ISBN 0-7923-4212-7
6. H.-J. Braun, F. Altay, W.E. Kronstad, S.P.S. Beniwal and A. McNab (eds.): *Wheat: Prospects for Global Improvement.* Proceedings of the 5th International Wheat Conference, Ankara, Turkey (June 10–14, 1996). 1997 ISBN 0-7923-4727-7
7. S.P. Singh (ed.): *Common Bean Improvement in the Twenty-First Century.* 1999
ISBN 0-7923-5887-2
8. G.T. Scarascia Mugnozza, E. Porceddu and M.A. Pagnotta (eds.): *Genetics and Breeding for Crop Quality and Resistance.* Proceedings of the XV EUCARPIA Congress, Viterbo, Italy (September 20–25, 1998). 1999 ISBN 0-7923-5844-9
9. Z. Bedö and L. Láng (eds.): *Wheat in a Global Environment.* Proceedings of the 6th International Wheat Conference, Budapest, Hungary (June 5–9, 2000). 2001
ISBN 0-7923-6722-7
10. G. Spangenberg (ed.): *Molecular Breeding of Forage Crops.* Proceedings of the 2nd International Symposium, Molecular Breeding of Forage Crops, Lorne and Hamilton, Victoria, Australia (November 19–24, 2000). 2001
ISBN 0-7923-6881-9
11. A. Hopkins, Z.-Y. Wang, R. Mian, M. Sledge and R.E. Barker (eds.): *Molecular Breeding of Forage and Turf.* Proceedings of the 3rd International Symposium, Molecular Breeding of Forage and Turf, Dallas, Texas, and Ardmore, Oklahoma, USA, (May 18–22, 2003). 2004 ISBN 1-4020-1867-3

KLUWER ACADEMIC PUBLISHERS – DORDRECHT / BOSTON / LONDON